AF897580

Philosophy and Medicine

Volume 122

Founding Co-Editor

Stuart F. Spicker

Senior Editor

H. Tristram Engelhardt, Jr., Department of Philosophy, Rice University, and Baylor College of Medicine, Houston, TX, USA

Series editor

Lisa M. Rasmussen, Department of Philosophy, University of North Carolina at Charlotte, Charlotte, NC, USA

Assistant Editor

Jeffrey P. Bishop, Gnaegi Center for Health Care Ethics, Saint Louis University, St. Louis, MO, USA

Editorial Board

George J. Agich, Department of Philosophy, Bowling Green State University, Bowling Green, OH, USA
Nicholas Capaldi, College of Business Administration, Loyola University, New Orleans, LA, USA
Edmund Erde, University of Medicine and Dentistry of New Jersey (Retired), Stratford, NJ, USA
Christopher Tollefsen, Department of Philosophy, University of South Carolina, Columbia, SC, USA
Kevin Wm. Wildes, S.J., President, Loyola University, New Orleans, LA, USA

The Philosophy and Medicine series is dedicated to publishing monographs and collections of essays that contribute importantly to scholarship in bioethics and the philosophy of medicine. The series addresses the full scope of issues in bioethics, from euthanasia to justice and solidarity in health care. The Philosophy and Medicine series places the scholarship of bioethics within studies of basic problems in the epistemology and metaphysics of medicine. The latter publications explore such issues as models of explanation in medicine, concepts of health and disease, clinical judgment, the meaning of human dignity, the definition of death, and the significance of beneficence, virtue, and consensus in health care. The series seeks to publish the best of philosophical work directed to health care and the biomedical sciences.

More information about this series at http://www.springer.com/series/6414

Dien Ho
Editor

Philosophical Issues in Pharmaceutics

Development, Dispensing, and Use

Springer

Editor
Dien Ho
School of Arts and Sciences
MCPHS University
Boston, MA, USA

ISSN 0376-7418 ISSN 2215-0080 (electronic)
Philosophy and Medicine
ISBN 978-94-024-0977-2 ISBN 978-94-024-0979-6 (eBook)
DOI 10.1007/978-94-024-0979-6

Library of Congress Control Number: 2017930927

© Springer Science+Business Media B.V. 2017
This work is subject to copyright. All rights are reserved by the Publisher, whether the whole or part of the material is concerned, specifically the rights of translation, reprinting, reuse of illustrations, recitation, broadcasting, reproduction on microfilms or in any other physical way, and transmission or information storage and retrieval, electronic adaptation, computer software, or by similar or dissimilar methodology now known or hereafter developed.
The use of general descriptive names, registered names, trademarks, service marks, etc. in this publication does not imply, even in the absence of a specific statement, that such names are exempt from the relevant protective laws and regulations and therefore free for general use.
The publisher, the authors and the editors are safe to assume that the advice and information in this book are believed to be true and accurate at the date of publication. Neither the publisher nor the authors or the editors give a warranty, express or implied, with respect to the material contained herein or for any errors or omissions that may have been made. The publisher remains neutral with regard to jurisdictional claims in published maps and institutional affiliations.

Printed on acid-free paper

This Springer imprint is published by Springer Nature
The registered company is Springer Science+Business Media B.V.
The registered company address is: Van Godewijckstraat 30, 3311 GX Dordrecht, The Netherlands

Contents

Introduction

Dien Ho

The ubiquitous presence of pharmaceuticals in our lives is underappreciated. In the United States between 2009 and 2012, almost half the population used at least one prescription drug and more than one in ten Americans used five or more prescription drugs within a 30-day period (Centers for Disease Control and Prevention 2015). The use of pharmaceuticals is so widespread that runoffs from incorrect disposal of drugs have become a pollutant in our drinking water. In 2009, researchers found 51 different pharmaceuticals from beta-blockers to antianxiety medications to anticonvulsants in 19 water treatment plants (Benotti et al. 2009). Environmental impacts are worrisome for sure, but the effects of drugs on us raise far deeper concerns.

The line between therapies and enhancements has been blurred. Consider the recent class of medications that aim to "treat" erectile dysfunction. Are these drugs treatments or are they enhancers? The answer depends on our understanding of what is healthy or biologically normal. If we believe that as men age, their sexual function naturally declines, then drugs like Viagra are enhancers—they allow men to do what they naturally wouldn't be able to. On the other hand, if we believe declining sexual function is not a necessary comorbidity of senescence, then these drugs are treatments. If we take "normal" and "healthy" as normatively loaded, then drawing the line between treatments and enhancements essentially involves defining who we are. To be sure, confronting the line between treatments and enhancements is not unique to pharmaceuticals. All of medicine, indeed, all human modifications (from education to cyberware implants), challenge our concepts of ourselves. Yet drugs offer the easiest and most socially accepted ways of self-modification. We take dietary supplements in the hope of preventing diseases and slowing down aging. Millions of Americans take antidepressants to "feel better than well" (to borrow a phrase from Peter Kramer). Diabetics use insulin pumps to

D. Ho (✉)
School of Arts and Sciences, MCPHS University, Boston, MA, USA
e-mail: dien.ho@mcphs.edu

© Springer Science+Business Media B.V. 2017

D. Ho (ed.), *Philosophical Issues in Pharmaceutics*, Philosophy and Medicine 122,
DOI 10.1007/978-94-024-0979-6_1

regulate their blood sugar levels. Chemotherapies continue to be a mainstay in our fight against cancer. Cholesterol-lowering drugs like statins claim to decrease the odds of someone suffering a stroke or a heart attack. Given the fact that cancer and heart diseases are the two leading causes of death in the USA, successes in drug therapies contribute significantly to the life expectancy of the average American. In other words, how long we *normally* live has been altered, and, with it, the range of activities possible for an average human has changed. Drugs, in this sense, transform the human experience.

At the same time, the proliferation of drug use has also created serious social and medical problems. The abuse of prescription painkillers is a national epidemic. Addiction to painkillers often begins with prescribed use, and patients simply continue to take them long after medically indicated need for pain management. Likewise, in a recent study, almost 7% of college students surveyed have used prescription stimulants for nonmedical purposes at some point in their lives (McCabe et al. 2005). These "spillovers" from prescription to nonprescription use point out one of the unique properties of pharmaceuticals. Unlike surgeries, for example, nonmedically indicated uses of prescription drugs are easily accessible, resulting in unsupervised medical treatments. Drugs are easy to abuse, misuse, buy and sell, and transport to and from faraway places. They can be as powerful as any other medical interventions but as portable and accessible as a bag of candies.

More worrisome, however, is the fact that drugs and their development can shape who we are in a far more conceptual sense. As with any product, drug manufacturers can only make money if there is a demand creating a financial incentive for the drug tail to wag the disease dog. Medicalization of human differences and conditions makes money. The combination of capitalism and drugs means that there will always be financial pressure to look for biological traits that can be fixed and controlled by drugs. In other words, our human imperfection is the source of tremendous profits, and our future identities are the results of scientific progress shaped by market forces.

Given the impact pharmaceuticals have on our social, political, biological, and philosophical identities, it is alarming that there has been a lack of a coherent treatment of philosophy and pharmaceutics. This volume aims to fill that gap and identify some of the major areas of philosophical interest. The book proceeds by tracing the flow of drugs from development through dispensing to their eventual use. At each point, there are fascinating and urgent philosophical issues that need to be addressed. Not only do millions of lives hang in the balance (e.g., how should antibiotics be developed and distributed in developing nations?), the very conception of who we are and who we will be as a species depends on our capacity and willingness to modify ourselves with drugs.

The book begins with a section on philosophical issues relating to the development of drugs. Jessica Flanigan begins by challenging the conventional model of drug development. Flanigan draws on our experience with a patient-driven approach to drug development for rare diseases, such as Niemann-Pick disease, type C. She argues that our respect for patient autonomy, the potential clinical benefits, and the advances in research all support her conclusion that patients should be permitted to undertake a more active role in drug development, including experimenting with not-yet-approved drugs. Anita Ho tackles the issue of pharmaceutical companies' responsibility to develop and provide affordable drugs for the least developed nations. Focusing on the current HIV/AIDS epidemic, she examines the arguments in favor of holding manufacturers responsible for easing the disproportionate disease burden placed on these nations. Her conclusion is that these arguments are at once too weak and too strong. The fact that many of the least developed nations assisted in the development of HIV/AIDS treatments means that pharmaceutical developers have a special reciprocity-based duty to help provide affordable drugs to them.

Sergio Sismondo's essay takes a close look at how pharmaceutical companies have enormous influence on the development of scientific knowledge. Framing his investigation within a political economy of knowledge in which knowledge is produced and justified not by individuals but by social institutions and forces (e.g., pharmaceutical companies), Sismondo traces how pharmaceutical companies sponsor clinical trials, pay scientists to sign on as authors of the trials, disseminate the information to clinicians at continual medical education courses, and plant key opinion leaders who reinforce the scientific legitimacy of these trials. The picture Sismondo paints is one in which the pharmaceutical industry has a great deal of control over the scientific landscape, much to their financial benefit. Lance Stell, on the other hand, challenges the notion that the close (and often, mutually beneficial) relationship between pharmaceutical companies and clinicians in fact poses a moral problem. Specifically, Stell examines the effect strict anti-conflict of interest policies has on patients' care. He concludes that there is simply no evidence to warrant a hypervigilant stance against close financial relationship between drug developers and clinicians.

To complete our section on the philosophical issues relating to the development of drugs, Leah McClimans explores the use of Patient-Reported Outcome Measures (PROMs) as endpoint measurements for clinical trials. McClimans argues that recent attempts to improve PROMs contain hidden values that may make positive outcomes more likely for a clinical trial. Although this consequence might align with the interests of pharmaceutical developers, it is not necessary in patients' best interests.

In the second section of our volume, we explore philosophical issues that arise in the distribution and dispensing of pharmaceuticals. We begin with an essay by Howard Brody, who argues that pharmaceutical companies have a financial incentive to increase the prescription rates of their products. They do so by heavy-handed marketing strategies that bombard clinicians and patients with drug advertisements. Moreover, they also create new clinical avenues and justifications for prescribing medications. Using antidepressants and diabetes as case studies, Brody argues that pharmaceutical companies subtly influence prescription habits to increase their market share. He calls this "drug-centered care." Kenneth Richman follows Brody's essay with a reexamination of drug-centered care. Richman argues that searching for additional uses of a product is not entirely unique to drugs; for example, we make meals on the basis of what ingredients we have available. If we reconceptualize drug-centered care, he argues that it is not obviously at odds with the goals of medicine.

Robert Veatch follows the discussion of drug-centered care with a look at our current laws regulating patients' access to prescription drugs. Echoing Flanigan's discussion, Veatch argues that respecting patients' autonomy means that we should permit certain patients to have access to drugs without prescriptions. There are simply no convincing reasons, according to Veatch, to justify the paternalism that grounds our current policies regulating the dispensing of prescription drugs. Zuzana Deans continues the focus on pharmacies and dispensing policies by examining the ethics of pharmacists' refusals to dispense certain medications on the basis of their personal moral beliefs. Deans surveys a broad range of key issues in conscientious refusals from philosophical concepts such as conscience and integrity to the application of the conventional compromise (often used to justify physicians' conscientious refusals) to the pharmacist setting.

In the third section of our volume, we look at how drug use affects individuals from a philosophical point of view. David DeGrazia begins with an examination of the ethics of human enhancements via the use of pharmaceuticals. Specifically, DeGrazia responds to the objections that enhancements threaten individual identity and authenticity. A close examination of these concepts, he concludes, shows that some enhancive self-transformations using pharmaceuticals are not morally problematic.

In their piece "The Wisdom of Nature: An Evolutionary Heuristic for Human Enhancement," Nick Bostrom and Anders Sandberg further explore the ethics of human enhancements. Bostrom and Sandberg critically explore arguments against human enhancement that appeal to nature's wisdom. To wit, these arguments, according to Bostrom and Sandberg, can draw from a number of different intuitions: from favoring the status quo to the superiority of nature. They develop a heuristic based on evolutionary medicine that captures some of the intuitions behind nature's wisdom while showing that it is morally permissible for us to improve on nature. Şerife Tekin, Owen Flanagan, and George Graham's essay nicely contrasts the two entries before. Tekin et al. highlight the limits of pharmaceutical interventions for medical conditions like addictions. Taking a new look at the nature of addiction, they argue that additive behaviors involve not just neurochemistry but also social and physical relationships. A reliance on pharmaceuticals to cure or improve a person's addiction is not scientifically justified.

Jennifer Radden follows with a discussion of the history of melancholy. The seventeenth-century writer Robert Burton argued that treatments of melancholy must be broadly based to include not only medicinal remedies like herbs but also proper exercises, fresh air, good diet, and so on. The principle responsibility of adhering to one's treatment falls on the patients. Radden draws from this holistic (and essentially self-help) approach and argues that it resonates well with contemporary treatments of mood disorders, such as network-based disorders consisting of diverse causes and requiring multimodal treatments (i.e., more than just the use of pharmaceuticals).

Dien Ho finishes the anthology with a look at the growing problem of antibiotic-resistant bacteria. Ho argues that slowing the emergence of antibiotic resistance requires solving a game of n-person prisoner's dilemma at a number of different levels: from the personal to the global. The standard solutions all assume participants are acting out of rational self-interest. As such, in order to convince participants to cooperate, we need to either change their payoffs (via fines or incentives) or remove the element of choice. Given the slim chance that such an international overseeing body can exist and function effectively, we should instead try to convince participants not to act out of self-interest. It is perhaps a fitting end to the anthology. Philosophical problems pepper the flow of drugs from development to usage. They are not only of deep intellectual interests but they raise some fundamental questions about who we are and what we want to be.

Acknowledgments I want to extend my deep gratitude to Provost Doug Pisano and Dean Delia Anderson of MCPHS University, who helped secure my sabbatical. Without their assistance, I would not have had the necessary time to undertake this project. I would also like to thank the Institute of Medical Humanities at the University of Texas Medical Branch, who awarded me a visiting fellowship in the spring of 2015. Their support was indispensable. In addition, I am grateful for the support I received as a visiting scholar at University of Akureyri in Iceland. Professor Sigurður Kristinsson was not only a great host and helpful colleague, but he has also become a dear friend.

The editing help I received from Peter Leo made finishing this project possible. His editorial instincts and sharp mind spotted improvements that I would have missed. My assistant Amanda Braconnier took care of some of the most demanding and tedious tasks with grace and aplomb. I am deeply grateful for her help. Finally, I dedicate this book to my wife Jane Leo whose comments on my essay and the introduction helped sharpen my thoughts. I also thank her for taking care of our feline daughters Sophie and Wilma while I was away for my visiting posts.

References

Benotti, M.J., R.A. Trenholm, B.J. Vanderford, J.C. Holady, B.D. Stanford, and S.A. Snyder. 2009. Pharmaceuticals and Endocrine Disrupting Compounds in U.S. Drinking Water. *Environmental Science & Technology* 43(3): 597–603.

Centers for Disease Control and Prevention. 2015. *Therapeutic Drug Use.* http://www.cdc.gov/nchs/fastats/drug-use-therapeutic.htm. Accessed 23 May 2015.

McCabe, S.E., J.R. Knight, C.J. Teter, and H. Wechsler. 2005. Non-medical Use of Prescription Stimulants among US College Students: Prevalence and Correlates from a National Survey. *Addiction* 100(1): 96–106.

Part I
Development

Patient-Driven Drug Development

Jessica Flanigan

Autonomy is a foundational principle of medical ethics. Patients are entitled to make their own treatment decisions, even if they don't make the medically advisable choice. Patients are also entitled to know their options. Health workers must inform patients of all available treatments that a reasonable person would consider. Patients also have the right to seek out approved alternative therapies as long as they can find a willing provider. Even when all existing treatments are disclosed and made available, though, some patients have very few options for fighting their diseases. These patients encounter the limits of medical autonomy—one can have the freedom to choose and refuse treatment but that freedom is of little value when effective treatments do not exist or are unknown.

For example, patients with rare diseases may not have effective treatment options because manufacturers do not typically invest in therapies that treat relatively small patient populations. Even when existing therapies can be used for rare-disease treatment, physicians may not be equipped to inform patients about off-label treatment options because they are not familiar with the latest research about their patient's rare condition. Patients who suffer from emerging diseases face similar hurdles. When patients' medical needs outpace the scientific community's medical knowledge, medical autonomy really amounts to the right to wait for the scientific community to develop an effective treatment.

In this essay, I discuss an emerging approach to drug development that is forged in rare-disease communities and patient advocacy networks. Drug development has historically advanced within the context of research institutions, such as universities and pharmaceutical companies. Within and outside of this framework, some patients are now taking greater control of their treatment options even when few options exist—not only as clinical trial participants but also as researchers. These citizen-scientists refuse to wait, even when there are few available treatments.

J. Flanigan (✉)
The Jepson School of Leadership Studies, The University of Richmond, Richmond, VA, USA
e-mail: jflaniga@richmond.edu

© Springer Science+Business Media B.V. 2017

D. Ho (ed.), *Philosophical Issues in Pharmaceutics*, Philosophy and Medicine 122,
DOI 10.1007/978-94-024-0979-6_2

Yet the patient-driven approach to drug development is controversial. Patients are not medical experts, so when citizen-scientists with rare diseases publicize information about untested and experimental treatments to rare-disease communities, their example could cause some people to harm themselves. Patient-driven drug development could also compromise the clinical trial system, thereby making it harder for other patients to know if a treatment is working. Another concern is that citizen-scientists will threaten existing pathways of drug development if patients' adverse reactions to experimental treatments ultimately discourage manufacturers from developing the therapy. On the other hand, patient-driven drug development could also divert resources from more promising therapies if resources for medical resources are very limited. Or patient-driven development could slow progress toward effective treatment insofar as including laypeople's priorities or judgments makes medical research less efficient. Citizen-scientists are not impartial. They mix research with advocacy in ways that undermine their credibility with academics. Outside rare-disease communities, some people wonder whether wealthy patients should set our priorities in medicine, when research funding would otherwise help a greater number of patients (Salmon 2014).

Despite these concerns, I argue that patient-driven development is best understood as a legitimate method of drug discovery and medical treatment that should be valued as a complement to the conventional development pipeline. The main reason in favor of including patients as partners in the development process is that patients are entitled to pursue treatment options. Including patients in the development process can also promote drug discovery and help manufacturers and physicians better understand emerging therapies.

I first explain some of the policies and practices that currently limit patients' treatment options. The current development pathway is often harmful and should be reformed to give patients greater access to experimental drugs, but I will not argue for that claim in this essay. Rather, I am concerned with the ethics of some patients' response to the current system, when they become partners in research and development. Next, I describe some forms of patient-driven development. I then argue that patient-driven development is morally permissible, even if it causes some patients to make risky treatment decisions. I also respond to concerns that patient-driven development would threaten drug development on balance and I make the case for patient-driven drug development, despite concerns about political bias and inequality.

Choosing Treatment, Creating Treatment

Governments in most developed countries require that drugs receive regulatory approval before they can be sold to patients. This process takes years, even for drugs that have been given expedited approval. The approval process is also extraordinarily expensive (DiMasi et al. 2003). It is risky for pharmaceutical companies and researchers to invest in drug development in light of the time and money required, because some drugs may be denied approval midway through the process. Approval requirements therefore discourage pharmaceutical innovation—a phenomenon known as drug loss.

Though it is difficult to estimate how many drugs would have been developed in the twentieth century in the absence of expensive approval policies, there is extensive evidence that costly requirements do discourage innovation. After the USA passed efficacy testing requirements, the average number of new drug applications per year significantly declined (Peltzman 1973; Grabowski and Vernon 1983). Subsequent studies confirmed that approval requirements had an adverse effect on research productivity (Wiggins 1981). International comparisons find that research productivity declines more in countries with higher barriers to approval (Grabowski et al. 1978). More recent economic analyses confirm that regulatory stringency is correlated with fewer drug discoveries (Jensen 1987; Scherer 2000).

Drug loss is particularly harmful to patients with rare diseases because researchers and manufacturers are especially unlikely to make risky investments in drugs that treat only a small group of patients. In 1983, the US Congress passed the Orphan Drug Act (ODA) to encourage drug development for rare conditions. The ODA established tax incentives and patent protections for drugs that treated rare diseases, and it effectively encouraged drug development relative to the previous system (Lichtenberg and Waldfogel 2008). Specifically, the ODA caused "a significant and sustained increase in new trials" for relatively prevalent rare diseases and increased the stock of drugs for less prevalent rare diseases (Yin 2008, 1060). Today, approximately a quarter of new drug are approved as part of the orphan drug program (Kesselheim and Darrow 2015).Yet drug development for rare and emerging diseases is still risky, and whatever limited market incentives exist for the treatment of rare diseases are offset by the risk of investing in a drug that could be denied approval. Public officials should change existing approval requirements to further lower the cost of drug development and raise the number of treatments available to all patients—especially patients with rare and emerging diseases. Policy changes like the ODA are steps in the right direction. The success of the ODA and other reforms that expedited the approval process and reduced development costs have shown that deregulation can encourage development without threatening patient safety.

Another advantage of shorter and less costly approval requirements is that fewer patients would die waiting for access to drugs. The time between initial development and approval is called the "drug lag." This delay causes patients to suffer and die waiting for drugs that exist but are not yet approved. Because of drug lag, approval requirements stand in the way of people accessing the means to save their own lives. Patients with rare diseases are also likely to be affected by drug lag because clinical trials are generally designed to treat the largest patient population, where the market for the drug will be strongest, so rare-disease patients who could benefit from access to experimental drugs cannot access treatment in the context of a trial. Most states have compassionate access provisions to facilitate access for patients with an especially urgent need for experimental treatment, but compassionate-use provisions do not meet the full demand for experimental drugs.

The current development pathway threatens rare-disease patients' health by contributing to drug lag and drug loss. It also violates patients' rights to make treatment decisions by barring access to experimental and unapproved drugs (Volokh 2007). The development pathway restricts access to drugs based on regulators' judgments about whether a drug is acceptably risky. But regulators are not experts on whether using a drug is "worth it" for a patient. Whether a drug is unacceptably risky or worth taking is a normative judgment that rare-disease patients are qualified to make for themselves, not a judgment for a distant regulator to make for an entire patient population. In this way, pharmaceutical regulations that are based on a public official's judgments of "acceptable risk" are fundamentally paternalistic.

Patients' rights advocates have recently advocated for further political reforms in light of the injustices and harms that are inflicted by the current system of drug development, including "right to try" laws and calls for a safety-only standard. Other patient groups work to facilitate greater access across regulatory systems for patients with rare and terminal diseases. For example, myTomorrows is an online patient platform that helps patients find drugs in development through a network of researchers and manufacturers. The platform connects patients to people who can provide experimental drugs, and it helps patients navigate compassionate access provisions and clinical trial opportunities across regulatory systems and borders. The company charges a commission from drug developers who provide development-stage treatments to myTomorrows patients. In some cases, manufacturers also offer royalties to myTomorrows for its role in providing valuable observational data about the effects of new drugs.

Historically, patients and researchers also protested regulatory barriers to access and facilitated illegal access to experimental drugs. For example, in 1988 police arrested at least 120 AIDS activists who lead a protest that shut down the FDA headquarters to draw attention to approval delays. AIDS activists also formed "buyers clubs" that distributed unapproved therapies to AIDS patients outside of clinical trials.

Political reform may ultimately be the most promising solution for patients with rare and emerging diseases who seek greater access to drugs, but it is not the only solution. Where regulatory burdens currently discourage drug development and access to experimental drugs, some patients are also playing a more active role in drug development. This phenomenon, which I call patient-driven drug development, consists in patients conducting and overseeing drug research, patients financing drug development, and patients participating in existing research networks despite their lack of formal training. Through the clinical trial system, many patients have participated in drug research. Patient-driven development differs from the trial system, however, in that patients not only participate in development as research subjects but also initiate and interpret research.

Patient-Driven Development

Patient-driven drug development refers to cases where patients and patient advocacy groups research potential treatment options and try new therapies outside the conventional drug development pipeline. Citizen-scientists may cooperate with researchers or they may act outside of the traditional research community. They may focus specifically on their own care or attempt to benefit a broader group of patients. Some citizen-scientists shape drug development by deciding to finance or invest in certain areas of research, whereas others play a more active role in conducting research. Some examples will help to illustrate the nature of citizen science.

Chris and Hugh Hempel exemplify the citizen-scientist approach. The Hempels' twin daughters, Addison and Cassidy, suffer from Niemann-Pick disease, type C (NPC), a rare and fatal progressive disease that causes degeneration of the nervous system and dementia. Most children with NPC die before they reach adulthood. Patients require considerable care as they lose the ability to walk, talk, or feed themselves. When Chris Hempel's children were diagnosed, she began researching all potential therapies for the treatment-resistant disease. She discovered very limited evidence that cyclodextrin, a substance that is used to dissolve other drugs, extended the lives of mice with NPC. Hempel ordered cyclodextrin and tested the substance on herself for a few weeks before giving it to her twin daughters. Yet cyclodextrin required intravenous infusion in order to effectively treat NPC, so the Hempels petitioned the FDA and were granted permission to use the unapproved drug to treat their daughters' rare disease.

Cyclodextrin did not cure the Hempel twins, but they did seem to improve after treatment and it seemed to slow the progression of their disease. Of course, the Hempels cannot be sure of the effects because the twins used many other drugs while they were using cyclodextrin, and their progress was not monitored in comparison to a control group that received the standard treatments. Nevertheless, Chris Hempel's discovery of cyclodextrin as a potential treatment for NPC did prompt physicians and researchers at the National Institutes of Health (NIH) to fast-track Phase I clinical trials for the drug.

Social media can help some patients access drugs that they would not have otherwise received. In 2014, a pediatric patient named Josh Hardy suffered from a life-threatening adenovirus infection. Hardy's physicians at St. Jude Hospital requested access to brincidofovir, an experimental drug that was being tested as a treatment for a different virus in immunocompromised adults, but which showed promise as a treatment for adenovirus as well. Yet the manufacturer, Chimerix, was initially reluctant to provide the drug. Only after an extensive social media campaign on Hardy's behalf did the company establish a new clinical trial for adenovirus that included Hardy as the first patient (Morrow 2015). Hardy recovered from the infection.

Similarly, David deBronkart's experience as a citizen-scientist illustrates how patients can use social media to influence drug development after approval as well. After being diagnosed with stage IV renal cell carcinoma, deBronkart was actively involved in ACOR, an online community for patients with cancer. Through ACOR, deBronkart learned the latest information about the state of the art for treatment, which was not easily accessible through peer-reviewed publications. ACOR's knowledge of renal cell carcinoma was continuously updated to reflect the latest medical knowledge. He attributes the success of his treatment to the fact that he incorporated this community knowledge into his own treatment plan. He has since become an advocate for including patients in treatment and providing people with greater access to their medical records. He recently wrote in the BMJ, "The value delivered by skilled clinicians is still there, but now we can see that it's no longer the only source [of valuable knowledge]—and sometimes it's not even the best" (deBronkart 2013: 346).

Patient-driven drug development can also consist in financing research. Venture philanthropy refers to a system where patients finance early-stage research for drugs that treat their own diseases (Marcus 2010). Some patients even purchase places in clinical trials for a chance at receiving an effective treatment (Masters 2015). The price of a place in a glioblastoma trial is $2 million. In other cases, patient advocacy groups profited from early investment in treatments. For example, the Cystic Fibrosis Foundation recently received a $3.3 billion payout, which was a financial return for its early investment in a drug that was subsequently approved by the FDA as a treatment for the lung disease.

Citizen-scientists also form clinical trial networks, establish tissue and specimen banks, and create patient registries (Frydman 2009). Though they acknowledge the limits of their approach, citizen-scientists are reluctant to wait for drug approval when a potentially effective treatment is available off-label, in the context of a clinical trial or through a compassionate-use program. Implicit in the patient-driven approach to research is the assumption that existing regulatory frameworks are inadequate for addressing patients' needs. Instead, patients are opting out of the conventional development process and taking drug development into their own hands.

The citizen-scientist movement began in the 1980s and 1990s at the start of the AIDS epidemic. At the time, scientists did not know a lot about AIDS, and patients were dying from a lack of available therapies. When experimental drugs and off-label therapies were suggested as potential treatments, patients formed buyers clubs that imported and distributed unapproved therapies to members. At the time, critics worried that buyers clubs would undermine the testing process and threaten approval for promising drugs. However, some AIDS activists ultimately established themselves as credible members of the research community, and scientists came to rely on activists for recruiting trial participants and educating patients. For example, Mark Harrington forged strong ties with the research community as an AIDS activist and eventually became a published AIDS researcher despite having no scientific training (Epstein 1995). The lessons of the AIDS movement are that citizen science is not incompatible with science and that patient communities may recognize the importance of testing drugs while rejecting the validity of the approval process.

When patients become citizen-scientists, when they develop their own therapies and form buyers clubs and scientific communities, it may seem that they are opting out of the regulatory framework entirely. Yet self-medication does not consist in a rejection of scientific authority. Patients like the Hempels, Josh Hardy, Dave deBronkart, Mark Harrington, and anyone who purchases spots in clinical trials all recognize the importance of scientific expertise and willingly share their experiences with researchers and other patients so that it may be put to greater use. Patients who take their medical choices in their own hands are extremely interested in scientific advances concerning their diseases—they do not deny the importance of medical knowledge. But citizen-scientists reject the scientific community's insistence that drugs be withheld until there is evidence that the risks outweigh the benefits. Activists and members of buyers clubs reject policies that prevent them from accessing potentially effective treatment when all other treatments have failed.

The Risks of Citizen Science

Patient-driven drug development gives patients more control over their treatment options than they would otherwise have. Yet it is not uncontroversial that patients should have more control. When patients play a role in development, they expose themselves (or their children) to the risks associated with using an unapproved drug. Patients are not medical experts. Though their experiences with their own diseases and the collective knowledge of disease communities are valuable forms of medical data, patients still do not generally have medical degrees or PhDs in the relevant disciplines. Patients also are not held to the professional codes, regulations, and heightened standards of liability that prevent medical professionals from taking undue risks or exposing people to unnecessary harm. Historically, these codes and regulations evolved to protect patients from being deceived or exploited by untrained medical practitioners. A potential risk of legitimizing patient-driven development is that it would more generally legitimize laypersons role in medicine to the detriment of patients.

In this section I argue that the risks of patient-driven development cannot justify restrictions on patients who seek a greater role in their treatment. Still, patients should consider the risks of participating in the development process when they make treatment choices.

Four kinds of risks merit consideration. First, patients take risks when they access experimental medicines outside the context of the clinical trial. In these cases, I argue that adult patients are generally in the best position to decide whether the risks of an experimental drug are worth taking. Pediatricians, parents, and patients should collaboratively make treatment choices in the interest of pediatric patients, but in some cases, experimental treatment is a child's interests. Second, patients may put other patients at risk insofar as patient advocates threaten the conventional development pipeline. Yet patient advocates are not obligated to restrain their efforts on the grounds that regulators and manufacturers may impermissibly slow or halt development in response to a patient's reaction to an experimental drug. Third, patients may put other patients at risk by publicizing dangerous or misleading information. Patients who play a role in drug development are obligated to disclose their level of expertise, and if they do, it is not impermissible to encourage other patients to participate in the development process. The fourth risk, which I address in the next section, is that patients could undermine the clinical trial process by accessing experimental treatments.

Consider first the concern that adult patients could harm themselves by making dangerous choices to use experimental drugs. Even if some drug choices are dangerous, the judgment about whether a drug is acceptably risky is one that each patient is uniquely qualified to make. Though physicians, researchers, manufacturers, and regulators all have expert knowledge about the pharmacological properties of a drug, each patient is an expert about whether the medical risks are worth it, given his or her other values and overall prognosis. Just as patients are entitled to refuse treatment against a medical expert's advice, so too do patients have rights against interference when they choose to try a risky treatment, even if medical experts recommend against it. When public officials prohibit patients from accessing treatment, they interfere with patients' rights to make treatment decisions (Flanigan 2012). The general principle of respect for medical autonomy also supports deferring to patients who choose to use experimental drugs, even if that choice proves harmful to the patient.

Unlike competent adults, pediatric patients are not entitled to make their own treatment decisions. Rather, treatment choices for children should be made in the interest of the child. The American Academy of Pediatrics recommends that pediatricians seek assent from patients who are not capable of giving consent (e.g., children under age 14) and permission from their parents (Committee on Bioethics 1995). This means that pediatricians and parents of children with rare diseases should work together to make treatment choices in the interest of the child. These standards are not in tension with patient-driven drug development. Josh Hardy's medical team supported his petition for access to brincidofovir. The Hempels worked with their pediatrician, Dr. Caroline Hastings, to administer experimental cyclodextrin to the twins through a catheter (Marcus 2015b; Thomas 2014).

Still, experimental drugs are more risky than standard treatments because researchers do not know what they will do. It is not always clear to parents or physicians if an experimental drug is in a patient's interest. Some parents of children with NPC criticized the Hempels' decision to provide the experimental drug on the grounds that they were taking unnecessary risks with their children (Marcus 2015a, b). One of the twins, Cassi, suffered complications related to the catheter implantation, even though it is typically a low-risk procedure. Cassi developed a blood clot at the site of the catheter that paralyzed the left side of her body.

Yet concerns about the risks of experimental therapy are not unique to patient-driven development. Pediatric patients in clinical trials face similar risks. Two other NPC patients who were enrolled in an NIH clinical trial of cyclodextrin also suffered complications from their catheter insertion. Even when the treatment does not encounter complications, there are risks associated with the new drug. Recent trials found that cyclodextrin infusion causes deafness for NPC patients, though it may also extend their lives (Marcus 2015a). Despite these significant risks, though, parents, patients, and pediatricians may still judge that the risks of failing to treat a progressive and fatal disease are greater. Policies that prohibit patients from making treatment decisions bear an even higher justificatory burden than other forms of paternalism because people's interests in their health and treatment choices are especially strong.

Critics of patient-driven development also object that activist patients could undermine drug development and thereby harm other patients. They point out that patients are not the only ones who bear the risks of using experimental therapies or publicizing information about breakthrough treatments. In some circumstances, patient-driven development could slow or halt drug development. Parents in the NPC community worried that patients who used cyclodextrin outside the context of a clinical trial were legally required to report any adverse reactions, which could have caused the FDA to prohibit all further testing or to ban anyone with NPC from using the drug ever again. (Marcus 2015b).

These concerns are not unwarranted. In 2014 the FDA ordered the drug manufacturer CytRx to halt Phase II clinical trials of an experimental cancer drug after a patient died while using the drug outside the context of a clinical trial. Though CytRx later modified the trials and the hold was lifted, the incident illustrates how patients can disrupt the traditional development pathway by gaining compassionate access to an experimental drug (Carroll 2015). If so, patient-driven development may harm other patients by depriving them of the opportunity to use a promising drug that would have been developed were it not for a compassionate user's adverse reaction.

Bioethicist Art Caplan expressed related concerns about creating a trial to accommodate Josh Hardy's case. Caplan described the balance between promoting a drug's approval chances and providing expanded access as "a tradeoff between the public good and self-interest" (Cha 2014). However, Caplan's characterization assumes a false dichotomy. Public officials and researchers can accommodate patient-driven drug development alongside the traditional pipeline, though they may need to reform the current system to do so. Insofar as expanded access to experimental drugs threatens development, it is because (a) regulatory agencies will be reluctant to approve a drug if there were adverse effects outside the clinical trials, (b) manufacturers are held liable for adverse effects outside of clinical trials, (c) manufacturers will suffer reputational harms if they provide experimental drugs and patients have adverse reactions, or (d) refusing to participate in a clinical trial would compromise existing trials (Darrow et al. 2015).

The first threat to development, which regulators would halt development based on the experience of patients who access a drug outside of a trial, assumes that regulators do not trust their approval requirements. If the current approval system is justified on the grounds that clinical trial data is sufficient for establishing safety and efficacy, then regulatory agencies should not be reluctant to approve a drug based on patients' experiences outside the context of a clinical trial. If data obtained outside the context of a clinical trial were grounds for halting a trial, then regulators should also rethink whether the existing trial system is sufficient to establish safety and efficacy.

The second threat to development is liability. Despite the recent passage of "right to try" legislation, most US states do not limit tort liability for manufacturers who grant patient's requests for unapproved drugs. Manufacturers cannot require releases of liability as a condition for providing a patient with a drug under an expanded access program. Even if a manufacturer tried to secure a release of liability, courts might not uphold it if they judge that a patient who released the manufacturer from liability was coerced to do so out of desperation. In these cases, patient-driven drug development does not directly threaten clinical trials, but it could undermine development indirectly, if a patient's adverse reaction requires that the manufacturer pay a costly settlement.

Manufacturers also expose themselves to potential reputational harms if they allow patients to access experimental drugs during the development phase. Bad publicity from a compassionate user's adverse reaction could shake investor confidence or undermine the drug's commercial prospects. Manufacturers are not required to support patients' calls for greater participation in the drug development process, just as they are not more generally required to provide drugs to all patients who could benefit (Huebner 2013). On the other hand, as Josh Hardy's example illustrated, manufacturers also encounter extremely bad publicity when they exclude patients from the development process by withholding access to experimental drugs.

Third, patient-driven development may lead some people to take undue risks and harm themselves. The online forums that are frequented by cancer patients like Dave deBronkart cite data and anecdotes that are not subject to rigorous peer review. Chris Hempel collected data from other parents whose children were using experimental cyclodextrin and posted it online. Though members of online communities generally disclose their lack of scientific expertise and the limits of the evidence, other patients may defer to unsubstantiated reports out of desperation. There are no guarantees that citizen-scientists will take appropriate measures to respect people's privacy or to avoid creating unnecessary risks when they share and analyze medical data. Though patient-driven development should abide by ethical standards such as respect for privacy and standards of truthfulness, there are few mechanisms to enforce these standards on online forums.

Whether it is permissible for citizen-scientists to put others at risk by sharing information depends on the kind of information they share. Surely it is permissible for a single patient to discuss her experiences on an online forum. And when members of patient forums aggregate data and anecdotes alongside peer-reviewed studies, as long as they do not deceive readers about the credibility of each source, they do not wrongfully mislead anyone. Patients who are concerned about the quality of online information can consult their physicians or disregard whatever they find. Sharing truthful information about one's own experience or providing access to information that is provided elsewhere is permissible. In these cases, as long as members of online communities share information in good faith and disclose their lack of expertise, they do not act impermissibly.

One may object to this argument on the grounds that laypeople are not capable of interpreting complex medical data, especially if they are in a vulnerable position (Darrow et al. 2015). Yet physicians, policy-makers, and researchers ought to presume that laypeople are capable of understanding information about their treatment options. Laypeople are required to interpret and educate themselves about complex medical data in order to give informed consent to approved therapies or to participate in clinical trials. Since there is little reason to think their capacity to interpret and understand information would be adequate in these circumstances but not in others, laypeople should be treated as if they are capable of understanding. Even if laypeople are not capable of understanding complex medical information and statistics, that lack of understanding may be a kind of learned helplessness that is caused by a lack of autonomy over their treatment options. Moreover, as Mark Harrington's story illustrates, patients with rare and emerging diseases may be relatively more educated than nonspecialist physicians and other kinds of patients because they develop specialized knowledge that others have little incentive to acquire.

On the other hand, sharing medical information online is impermissible when it is deceptive or when sharing violates privacy rights. It is generally wrong to deceive people about potential treatments. It is especially immoral to mislead people who are vulnerable, as desperate and dying patients may be. It is also wrong to share personal information about someone without his or her consent, unless she was not entitled to control that information in the first place. Without delving into the justification for privacy rights here, whatever standards and privacy protections ought to inform physicians' and parents' disclosure of information about patients should also serve as guidelines for citizen-scientists who post information online. This may require that they remove identifying information about some cases or that patient communities refrain from publicizing information about individual genomes.

Can Drug Development Make Room for Patients?

In addition to concerns about adverse reactions compromising support for clinical trials, patient-driven drug development can also potentially undermine the clinical trial system by making it harder for manufacturers to recruit trial participants. Regulatory agencies require evidence from clinical trials before they approve a drug, so manufacturers rely on willing patients to participate in trials. Patients-driven drug development could threaten development on balance if patients access drugs outside of the traditional pipeline and independently develop and use off-label therapies.

Patient advocates and government officials cited this concern as a justification for preventing patients from accessing drugs outside of the context of a clinical trial. In 2001, 21-year-old Abigail Burroughs died while waiting for access to an experimental drug that was subsequently approved to treat her form of cancer. That year, her family formed the Abigail Alliance, which sued the FDA for better access to unapproved drugs. Yet an appellate court ultimately ruled in 2007 that terminally ill patients were not entitled to access unapproved therapies, partly on the grounds that expanded access would undermine the clinical trial system.

Prominent officials echoed this concern. William Schulz, a former deputy commissioner of the FDA, submitted a brief on behalf of the National Organization of Rare Diseases that argued that greater access to experimental drugs would undermine the entire approval process (Pollack 2007). When the Abigail Alliance was denied a final appeal in 2008, the FDA expressed concerns that even compassionate access could undermine clinical trials (Greenhouse 2008). These concerns about clinical trial participation are especially salient for people with rare diseases because each trial participant substantially increases the statistical power of a trial, which means that each participant makes it more likely that the trial will find statistically significant effects for an experimental treatment. In this way, opting out of a clinical trial could threaten a drug's chances for subsequent approval, thereby harming the prospects of everyone else with that disease.

However, this concern about patient involvement in drug development assumes that the rest of the development process must stay the same while patients gain expanded access to experimental therapies. Regulatory agencies could change their standards for approval. Or, given the current standards of approval, patient-driven development could compliment and improve upon the existing research and development pipeline. Legislators, regulators, and manufacturers could consider offering other incentives for patients to participate in clinical trials. Already, patients continue to participate in clinical trials for drugs they could otherwise access off-label because clinical trials often offer high-quality affordable treatment to patients who are reasonably uncertain about whether the standard of care or experimental treatment is better.

There are also other ways to measure whether a drug is effective outside the context of a clinical trial, such as well-designed observational studies. Though randomized clinical trials are the gold standard for drug research, they needn't be the only standard. Even if observational studies are not as effective as clinical trials at estimating the effects of a treatment, they can be more effective at identifying adverse events because they include more participants. Moreover, though clinical trials are designed to eliminate selection bias, they face two competing risks in their selection criteria. On one hand, if a trial is designed for a very narrow patient population, researchers may miss an opportunity to learn about the potential uses of a drug for different patient types. If a drug fails to establish efficacy for its patient population, researchers may abandon what would be a promising treatment for another patient group. On the other hand, if a trial is designed for a very broad patient population, what is sometimes referred to as an "all comers trial," the trial may not be designed in a way that can establish the efficacy of a drug even if the drug does effectively treat a subset of the trial participants.

I do not mean to suggest that manufacturers should abandon clinical trials nor to discredit their usefulness by acknowledging their limits. However, observational approaches can potentially improve upon clinical trials alone by increasing the amount of useful data about a drug. Were patients further empowered to access experimental drugs outside of clinical trials, researchers could also collect observational data about the risks and benefits of an experimental drug for different patient types and in typical use conditions. Additionally, if the possibility of patients opting out of the clinical trial system is truly of concern to legislators and regulators, then that concern may indicate that existing trials do not meet standards of clinical equipoise. Currently, researchers are encouraged to design trials so that they are genuinely uncertain whether the treatment is better or worse than the standard of care. If patients would overwhelmingly opt for treatment over the standard of care if given the choice to access the treatment outside of a trial, then those patient choices should indicate that a trial did not meet standards of equipoise.

For patients with rare diseases, observational studies are also valuable because it is more difficult to conduct a well-designed clinical trial with a small patient population (Thadhani 2006). As long as randomized controlled trials are required for approval, though, patients with rare diseases could compromise orphan drug development by opting out of the clinical trial system. Returning to the example of NPC, the Hempels' decision to research and treat their daughters outside of the context of a scientific study was controversial in part because researchers struggle to recruit enough participants for randomized controlled trials when a rare disease like NPC effects only a few hundred families, and not all will qualify. In these cases each qualifying participant is incredibly valuable to the overall community of NPC patients because his or her experience can establish grounds for FDA approval. Trial participants can also help all rare-disease patients learn whether a drug's benefits are statistically significant or produced by chance, whereas patients who use experimental drugs outside of a trial cannot.

In these cases, the moral considerations against threatening the statistical power of a clinical trial do weigh against allowing patients to use experimental drugs that are still in development. Therefore, Caplan's concern about trade-offs between individual interests and public goods is more apt for patient populations with less prevalent rare diseases. Yet even in these circumstances, I am skeptical that citizen-scientists are obligated to rare-disease communities simply because they too suffer from a rare disease. Patients are under no obligation to endanger their own lives or their children's lives so that their experience can generate better data for subsequent patients.

For each patient who opts out of the approval process and potentially compromises the clinical trial system, there are other patients like Abigail who are put at risk by the clinical trial system. Patients who do not qualify for participation in trials are denied access to experimental drugs during drug development even if they could benefit. Or, even if a patient does qualify to participate in a trial, researchers might require that he receive the standard of care even when he has not responded to existing therapies. For example, Gideon Sofer suffered from severe Crohn's disease. In 2007 he enrolled in a clinical trial of an adult stem-cell therapy that aimed to regenerate damaged intestinal tissue. All Phase II trial participants who received the therapy demonstrated improvement, but the FDA required double-blind randomized Phase III testing that included a treatment group, a group at a lower dosage, and a control group that received the standard of care but no adult stem-cell therapy. Sofer enrolled in the Phase III trial but seemingly received only the standard of care because his condition deteriorated. After his participation in the trial ended, he still could not access the treatment because the drug had not yet been approved. Sofer died from complications relating to his disease in 2011.

Sofer's experience as a clinical trial participant illustrates a potential hazard of limiting the patient's role in drug development. Reflecting on his experience, he argued, "If trial patients are treated like lab rats, they won't feel obliged to cooperate unconditionally and report accurate data" (Sofer 2008). By giving patients a greater role in drug development, they may cooperate with researchers and provide better information about experimental drugs. More generally, patient-driven development may also spark more support for clinical trials, if only because patients learn about opportunities for trial participation through patient networks. Patient-driven development also generates support and funding for additional trials and encourages people who might otherwise have opted for the standard of care to enroll in a trial. These are empirical questions, but the burden of proof should rest with those who oppose expanding patient's choices to establish that allowing additional pathways to developmental drugs would cause more harm than good.

A related objection to patient-driven drug development is that patients who petition for expanded access or pursue off-label treatments could divert resources from clinical trials and expert-driven drug development. And in these cases, a similar reply is warranted. Patients do not have special obligations to generate high-quality medical data simply because they are well-placed to do so. If a patient judges that an expert-recommended experimental treatment isn't the right choice for him, then he may refuse it in favor of another experimental treatment or an off-label treatment.

Clinical trials are governed by standards of informed consent. Patients must be told the risks and benefits of participation in a trial rather than receiving the standard of care, and patients always have the option to choose the standard of care over participation. It is always permissible for patients to do so, despite the potential benefits of trial participation to the broader patient community. So given that patients are not obligated to participate in a trial when their only other choice is the standard treatment, why would it be impermissible for patients to decline to participate in a trial in favor of an experimental treatment? A patient's reasons for refusing or choosing to make a risky medical choice do not bear on the permissibility of that choice, so it is not wrong for a person to opt out of the clinical trial system in order to use an experimental drug as part of a compassionate-use program. Therefore, patients who are desperate for experimental treatments should be permitted to use them even if they qualified for clinical trials.

The Politics of Patient-Driven Development

Consider one more story of patient-driven drug development. The English poet Sarah Broom was diagnosed with lung cancer at age 35, and her doctors predicted she had only a few months to live. Desperate for a treatment, she contacted everyone she knew. Eventually, Sarah was introduced to a leading cancer researcher who informed her that her cancer had a mutation that had only been discovered in the past year (Sanghavi 2013). Broom quickly enrolled in a trial for crizotinib, an experimental drug that might have treated her mutation. The drug seemingly gave Sarah two years of normal life before the cancer returned. She then petitioned Novartis for access to an experimental compound, and she moved from New Zealand to Boston to join the trial. When that compound stopped working, she petitioned Novartis for compassionate access to another compound, and after she was initially denied, she eventually gained access to that drug as well. Sarah lived four years past her initial prognosis. Her story is unusual because her ability to access promising experimental treatments relied on multiple advantages, including money, connections, and an educational background that many patients lack.

Not all patients have the resources to participate in the development process as Sarah Broom did. The high cost of patient-driven drug development has raised concerns about political bias and inequality. Should the patients with the loudest voices gain access first? Should the deepest pockets drive drug development? Oftentimes, patient-driven drug development channels resources toward marketable and sympathetic patients, such as children and people with genetic diseases, even if other diseases are more serious or prevalent. Should drug development be held to standards of effective altruism?

It may strike egalitarians as unjust that multiply advantaged patients can access promising experimental drugs before others. This form of inequality is partly due to the case-by-case nature of compassionate access, which favors patients who are antecedently empowered to petition for an experimental drug. Expanded access systems require information, connections, and funds. Patient advocacy networks provide these resources to an extent, but patients who are multiply advantaged can deploy them more easily. The most extreme example is the sale of spots in clinical trials, which amounts to an explicit preference for treating patients on the basis of their economic status (Masters 2015).

Egalitarian challenges to the distribution of healthcare are not unique to patient-driven drug development. In all healthcare systems, multiply advantaged people have more and better treatment options than others. Attempts to limit health inequality risk "leveling down" some people's access to healthcare for the sake of equal treatment, when the egalitarian response should be empower disadvantaged patients to access the same treatment options. Limits on patient-driven development for the sake of equality could also risk limiting markets in healthcare at the expense of all patients (Daniels 2001). Furthermore, researchers and patients are uncertain whether experimental therapies are helpful—they may not be in some cases. So even if patients with advantages do enjoy more options, these options may not translate to better health. Wealthy patients create markets for experimental drugs that otherwise would not progress in the development pipeline at all. For these reasons, it would be harmful to all patients to limit patient-driven development out of concerns about unequal access.

A related objection to patient-driven development is that it is inefficient because it diverts resources from those who need it most (Salmon 2014). However, people who advocate for better access to drugs and those who finance drug development for disease-specific charities are not obligated to allocate resources as efficiently as possible. Even if patient-driven drug development benefits only members of a small patient community, patient advocates and philanthropists are not morally required to be as efficient in their donations as possible. Though it may be *more* praiseworthy to support general medical research or research for conditions that effect a larger population, it is still praiseworthy for patients to promote research that cures diseases.

Fears about inefficiency also assume that there is a fixed pool of resources for medical research and that the benefits of disease-specific drug development will only accrue to members of that disease community. But some people who donate to disease-specific charities would not otherwise donate at all. Patient-driven approaches to development mobilize networks to generate funding that was inaccessible to researchers beforehand, such as the iCancer fundraising efforts that restarted a study of an experimental cancer treatment when researchers were unable to find funding through public or industry channels ("About | iCancer" 2015). Moreover, research in one disease area can improve physicians' and patients' understanding of other diseases as well. In this way, disease-specific, patient-driven drug development can generate research with multiple applications.

Commentators also criticize venture philanthropy on the grounds that it contributes to economic injustice in healthcare. Disease-specific charities such as the Cystic Fibrosis Foundation develop conflicting goals when they invest in new treatments because they are also charged with advocating for patients. Their goals therefore compete when foundations have an incentive to support very high drug prices to secure a return on their investment, while they are also directed to support very low drug prices so that patients can access affordable treatment. Nonetheless, nonprofits that advocate on behalf of patients in general must weigh these competing goals, but striking a proper balance is not impossible. Nonprofit hospitals, for instance, must decide how much care to provide for free and how much to charge, so that the institution remains financially sustainable. Nonprofit blood-product providers also have an interest in the price of blood products being high, but they also aim to ensure that everyone can use blood products regardless of cost. Disease-specific nonprofits face similar challenges in weighing their goals, but they can still reconcile investment and fiscal sustainability with patient advocacy. In this respect, it is not obvious that patient-driven developers face unique moral and institutional conflicts of interest that other nonprofit healthcare organizations have not successfully reconciled.

Another objection to venture philanthropy is that it is "essentially, another term for privatizing scientific research. Instead of decisions about the fate of scientific funding being made by publicly oriented institutions, those decisions are being put in the hands of anonymous philanthropists and ostensibly benevolent nonprofits" (Hinkes-jones 2015: A23). This objection assumes that publicly oriented institutions ought to make the bulk of the decisions about funding for disease research, even though public institutions to date have not provided sufficient funding for medical research. One explanation for the inadequacy of public funding for medical research is that there are few incentives to publicly fund research on cures for specific diseases. Public officials are accountable to large heterogeneous populations, and the average citizen is more concerned with issues like drug prices and reimbursement than potential cures for diseases they are unlikely to get.

I agree that public officials should allocate more funding to disease research. But in the absence of public funding, private philanthropists should not be barred from drug development. Philanthropists and nonprofits are better placed to allocate resources toward potential cures because they are accountable to specific patient communities. This accountability creates problems of equity, as I discussed, but it also creates incentives to develop lifesaving drugs for their patients. Consider an analogy to education. In the absence of adequate public schools, it would be impermissible to ban private schooling or to forbid parents from reading to their children. Support for private investment in some children's education does not imply that public institutions are not obligated to ensure that every child has an adequate education. But even if states did give every child an adequate education, children and communities could still benefit from private investment in education in addition to what is publically provided. Whether this argument for private investment in education succeeds, the case for private investment in health is even stronger because, unlike education, health is not a positional good, meaning that one person's improved health does not reduce the value of health for others.

Conclusion

Patient-driven drug development is part of a larger trend in medicine toward greater patient involvement in medical treatment (Dwamena et al. 2012). Technology is giving patients ever more opportunities to collect data, collaborate, and measure the effectiveness of treatment alongside researchers and physicians (Topol 2015). Physicians and researchers might resist this trend out of a fear that they will be inundated with time-consuming requests from patients who are not medical experts, as more patients demand a role in the development of new therapies. The worry that patient-driven development would increase the workload of physicians and researchers might be legitimate. Nonetheless, just as physicians required longer conversations with patients after the introduction of informed consent requirements, the moral and therapeutic gain might warrant inclusion of patients as research partners. Patients may take a more active role in their own care and they may make significant contributions to our understanding of diseases and of quality care more generally.

Finally, as pharmaceutical research is increasingly focused on personalized medicine, new advances in technology may also complement patient-driven approaches to drug development and vice versa. As patients continue to claim greater control over their treatment, calls for compassionate access and crowd-sourced funding will play a greater role in the development process. Though there are some risks to the patient-driven approach, patients can offer manufactures and researchers resources that promote better drug discovery and better treatment. Policy-makers, researchers, and manufacturers should therefore include patients in the development process as much as possible, even if doing so requires changes to the clinical trial system, expanded access programs, or the current approval process.

References

"About | iCancer". 2015. http://icancer.org.uk/about/. Accessed 27 Apr 2015.

Carroll, J. 2015, January 20. CytRx Surges after FDA Lifts a Partial Hold on Late-Stage Cancer Drug. *Fierce Biotech*. http://www.fiercebiotech.com/story/cytrx-surges-after-fda-lifts-partial-hold-late-stage-cancer-drug/2015-01-20. Accessed 29 Apr 2015.

Cha, A.E. 2014, March 23. Crowdsourcing Medical Decisions: Ethicists Worry Josh Hardy Case May Set Bad Precedent. *The Washington Post*.

Committee on Bioethics. 1995. Informed Consent, Parental Permission, and Assent in Pediatric Practice. *Pediatrics* 95(2): 314–317.

Daniels, N. 2001. Justice, Health, and Healthcare. *The American Journal of Bioethics* 1(2): 2–16.

Darrow, J.J., A. Sarpatwari, J. Avorn, and A.S. Kesselheim. 2015. Practical, Legal, and Ethical Issues in Expanded Access to Investigational Drugs. *The New England Journal of Medicine* 372(3): 279–286. doi:10.1056/NEJMhle1409465.

deBronkart, D. 2013. How the E-Patient Community Helped Save My Life: An Essay by Dave deBronkart. *British Medical Journal* 346: f1990.

DiMasi, J.A., R.W. Hansen, and H.G. Grabowski. 2003. The Price of Innovation: New Estimates of Drug Development Costs. *Journal of Health Economics* 22(2): 151–185.

Dwamena, F., M. Holmes-Rovner, C.M. Gaulden, S. Jorgenson, G. Sadigh, A. Sikorskii, S. Lewin et al. 2012. Interventions for Providers to Promote a Patient-Centred Approach in Clinical Consultations. In *Cochrane database of systematic reviews*, ed. The Cochrane Collaboration. Chichester: Wiley. http://doi.wiley.com/10.1002/14651858.CD003267.pub2. Accessed 11 Aug 2016.

Epstein, S. 1995. The Construction of Lay Expertise: AIDS Activism and the Forging of Credibility in the Reform of Clinical Trials. *Science, Technology & Human Values* 20(4): 408–437.

Flanigan, J. 2012. Three Arguments against Prescription Requirements. *Journal of Medical Ethics* 38(10): 579–586.

Frydman, G.J. 2009. Patient-Driven Research: Rich Opportunities and Real Risks. *Journal of Participatory Medicine* 1(1): e12.

Grabowski, H.G., and J.M. Vernon. 1983. *The Regulation of Pharmaceuticals: Balancing the Benefits and Risks*. Washington, DC: American Enterprise Institute for Public Policy Research.

Grabowski, H.G., J.M. Vernon, and L.G. Thomas. 1978. Estimating the Effects of Regulation on Innovation: An International Comparative Analysis of the Pharmaceutical Industry. *Journal of Law and Economics* 21(1): 133–163.

Greenhouse, L. 2008, January 15. Justices Won't Hear Appeal on Drugs for Terminally Ill. *The New York Times*.

Hinkes-Jones, L. 2015, January 5. Stop Subsidizing Big Pharma. *The New York Times*.

Huebner, J.M. 2013. Moral Psychology and the Intuition That Pharmaceutical Companies Have a 'Special' Obligation to Society. *Journal of Business Ethics* 122(3): 501–510.
Jensen, E.J. 1987. Research Expenditures and the Discovery of New Drugs. *The Journal of Industrial Economics* 36(1): 83–95.
Kesselheim, A.S., and J.J. Darrow. 2015. FDA Designations for Therapeutics and Their Impact on Drug Development and Regulatory Review Outcomes. *Clinical Pharmacology and Therapeutics* 97(1): 29–36.
Lichtenberg, F.R., and J. Waldfogel. 2008. Does Misery Love Company–Evidence from Pharmaceutical Markets before and after the Orphan Drug Act. *Michigan Telecommunications and Technology Law Review* 15: 335.
Marcus, A.D. 2010. The Loneliness Of Fighting A Rare Cancer. *Health Affairs* 29(1): 203–206.
———. 2015a, March 2. Deaf or Death? In Drug Trial, Parents Weigh Life vs. Hearing Loss. *Wall Street Journal.*
———. 2015b. Niemann-Pick Type C: A Fight to Save Children with a Drug. *The Wall Street Journal*. http://on.wsj.com/1due624. Accessed 10 Mar 2015.
Masters, A. 2015. A Plutocratic Proposal. Mosaic. http://mosaicscience.com/story/plutocratic-proposal. Accessed 7 Apr 2015.
Morrow, K.J. 2015, April 25. Drug Development in the Age of Social Media. Life Science Leader.
Peltzman, S. 1973. An Evaluation of Consumer Protection Legislation: The 1962 Drug Amendments. *Journal of Political Economy* 81(5): 1049–1091.
Pollack, A. 2007, August 8. Court Rejects the Right to Use Drugs Being Tested. *The New York Times.*
Salmon, F. 2014, August 24. The Ice Bucket Challenge Isn't Going Away: But Giving Money to Disease-Specific Charities Is Still a Bad Idea. Slate. http://www.slate.com/articles/business/moneybox/2014/08/als_ice_bucket_challenge_giving_money_to_disease_specific_charities_is_a.html. Accessed 29 Apr 2015.
Sanghavi, D.M. 2013, October 31. How Dying Patients Get Access to Experimental Drugs. *The New York Times.*
Scherer, F.M. 2000. The Pharmaceutical Industry. In *Handbook of Health Economics*, ed. A.J. Culyer and J.P. Newhouse, 1297–1336. 1st edition. Elsevier. http://www.sciencedirect.com/science/article/pii/S1574006400800384. Accessed 11 Aug 2016.
Sofer, G.J. 2008, December 30. The FDA Is Killing Crohn's Patients. *Wall Street Journal.*
Thadhani, R. 2006. Formal Trials versus Observational Studies. In *Fabry Disease: Perspectives from 5 Years of FOS*, ed. A. Mehta, M. Beck, and G. Sunder-Plassmann. Oxford: Oxford PharmaGenesis.
Thomas, G. 2014, February 7. Oakland Pediatrician Opens Her Home To Patients. *NBC Bay Area.*
Topol, E. 2015. *The Patient Will See You Now: The Future of Medicine Is in Your Hands*, 1st ed. Basic Books, New York.
Volokh, E. 2007. Medical Self-Defense, Prohibited Experimental Therapies, and Payment for Organs. *Harvard Law Review* 20: 1813–1846.
Wiggins, S.N. 1981. Product Quality Regulation and New Drug Introductions: Some New Evidence from the 1970s. *The Review of Economics and Statistics* 63(4): 615–619.
Yin, W. 2008. Market Incentives and Pharmaceutical Innovation. *Journal of Health Economics* 27(4): 1060–1077.

Global Health Disparity and Pharmaceutical Companies' Obligation to Assist

Anita Ho

Introduction

In recent decades, in response to the ongoing global health disparity and the relative inaccessibility of pharmaceutical products for people in the least developed countries (LDCs), various international organizations, including the World Health Organization (WHO), have been exploring ways to increase access and affordability to essential treatments. As healthcare is often considered chief among all goods, many are claiming that pharmaceutical companies have a moral obligation to assist the afflicted in LDCs. To ease the dire situation in these impoverished regions, many have advocated for tiered or differential pricing, parallel imports, drug donation, and waiver of patent rights by resourceful companies for these regions (Spinello 1992; Resnik 2001; De George 2005a). Tiered or differential pricing is the idea that drug prices should be adjusted to reflect what different patient populations can afford, whereas parallel imports involve importing drugs retailed for lower prices in one country for resale in another.

This chapter critically explores the extent to which pharmaceutical companies have a moral obligation to assist poor patients in LDCs who currently have no or inadequate access to lifesaving medications. Focusing on the ongoing HIV/AIDS epidemic in LDCs, the first section of this essay will begin with some background information of the disproportionate burden of HIV/AIDS in LDCs. The second section will provide a brief overview of some of the salient arguments for holding multinational antiretroviral treatment (ART) manufacturers as morally responsible for easing the disproportionate global disease burden. The third section explains that these arguments regarding pharmaceutical companies' duty to assist are going

A. Ho (✉)
Centre for Biomedical Ethics, National University of Singapore, Singapore, Singapore

Centre for Applied Ethics, University of British Columbia, Vancouver, BC, Canada
e-mail: anitaho.ethics@gmail.com

© Springer Science+Business Media B.V. 2017
D. Ho (ed.), *Philosophical Issues in Pharmaceutics*, Philosophy and Medicine 122,
DOI 10.1007/978-94-024-0979-6_3

too far on the one hand by downplaying other non-pharmacological contributors to the slow response, but are also not going far enough on the other hand in upholding reciprocity-based duties. As the international community and researchers battle the reemergence of Ebola in Africa, the fourth section will explore how lessons from the HIV/AIDS situation can help to address what pharmaceutical companies may subsequently owe patients in this region and how the international community should respond to ongoing unequal disease burden.

Global Disparity in HIV/AIDS

In 2013, the UNAIDS estimated that 35 million people globally were living with HIV/AIDS (UNAIDS 2014a). While sub-Saharan Africa has less than 13% of the world's population (Population Reference Bureau 2014a), it is home to 71% (24.7 million) of all people living with HIV/AIDS and 91% of children with this condition (UNAIDS 2014b). Despite the introduction of ART in 1996 and the concerted efforts by various international agencies such as The Global Fund to Fight AIDS, Tuberculosis and Malaria (GFATM), the Joint United Nations Programme on HIV and AIDS (UNAIDS), and the Bill & Melinda Gates Foundation to increase treatment availability in the last decade, the 2013 data showed that only 37% of sufferers in this region had access to ART (UNAIDS 2014a). In some countries such as Nigeria, 80% of people did not have access to treatment (UNAIDS 2014a). Other countries in the region, including Central African Republic, Democratic Republic of Congo, and South Sudan, faced not only high HIV burden and low treatment coverage but also no or little decline in new infection.

Because of HIV/AIDS infection, life expectancy in sub-Saharan Africa is only at 54.9 years (UNDP 2013), which is more than 20 years lower than that in many developed countries, such as the United States (78.9 years), the United Kingdom (80.5 years), and Canada (81.5 years) (UNDP 2013). Swaziland, which has the highest HIV prevalence in the world (27.4%), has a life expectancy of just 48.9 (UNDP 2013). Botswana, despite having already seen a dramatic decline in AIDS-related deaths (58%) due to the rapid increase of placing patients on ART, maintains a high HIV prevalence (23%) and low life expectancy of 53 years (UNDP 2013, UNAIDS 2014b).

HIV/AIDS infection also perpetuates the cycle of poverty because of its serious impact upon a household's ability to generate income. When the income earners die or become too sick to work, medical costs can mount, children can be forced to abandon their education, and family members may have to forego any paid employment to care for the sick. In some cases, people have no other viable income-generating opportunities but to engage in sex work, which subsequently increases the risk of HIV/AIDS transmission and repeats the cycle (Miller et al. 2011). As HIV infection weakens the immune system, it can increase one's susceptibility to other infectious diseases and illnesses. In low-resource countries, tuberculosis (TB) is the most common opportunistic infection associated with HIV and a leading cause of death among people with HIV/AIDS, who have 26–31 times greater chance of developing TB than people without HIV infection (WHO 2014).

Holding Pharmaceutical Companies Morally Responsible

Many would agree that the disproportionate burden and grossly unequal access to ARTs are troubling, if we accept the premises of equal worth and universal dignity of all persons. Article 25.1 of the Universal Declaration of Human Rights states that every person has the right to a standard of living adequate for the health and well-being of oneself and one's family, including medical care. The WHO, which also recognizes the enjoyment of the highest attainable standard of health as a fundamental right of every human being (WHO 2006), makes the normative claim that "all segments of society" should have access to essential drugs, which are defined as drugs "considered to be of the utmost importance and hence basic, indispensable, and necessary for the health needs of the population" (Reich 1993). The WHO recognizes the concept as globally applicable, even though the specific list of essential drugs may differ among countries, depending on disease prevalence, evidence of efficacy and safety, and comparative cost-effectiveness. ART, which has been shown to be effective in controlling HIV and essential for maintaining the health of many infected patients, would fit the criteria of essential drugs.

In recent decades, many moral philosophers have argued that people in affluent countries have a positive obligation to help those who are vulnerable and in need, especially in cases where there is grave resource disparity (Singer 1972; Goodin 1985; Scanlon 1998). Two general and intersecting arguments are salient. First, it has been argued that, when some people are in dire need, if we can help them without sacrificing anything morally significant, we have a moral duty to do so, regardless of their geographical location. This argument, based on what Scanlon calls the principle of rescue (Scanlon 1998: 224–228), has been used to support moral duties to combat world hunger and help global victims of natural disasters. Scanlon believes that this principle is highly plausible and perhaps could not reasonably be rejected, at least not if the threshold of sacrifice considers previous contribution and does not demand unlimited contributions of small increments (Scanlon 1998: 224). If we accept that the main criterion for attributing a moral duty to aid is the ability to do so without significant sacrifice, many pharmaceutical corporations that sell ARTs can be good candidates for such a duty, as long as the effort required does not impose significant burden. The stakeholder theory reminds us that corporations are not purely economic agents that promote only shareholder value—corporate actions may have impact on many parties, including suppliers, employees, customers, and community members. With such recognition, the courts have repeatedly ruled that corporations are entities created by law that derive their powers and capacities from the law, with the purpose of serving the public or common good (Kelly 2001). Corporations are protected as legal persons under the US Constitution, and it seems reasonable to hold them accountable as we would with individuals. As DeGeorge points out, since corporations act according to a rational decision-making procedure and that their rational actions affect people, they are moral actors, and their decisions and operations can be evaluated from a moral point of view (De George 2005b: 184, 188).

In the context of the HIV/AIDS epidemic in LDCs, not only are ART manufacturers uniquely positioned to help by virtue of their therapeutic products; they also can control the supply and pricing of these products. It is thus unsurprising that many expect the pharmaceutical industry to pay attention to more than shareholder values—they also expect the industry to do its share in creating better access to medicines (Leisinger 2005). A poor host country's government and similarly impoverished neighboring nation-states may be incapable of providing enough aid, making corporations with the ability to provide these drugs or adjust their pricing or patent schemes plausible candidates for rendering assistance (Jackson 1993: 549). Even in cases where nongovernment organizations (NGOs) and governments of wealthier countries are willing to render assistance to LDCs, their ability to help still depends largely on access to affordable drugs. The technology to make such drugs is available, and the market for ARTs in developed countries provides a conducive business environment. Given that LDCs would not have been a lucrative market despite its high population of HIV sufferers due to their economic plight and lack of infrastructure, pharmaceutical companies can allow generic versions of their drugs to be manufactured in lower-cost countries (e.g., India) and then parallel-exported to LDCs without significantly affecting their bottom line. Richard Spinello (1992), for example, argues that there is little room for unequal distribution of a vital commodity such as essential medicines, defined as "those that satisfy the priority healthcare needs of the population" and are "selected with due regard to disease prevalence, evidence on efficacy and safety, and comparative cost-effectiveness" (WHO 2015a). Spinello recognizes that pharmaceutical companies are entitled to a reasonable profit and thus are not obligated to become charities by distributing essential drugs free of charge or at prices so low that they may sustain meager profits or even losses. Nonetheless, he argues that entitlement to sustainable profit does not justify premium prices for medicines that are truly essential, if that would prohibit people from accessing medicines that are vital for their well-being and life pursuits. In balancing goals of justice and considerations of economic viability, pharmaceutical companies have a duty to help those in LDCs by engaging in practices that will allow the widest possible distribution. If people in LDCs are in dire need and pharmaceutical corporations can assist without much sacrifice, there is a *prima facie* duty for them to do so.

Second, it is often argued that pharmaceutical companies' profit margin is excessive and that the high costs for essential medicines make the pricing strategy exploitative (Sterckx 2004). There is a public perception that, particularly for LDCs under conditions of collective and individual poverty, the prices for lifesaving drugs are too high (Leisinger 2005). Such premium pricing is considered morally unjustified, especially since governments and state-funded institutions often heavily subsidize medical research and provide generous tax deductions (Angell 2004: 40; Sterckx 2004). Patent protections, subsidies, tax credit, and other price protection provisions also support pharmaceutical companies' financial success. While research and development of new compounds for lifesaving treatments are time-consuming and costly, the financial burden of such development is not always borne completely by pharmaceutical companies. Only a small percentage of all new drugs are developed entirely within drug companies. Publicly funded researchers often license a private pharmaceutical corporation to act as the exclusive marketer of a drug. In the case of HIV/AIDS drugs, even though a private pharmaceutical corporation's researchers first conceived of the idea of using azidothymidine (AZT) in ARTs, the compound was initially synthesized and clinically trialed by researchers at federally funded institutions. Even if the duty to aid in the global community generally falls within the sphere and responsibility of states (Badcott 2013), given that government support was involved in the research and development of AZT, the reciprocity argument suggests that among potential actors, pharmaceutical companies have a moral obligation to partner with public agencies in facilitating their humanitarian efforts in making these lifesaving drugs available to patients in LDCs. The developing world component of the total pharmaceutical market is relatively low due to its economic status. The North American HIV market occupies more than 50% of the global revenue generated from ARTs, followed by the European market at around 30% of the global share. The generic-inaccessible ART market only represents 5% of the patients in LDCs (Clinton Health Access Initiative 2014), such that patent waivers and differential pricing would not significantly affect the industry's market share. If we focus only on LDCs that are facing the HIV/AIDS epidemics and strategies that would not impose a significant economic burden on these corporations, it is unclear that attribution of a moral duty to rescue is unreasonable. The question thus becomes one of how to reify such obligation to promote access to AZT to people in LDCs in a fair, feasible, and structured manner.

Duty to Assist: How Far Can We Go?

The Universal Declaration of Human Rights explicitly refers to non-state human rights duties that belong to "every organ of society." As many multinational corporations (MNCs) gradually become more powerful in the global marketplace and global political economy, governments are said to be losing their effectiveness as the sole protector of people's rights (Wettstein 2012). Many within the international legal community are thus increasingly supportive of imposing human rights duties directly upon MNCs. Recognizing that the ability to enjoy the highest attainable standard of health is essential in realizing other human rights, and that coordinated efforts of all components of civil society are important in realizing these rights, the Committee on Economic, Social and Cultural Rights, for example, argues that non-state actors, including corporations, have responsibilities to assist regarding the realization of the right to health.

If we take these rights seriously and accept the principle of rescue, it seems that many pharmaceutical corporations that sell ARTs can be good candidates for such a duty to aid, as long as the effort required does not impose significant burden. But questions remain as to how to determine the extent of such a duty and ensure compliance. This section explains how common arguments regarding pharmaceutical companies' duty to assist either go too far in singling out pharmaceutical pricing as the culprit of the lack of access to treatment or do not go far enough in ensuring that pharmaceutical companies, particularly those with prior relationships with affected LDCs in need of ARTs, are doing their fair and required share. Any rights to health are meaningless unless practices and institutions are established to determine who is to respond to these claims and to enforce compliance (Badcott 2013).

Going Too Far: Downplaying the Broader Context

There is a widespread public perception that the pharmaceutical industry has been unfair in its dealings with governments, regulators, healthcare providers, and the public (Huebner 2014). In particular, some activist groups have argued that drug companies constantly violate health-related human rights, prioritizing patents and profits over people's access to essential medicines (Gruskin and Raad 2010). As human life and health are highly valuable, there exists a common perception that an individual's posited right to essential medicines should not be encroached upon by premium pricing (Huebner 2014). Given that access to ARTs is crucial in conquering the HIV epidemic and that the pharmaceutical industry's pricing and other business strategies have strong influence on access to these medicines and health systems more broadly, drug companies are often singled out as the villains in the fight against the HIV epidemic in LDCs. Sociological research has found that consumers are indignant toward profit-oriented justifications for essential therapeutics, even though they find similar pricing rationalizations for other products acceptable (McGraw et al. 2011).

Indeed, access to ARTs has been important in controlling the epidemic, adding support to the call for more affordable treatments for those in LDCs. In countries that witnessed a significant decline in AIDS-related deaths, such as South Africa (48%), Rwanda (76%), Eritrea (67%), Burkina Faso (58%), Ethiopia (63%), Kenya (60%), Zimbabwe (57%), Malawi (51%), and the United Republic of Tanzania (44%), the success was directly due to the rapid increase in the number of people on ARTs (UNAIDS 2014b), made possible by pricing and licensing strategies. According to the 2014 Access to Medicine Index, out of the 33 ARTs on the market, 26 have been licensed, 22 of which to five or more generic medicine manufacturers to make and distribute these essential medicines (Access to Medicine Index 2014). As studies have shown, early initiation of HIV treatments can reduce the viral load in bodily fluid to "undetectable" level, decreasing an individual's risk of HIV transmission to one's sexual partner by 96% (Cohen et al. 2011). As treatment also serves as a form of prevention, new HIV infections have also declined globally by 38% since 2001. Such success is particularly important in HIV-prevalent countries such as Botswana (Thigpen et al. 2012). With recent progress, the global prevalence rate for adults between 15 and 49 years old has leveled in recent years, at 0.8% in 2013 (UNAIDS 2014b). In the period 2002–2012, life expectancy in sub-Saharan Africa increased by 5.5 years due mainly to the dramatic scaling up of antiretroviral treatment (UNDP 2013).

However, the glimpse of hope attributed to increasing pharmaceutical access notwithstanding, the story of the HIV epidemic in LDCs is a more complex one. Even if pharmaceutical companies continue to allow parallel imports, donate ARTs, grant generic licenses, and adopt differential pricing in LDCs, the success of such strategies in improving treatment availability will depend greatly on other contextual factors that are also hindering treatment access. It is noteworthy that while only 37% of HIV patients in sub-Saharan Africa are getting ARTs, the percentage of those treated in some other regions such as Eastern Europe and Central Asia (21%) as well as Middle East and North Africa (11%) are reportedly even lower (UNAIDS 2014a), partly due to stigma, discrimination, and worries of unauthorized disclosure of HIV status (Setayesh et al. 2014). Even Latin America (44%) as well as Western/Central Europe and North America (51%), which are comparatively more affluent regions, have substantial unmet needs (UNAIDS 2014a).

Indeed, it is important to note that cultural practices, lack of access to credible medical information, government policies, and other geopolitical factors have contributed to the crises in many countries in sub-Saharan Africa. For example, in the case of South Africa, infant feeding practices, women's inability to refuse sexual advances, healthcare workers' limited training and capacity, and the reluctance of the former President Thabo Mbeki and the counter-epistemic community to acknowledge the causal relation between HIV and AIDS have all contributed to the epidemic in that country (Youde 2005). It is also noteworthy that while 90% of people in sub-Saharan Africa who know their HIV status get treatment, an estimated 19 million of the 35 million people living with HIV globally do not know their HIV-positive status (UNAIDS 2014c). In South Sudan, where many residents continue to be displaced due to ongoing fighting, there is not only a shortage of laboratory space required to diagnose patients and the equipment needed to monitor the virus but also a scarcity of community health outreach resources (UNAIDS 2014d). In 2012, South Sudan spent 9.3% of its GDP on military activities (World Bank 2015a) while only 2.6% on healthcare (World Bank 2015b). In Central African Republic, minimal involvement of civil society and ineffective financial management have contributed to the epidemic in that country (UNAIDS 2010). An absolute shortage of physicians (less than 5 physicians per 100,000 people in 2009) also exacerbated the desperate situation in this country (WHO 2015b). With high job vacancy rates within the public health sector in the most vulnerable parts of Africa (Schneider et al. 2006), there is an additional challenge of ensuring that potentially infected people are educated about the disease, tested, and treated accordingly. In other words, the broader contexts of poverty, malnutrition, and underdevelopment in this region, which are primarily the responsibilities of governments, have been framing people's general health status and illness experience (Hanson et al. 2003). Mere provision of ARTs to LDCs without the necessary infrastructure and education to promote patient adherence and safe administration of these medicines may risk suboptimal therapeutic success as well as resistance to ARTs and important antibiotics in the long run (Leisinger 2005).

In determining the extent of pharmaceutical companies' duty to aid, we thus need to consider the larger realities. While the call for drug manufacturers to negotiate with governments and NGOs to promote affordable distribution of ARTs is legitimate, singling out the industry's moral duty to assist ignores how sociocultural and geopolitical barriers may shroud the reality that many impoverished countries have not or are not able to develop the necessary infrastructure to deliver basic healthcare, let alone ARTs (Schüklenk and Ashcroft 2002). Even when ARTs are available, stigmatization may inhibit serostatus disclosure to sexual partners and potential treatment supporters, delay ART initiation, and contribute to treatment nonadherence (Tsai et al. 2013). As the WHO contends (2011), long-term strategies to combat the epidemic will require various countries to strengthen their health and community systems, tackle the social determinants of health that both drive the epidemic and hinder the response, and protect and promote human and gender rights. Some of the aforementioned countries such as Malawi and Tanzania, which have enjoyed sharp decline in HIV-related deaths, have also been experiencing better infrastructure, access to water/sanitation, and higher political stability (World Bank 2014). As effective distribution and administration of ARTs depend on states to set up infrastructure, train healthcare personnel, provide surveillance, coordinate various services, and promote prevention and testing, the level of pharmaceutical companies' moral obligation to aid depends partly on whether the local governments and other relevant actors are first and foremost upholding their responsibilities to promote the necessary infrastructure for such endeavors.

If we recognize that other infrastructure and social conditions are necessary for efficient and effective distribution of ARTs, and if we also accept that the main criterion supporting the duty of rescue is the ability to do so, it would seem that we cannot attribute all or even most of the responsibility to assist to the pharmaceutical industry. Other industries—be it labor, construction, water sanitation, electricity, automobile, or petroleum—that can help to build the necessary infrastructure for a responsive primary healthcare system that can facilitate safe and effective distribution of HIV treatments may also have a prior obligation to assist. Unless there are morally relevant differences between the pharmaceutical and other industries in the overall goal of curbing the HIV epidemic, which will be explored in the next section, holding only pharmaceutical companies responsible without calling upon other industries to assist under the duty of rescue is going too far and unfair to the drug companies (Chang 2006).

Not Far Enough: Rethinking Prior Relationships

The last point regarding any morally relevant differences between the pharmaceutical and other industries helps to explain why some arguments regarding drug companies' duty to assist are not going far enough. Many common arguments supporting pharmaceutical companies' duty to assist have yet to emphasize and incorporate pertinent justice-related considerations in designing a system that can promote and enforce such duty.

I agree that the moral duty to rescue applies to all capable actors, not simply to pharmaceutical companies, and thus it is inappropriate to impose the heaviest burden to assist on these entities, unless their profit margins indeed suggest that they have more ability to render help or can assist more efficiently than other actors. However, as Finegold and Moser (2006) claim, pharmaceutical companies make products that have life-and-death impact, such that it is inevitable that they will come under greater ethical scrutiny. Moreover, there is another important element of reciprocity that is often neglected in arguments supporting a duty to assist. Below, I explain how the reciprocity requirement based on prior relationships sets ART manufacturers apart from most companies in other industries that may also have the capacity to aid. I argue that relationship-based reciprocity makes the duty to assist one of justice, which can be incorporated into the social and legal frameworks within which pharmaceutical companies operate.

Recall that the funding-based reciprocity argument suggests that pharmaceutical companies should facilitate or sponsor government efforts in providing ARTs to those in LDCs because research and development of these treatments were supported by public funding and infrastructure. I agree that these prior investments help to establish a duty of reciprocity to assist, but this argument on its own would not explain why these companies might have a duty to help those in LDCs but not in their home country. Moreover, some have rightly pointed out that many other industries are also supported by publicly funded organizations or benefit from research conducted at universities, but we do not generally impose a duty of reciprocity to assist on these industries (Chang 2006).

I argue that pharmaceutical companies that manufacture HIV treatments have two different types of reciprocity-based duties. In addition to public funding, many early investigations of HIV pathology and disease pathways were carried out in Africa, and various clinical drug trials have been conducted in some of the LDCs in this region—a high incidence of new HIV infections helped to produce valid and timely results. As variable strains and other confounding local realities require candidate vaccines to be tested in different regions, the contribution of the afflicted in LDCs facilitated the development or research of various biological or chemical phenomena. This prior relationship paved the way for scientific and commercial successes for many multinational pharmaceutical companies. The involvement of those living in the most impoverished regions helped to generate findings necessary to advance knowledge and develop new medical interventions for not only people in the LDCs but also the international community, thereby creating a broad duty of reciprocity. From the controversial placebo-controlled trials on perinatal transmission in Africa in the 1990s to the recent preexposure prophylaxis trials in Zimbabwe, Uganda, and South Africa, HIV research trials in LDCs have led to better understanding of biologic and behavioral factors contributing to the epidemic and respective treatment strategies. Distributive justice requires that burdens and benefits of research be distributed fairly, and justice as reciprocity demands that people receive what they deserve as a function of their contribution to an enterprise, both at the micro- and macro-levels. In the context of the HIV epidemic in LDCs, this prior relationship adds to the duty of rescue and bolsters pharmaceutical companies' duty to assist. Such a duty goes beyond one of charity—the prior relationship renders the duty to assist as one of distributive justice, and the extent of this duty depends on the level of involvement and contribution from the local residents and research environment. Some may consider reciprocity a microlevel obligation, borne only by researchers directly involved in specific clinical trials and only toward identifiable participants. However, this viewpoint is myopic and also neglects the broader social history of international research involving disadvantaged populations as well as colonialism and neocolonialistic practices that continue to plague many in LDCs. The magnitude of the epidemic in LDCs also reminds us that various geopolitical conditions that have framed international relationships have had profound and enduring effects on the distribution of not only economic but health burdens and benefits around the world (Pogge 2005). In constructing a fair global framework within which pharmaceutical companies operate and accrue profits, these macro-level ethical and geopolitical concerns about how to achieve greater justice between developed and least developed countries need to be incorporated.

The recognition of the duty to assist as one of justice that arises from prior relationships helps to reframe the discussion of how (much) pharmaceutical companies ought to contribute in the global effort to curb the HIV epidemic in LDCs and how multilateral cooperation should be coordinated to ensure proper discharge of such duties. In recognizing that pharmaceutical companies are profit-driven entities, many who advocate for a duty to assist caution that such an obligation will need to be balanced against other obligations and commitments in light of economic, social, legal, political, and cultural conditions (Resnik 2001). In particular, the nature of pharmaceutical companies as for-profit entities is often highlighted. As Badcott (2013) argues, to ignore market imperatives and portray the pharmaceutical industry and its employees as if they were quasi-healthcare practitioners with primary motivations of beneficence would constitute an insurmountable barrier to business operation generally.

While I agree that a balance of all conditions is necessary, a recognition of reciprocity-based duties reminds us that attention to the intersecting contexts does not imply that the profit motive should be the *starting* point for consideration, pitted against all other obligations. Thinking of the operating structure as one of profit that frames or overrides all other considerations, or is in perpetual conflict with economic imperatives (Schüklenk and Ashcroft 2002), would reduce the obligation to assist to one of charity, to be fulfilled at the companies' discretion. Doing so will not only place distributive justice concerns as secondary but also can have profound implications for global support for HIV responses.

In the face of reciprocity-based duties, we need to think of these intersecting factors not as conflicting considerations but as part of a social framework within which all citizens, families, organizations, and corporations coexist and rely on each other. Public investment in research and development highlights the vast amount of effort and coordination among governments, academia, and private industries required to treat various conditions. The research contribution of those in LDCs also reminds us that residents in the region, despite being plagued with HIV, are not simply passive recipients of assistance. They are part of the global research enterprise that contributed to scientific progress and commercial success of these pharmaceutical companies in the first place. More broadly speaking, pharmaceutical companies, like any other for-profit businesses, are not independent of the social context or exempt from social and legal norms. Rather, they operate *within* a social and interconnecting framework that supports and facilitates each party's flourishing. Pharmaceutical (or other) industries, despite their profit motives, are not outside of these social and legal frameworks, which have been created and shaped by social events, forces, and history. As I have explained elsewhere (Ho 2005), the patent system and other understandings and provisions regarding property rights are developed through the complex interplay of law and custom according to a social context and could change depending on our collective goals, economic realities, cultural norms, and so on. Pharmaceutical research, production, and distribution are highly regulated to balance safety and social welfare needs. The monopolistic structure and the powers it automatically confers through patents are granted by recent governments of certain countries to specific persons or companies "because this construction is believed to have positive social benefits" (Holm 2000). These monopolies are socially created to advance not only the progress of science but also social welfare (Leventer 2002). The terms of these legal rights are determined by various criteria based on the social goals we want to achieve, and can change if our social goals and disease patterns change, or if we discover better ways to respond to the domestic and global health realities, such as to control the HIV epidemic in LDCs.

In other words, profit, or how it is to be acquired, is *one* of the considerations within this social and legal framework rather than the starting point that should set the background context for everything else. Obligations to those in LDCs should not be an afterthought matter of discretional charity but one of justice that can and ought to be built into the system within which pharmaceutical corporations operate. The principle of reciprocity, in particular, reminds us that corporations, and in our context pharmaceutical companies, often benefit from or even rely on the contributions of the state and other stakeholders for their financial success. Given the ongoing global inequities and unequal conditions of benefit exchange for various stakeholders, attention to macro-distributive justice issues is essential in ensuring a fair system. As such, systemic coordination with drug companies regarding their responsibilities via treaty and patent provisions in light of other conditions is important to clarify expectations and promote compliance. It would be up to governments and international bodies to determine with the industry regarding how to construct a fair social and legal system that would balance ethical and economic imperatives of relevant stakeholders in light of the plight of those in LDCs, especially if residents in these regions have contributed to the development of these marketable drugs.

To ensure appropriate adherence to their duty to assist and to protect intra-industry fair competition, pharmaceutical corporations' duties to assist can be built into the framework in which these companies operate, such that all companies whose drugs are of certain characteristics (e.g., lifesaving, developed with prior contribution from a LDC) will be expected to contribute accordingly. Discretionary contributions can lead to variable and inconsistent offerings based on market conditions and the goodwill of individual firms. While companies can certainly provide additional assistance after their required contribution—after all, individual citizens can also donate to charities after paying taxes—concerted efforts by governments and relevant parties can coordinate the baseline contribution across ART manufacturers.

One notable example that could serve as a model for coordination is the Medicines Patent Pool (MPP), a United Nations-backed organization whose goal is to improve access to HIV medicines in developing countries. The MPP recognizes that although patents protect and encourage innovation on the part of pharmaceutical companies, they can restrict access to ARTs in developing countries. Through agreements with pharmaceutical companies such as Bristol-Myers Squibb and Gilead Sciences, the MPP provides developing countries the opportunities to produce and distribute generic versions of patented ARTs, while providing patent holders fair compensation (MPP 2015). In other cases, countries such as Brazil have made deliberate efforts to use existing international agreements such as the World Trade Organization (WTO) Agreement on Trade-Related Aspects of Intellectual Property Rights (TRIPS) to negotiate with pharmaceutical companies to ensure availability of affordable drugs for its people. This country's policies and actions have changed norms regarding global health, human rights, and trade policies related to access to essential medicines (Nunn et al. 2009). Depending on the level of local and regional involvement in developing these drugs, tiered pricing or licensing terms for various ARTs can be considered accordingly.

Lessons for Ebola

In late 2014, the rampant outbreak of Ebola in some of the most impoverished areas in Africa led the director of the Centers for Disease Control and Prevention to urge that "we have to work now so that this is not the world's next AIDS" (Rettner 2014). Ebola first emerged in Sudan (now South Sudan) and Zaire (now Democratic Republic of Congo) in 1976. Since then, there have been approximately two dozen outbreaks in various rural areas in Africa. The size, spread, and virulence of the most recent outbreak in 2014–2015, which reported more than 23,000 cases and killed over 10,500 people in Guinea, Liberia, and Sierra Leone as of April 2015

(WHO 2015c), were the most severe on record (Sayburn 2014). As multiple cities were affected and that infected healthcare workers and others traveled to various parts of the world, there were also fears that the virus could quickly become a global epidemic if not managed properly.

The statement comparing HIV and Ebola was controversial, especially since there are important clinical differences between these two viruses. HIV has a long latency period that lasts up to 10 years, during which asymptomatic people can spread the virus to others. Ebola has a short incubation window of up to 21 days, and only symptomatic people can spread the virus. Nonetheless, the macro and international sociopolitical context within which the HIV epidemic evolved in the last few decades offers important lessons for the international community regarding how justice requires various parties—government agencies, NGOs, pharmaceutical companies, and researchers—to work together in combating the spread of Ebola and other tropical diseases affecting people in LDCs. Both viruses emerged in Africa with no vaccine, and both are fatal if left untreated.

HIV was initially considered a disease only affecting poor Africans or gay men, and the stigmatization and marginalization of these populations prevented domestic and international communities from acknowledging the macro conditions contributing to and exacerbating the circumstances. It was only when the disease became more widespread in heterosexual people even in wealthy countries that research and development activities ramped up. Ebola has also been among one of the neglected tropical diseases. Dozens of small outbreaks for the past 40 years in rural Africa did not prompt significant public health research effort. Despite killing 50–90% of those infected with the virus, the "market condition" for Ebola simply provided little to no incentive for pharmaceutical companies to invest in developing a treatment, making this an orphan condition. It was only the dramatic fear brought on by the high susceptibility of healthcare workers—including international professionals working in Africa and subsequently treated in the west—and the perception of the global spread that recently motivated international pressures on fast-tracking research efforts. A few individuals were granted emergency access to ZMapp, the anti-Ebola virus three-antibody mixture that has gone through animal studies but not randomized clinical trials (RCTs) to evaluate efficacy and safety in humans (Pollack 2014).

Much of the recent ethics literature regarding ZMapp has been focusing on the ethical considerations regarding informed consent and balancing risks and benefits in using experimental drugs (Rid and Emanuel 2014). Distributive justice discussions have mostly been around the issue of allocating very limited doses of unproven compounds (Donovan 2014; Joffe 2014). There are also debates regarding whether the use of RCTs would be most appropriate in dire and emergency circumstances. However, as we have seen in the HIV situation, economic, sociopolitical, and cultural contexts affect how an epidemic emerges, such that these factors need to be considered as government agencies, international organizations, and pharmaceutical companies work together in finding ways to curb the epidemic and prevent the next outbreak. Stigma, health illiteracy, cultural practices, the general lack of health infrastructure, and the unavailability of effective treatment for Ebola have prevented many infected individuals from seeking medical attention. For every 100,000 people, Sierra Leone and Liberia only had 2.2 physicians in 2010 and 1.4 physicians in 2008, respectively (WHO 2015b). With much of the already weak infrastructure in these countries further damaged due to civil wars or conflicts, the usual systems of disease control used in previous outbreaks could not contain the spread of the virus.

Many have wondered why pharmaceutical corporations have not developed effective drugs for Ebola. It is noteworthy that ZMapp, which was the research result of collaborating biodefence companies that have been financially supported by various American federal agencies and the Public Health Agency of Canada (Kroll 2014), was not developed primarily as a treatment for infected patients in Africa. This is despite the fact that the virus was initially discovered in the region in the mid-1970s and that much of the key discoveries of how the virus spread were first conducted there. Rather, the antibody cocktail was developed because Ebola was considered a top-level threat for bioterrorism in the post 9–11 era.

At the height of the latest outbreak, there was political pressure from the international community to fast-track and ramp up R&D for this disease. The National Institute of Allergy and Infectious Diseases (NIAID) announced in February 2015 that the agency would partner with relevant research units in Liberia to test two experimental vaccines in that country (NIAID 2015). As the epidemic begins to ease, the study investigators anticipate that flexibility in the conduct and design of the trial will be necessary to address the changing nature of the outbreak.

I contend that these considerations regarding how to prevent the next outbreak need to go beyond study design. Assuming scientific success, larger questions of how these compounds may be developed and made available to those in LDCs in the long term need to be considered. As the fear of an imminent global epidemic fades, drug developers and international agencies may once again question if investment in an orphan drug would be financially worthwhile, and who may have the moral obligation to develop such treatment, since the principle of rescue does not require potential actors to make substantial sacrifice. Detailed discussion of this issue is beyond the scope of this chapter, but it is worth noting that drug availability is only a small part of disease prevention and treatment. As the high recovery rate of Ebola patients in the west has shown, infrastructure is important, such that the international community should not only focus on drug development. It also needs to consider how to support local actors in building capacity for the long run. While the United States spent hundreds of millions of dollars and deployed around 3000 troops to build treatment centers in Liberia in late 2014 and early 2015 (Onishi 2015), many of these treatment centers now sit empty as the number of new cases decreased drastically. Since LDCs facing Ebola also battle other deadly infections that continue to claim even more lives, the international community must not abandon its effort in understanding the fundamental economic, cultural, and geopolitical issues that may have contributed not only to the latest outbreak but also people's daily hardship. Many residents in the aforementioned LDCs already face HIV, malaria, water-borne diseases, TB, and other health problems that have become part of their everyday life. To prevent another epidemic, whether of Ebola or other viral diseases, we need to ask fundamental questions regarding what macro global and local conditions may contribute to the spread of diseases, how to make sure that the necessary and integrated infrastructure is available to tackle basic health needs, what conditions and populations should be targeted for pharmaceutical research, how such research should be supported and funded, and how proprietary regulations should be constructed to ensure that those in the most disadvantaged positions will not be denied access to essential medicines. It is only when we attend to these broader contextual issues that we can prevent the world's next AIDS.

References

Access to Medicine Index. 2014. *Licensing Medicines for HIV/AIDS: Critical Mechanism for Ending the Epidemic*. http://www.accesstomedicineindex.org/licensing-medicines-hivaids-critical-mechanism-ending-epidemic. Accessed 18 Feb 2015.

Angell, M. 2004. *The Truth About the Drug Companies: How They Deceive Us and What to Do About It*. New York: Random House.

Badcott, D. 2013. Big Pharma: A Former Insider's View. *Medicine, Health Care and Philosophy* 16: 249–264.

Bill & Melinda Gates Foundation. 2015. www.gatesfoundation.org. Accessed 18 Feb 2015.

Chang, P.L. 2006. Who's in the Business of Saving Lives? *The Journal of Medicine and Philosophy* 31(5): 465–482.

Clinton Health Access Initiative. Dec 2014. *ARV Market Report: The State of the Antiretroviral Drug Market in Low- and Middle-income Countries*. Boston: USA.

Cohen, M.S., et al. 2011. Prevention of HIV-1 Infection with Early Antiretroviral Therapy. *The New England Journal of Medicine* 365(5): 493–505.

De George, R.T. 2005a. Intellectual Property and Pharmaceutical Drugs: An Ethical Analysis. *Business Ethics Quarterly* 15(4): 549–575.

———. 2005b. *Business Ethics*. 6th ed. Upper Saddle River: Prentice Hall.

Donovan, G. 2014. Ebola, Epidemics, and Ethics—What We Have Learned. *Philosophy, Ethics, and Humanities in Medicine* 9: 15.

Finegold, D., and A. Moser. 2006. Ethical Decision-making in Bioscience Firms. *Nature Biotechnology* 24(3): 285–290.

Goodin, R. 1985. *Protecting the Vulnerable: A Reanalysis of Our Social Responsibilities*. Chicago: University of Chicago Press.

Gruskin, S., and Z. Raad. 2010. Are Drug Companies Living Up to Their Human Rights Responsibilities? Moving Toward Assessment. *PLoS Medicine* 7(9): 1–3.

Hanson, K., et al. 2003. Expanding Access to Priority Health Interventions: A Framework for Understanding the Constraints to Scaling-Up. *Journal of International Development* 15: 1–14.

Ho, A. 2005. Pharmaceutical Corporations and the Duty to Aid in HIV/AIDS Epidemic. *Business & Professional Ethics Journal* 24(4): 51–81.

Holm, S. 2000. Three Reasons Why a Global Market in Pharmaceutical Products Is Inherently Unjust. *Journal of Social Philosophy* 31: 398.

Huebner, J.M. 2014. Moral Psychology and the Intuition that Pharmaceutical Companies Have a 'Special' Obligation to Society. *Journal of Business Ethics* 122: 501–510.

Jackson, K. 1993. Global Distributive Justice and the Corporate Duty to Aid. *Journal of Business Ethics* 12: 549.

Joffe, S. 2014. Evaluating Novel Therapies During the Ebola Epidemic. *Journal of the American Medical Association* 312(13): 1299–1300.

Kelly, M. 2001. Property Privileges (Not Rights): The Cipro case reminds us that property rights are subordinate to the public good. *Business Ethics* 15(6, Nov/Dec): 4.

Kroll, D. 2014. Ebola 'Secret Serum': Small Biopharma, The Army and Big Tobacco. *Forbes*. http://www.forbes.com/sites/davidkroll/2014/08/05/ebola-secret-serum-small-biopharma-the-army-and-big-tobacco/. Accessed 23 Feb 2015.

Leisinger, K.M. 2005. The Corporate Social Responsibility of the Pharmaceutical Industry: Idealism Without Illusion and Realism Without Resignation. *Business Ethics Quarterly* 15(4): 577–594.

Leventer, H. 2002. It Is Not Unethical, Though It Is Often Unwise, to Override Patents. *The American Journal of Bioethics* 2: 50.

McGraw, A.P., J.A. Schwartz, and P.E. Tetlock. 2011. From the Commercial to the Communal: Reframing Taboo Trade-offs in Religious and Pharmaceutical Marketing. *Journal of Consumer Research* 39: 157–173.

Medicine Patent Pool. 2015. *About the MPP*. http://www.medicinespatentpool.org/about/. Accessed 18 Feb 2015.

Miller, C.L., et al. 2011. Food Insecurity and Sexual Risk in an HIV Endemic Community in Uganda. *AIDS and Behavior* 15(7): 1512–1519.

National Institute of Allergy and Infectious Diseases (NIAID). 2015. *Ebola Vaccine Trial Opens in Liberia—Study Led by Liberia-NIH Partnership Will Test Two Experimental Vaccines*. http://www.niaid.nih.gov/news/newsreleases/2015/Pages/PREVAIL.aspx. Accessed 1 Mar 2015.

Nunn, A., E. Fonseca, and S. Gruskin. 2009. Changing Global Essential Medicines Norms to Improve Access to AIDS Treatment: Lessons from Brazil. *Global Public Health* 4(2): 131–149.

Onishi, N. 2015, April 11. Empty Ebola Clinics in Liberia are Seen as Misstep in U.S. Relief Effort. *New York Times*. http://www.nytimes.com/2015/04/12/world/africa/idle-ebola-clinics-in-liberia-are-seen-as-misstep-in-us-relief-effort.html. Accessed 16 Apr 2015.

Pogge, T.W. 2005. World Poverty and Human Rights. *Ethics International Affairs* 19: 4.

Pollack, A. 2014, August 8. Ebola Drug Could Save a Few Lives. But Whose? *New York Times*. http://www.nytimes.com/2014/08/09/health/in-ebola-outbreak-who-should-get-experimental-drug.html. Accessed 4 Mar 2015.

Population Reference Bureau. 2014 Africa—Sub-Saharan. http://www.prb.org/DataFinder/Geography/Data.aspx?loc=247. Accessed 18 Feb 2015.

Reich, M. 1993. Essential Drugs: Economics and Politics in International Health. In *The Pharmaceutical Corporate Presence in Developing Countries*, ed. Travis and Williams, 83–108. Notre Dame, Indiana: University of Notre Dame Press.

Resnik, D.B. 2001. Developing Drugs for the Developing World: An Economic, Legal, Moral, and Political Dilemma. *Developing World Bioethics* 1(1): 11–32.

Rettner, R. 2014. Could Ebola Really Be The 'Next AIDS'? *Lifescience*. http://www.livescience.com/48232-ebola-aids-comparison.html. Accessed 23 Feb 2015.

Rid, A., and E.J. Emanuel. 2014. Ethical Considerations of Experimental Interventions in the Ebola Outbreak. *Lancet* 384(9957): 1896–1899.

Sayburn, A. 2014. WHO Gives Go Ahead for Experimental Treatments to be Used in Ebola Outbreak. *British Medical Journal* 349: g5161.

Scanlon, T.M. 1998. *What We Owe to Each Other*. Cambridge, MA: Belknap Press of Harvard University Press.

Schneider, H., et al. 2006. Health Systems and Access to Antiretroviral Drugs for HIV in Southern Africa: Service Delivery and Human Resource Challenges. *Reproductive Health Matters* 14(27): 12–23.

Schüklenk, U., and R.E. Ashcroft. 2002. Affordable Access to Essential Medication in Developing Countries: Conflicts Between Ethical and Economic Imperatives. *The Journal of Medicine and Philosophy* 27(2): 179–195.

Setayesh, H., et al. 2014. *HIV and AIDS in the Middle East and North Africa*. Washington, DC: Population Reference Bureau.

Singer, P. 1972. Famine, Affluence, and Morality. *Philosophy and Public Affairs* 1: 229–243.

Spinello, R.A. 1992. Ethics, Pricing and the Pharmaceutical Industry. *Journal of Business Ethics* 11(8): 617–626.

Streckx, S. 2004. Patents and Access to Drugs in Developing Countries: An Ethical Analysis. *Developing World Bioethics* 4(1): 1471–8847.

The Global Fund to Fight AIDS, Tuberculosis and Malaria. n.d. http://www.theglobalfund.org/en/. Accessed 18 Feb 2015.

The World Bank. 2014. *Worldwide Governance Reports: Country Data Reports*. http://info.worldbank.org/governance/wgi/index.aspx#countryReports. Accessed 27 Feb 2015.

———. 2015a. *Military expenditure (% of GDP)*. http://data.worldbank.org/indicator/MS.MIL.XPND.GD.ZS/countries. Accessed 27 Feb 2015.

———. 2015b *Health expenditure, total (% of GDP)*. http://data.worldbank.org/indicator/SH.XPD.TOTL.ZS/countries. Accessed 27 Feb 2015.

Thigpen, M.C., et al. 2012. Antiretroviral Preexposure Prophylaxis for Heterosexual HIV Transmission in Botswana. *The New England Journal of Medicine* 367: 423–434.

Tsai, A.C., D.R. Bangsberg, and S.D. Weiser. 2013. Harnessing Poverty Alleviation to Reduce the Stigma of HIV in Sub-Saharan Africa. *PLoS Medicine* 10(11): e1001557.

UNAIDS. 2010. *Removing Bottle-necks for Universal Access in Central African Republic*. http://www.unaids.org/en/resources/presscentre/featurestories/2010/february/20100208car. Accessed 23 Feb 2015.

———. 2014a. *Fact sheet*. http://www.unaids.org/en/resources/campaigns/2014/2014gapreport/factsheet. Accessed 18 Feb 2015.

———. 2014b. *The Gap Report*. http://www.unaids.org/sites/default/files/media_asset/UNAIDS_Gap_report_en.pdf. Accessed 18 Feb 2015.

———. 2014c. *UNAIDS Report Shows that 19 Million of the 35 Million People Living with HIV Today Do Not Know that They Have the Virus*. http://www.unaids.org/en/resources/presscentre/pressreleaseandstatementarchive/2014/july/20140716prgapreport. Accessed 18 Feb 2015.

———. 2014d. *Raising Awareness of HIV Testing and Counselling in South Sudan*. http://www.unaids.org/en/resources/presscentre/featurestories/2014/june/20140623southsudantesting. Accessed 23 Feb 2015.

United Nations. Declaration of Universal Human Rights. http://www.un.org/Overview/rights.html. Accessed 18 Feb 2015.

United Nations Development Programme. 2013. *Human Development Report 2013. The Rise of the South: Human Progress in a Diverse World*. http://hdr.undp.org/sites/default/files/reports/14/hdr2013_en_complete.pdf. Accessed 18 Feb 2015.

Wettstein, F. 2012. CSR and the Debate on Business and Human Rights: Bridging the Great Divide. *Business Ethics Quarterly* 22(4): 739–770.

World Health Organization. 2006. *Constitution of the World Health Organization*. http://www.who.int/governance/eb/who_constitution_en.pdf. Accessed 18 Feb 2015.

———. 2011. *Global Health Sector Strategy on HIV/AIDS 2011–2015*. http://whqlibdoc.who.int/publications/2011/9789241501651_eng.pdf?ua=1. Accessed 23 Feb 2015.

———. 2014. *Fact Sheet No. 104 Tuberculosis*. http://www.who.int/mediacentre/factsheets/fs104/en/. Accessed 18 Feb 2015.

———. 2015a *Essential Medicines and Health Products*. http://www.who.int/medicines/services/essmedicines_def/en/. Accessed 18 Feb 2015.

———. 2015b *Health Workforce: Density of Physicians (Total Number per 1000 Population): Latest available year*. http://gamapserver.who.int/gho/interactive_charts/health_workforce/PhysiciansDensity_Total/tablet/atlas.html. Accessed 27 Feb 2015.

———. 2015c. *Ebola Situation Report–8 April 2015*. http://apps.who.int/ebola/current-situation/ebola-situation-report-8-april-2015. Accessed 16 April 2015.

Youde, J. 2005. The Development of a Counter-Epistemic Community: AIDS, South Africa, and International Regimes. *International Relations* 19: 421–439.

Hegemony of Knowledge and Pharmaceutical Industry Strategy

Sergio Sismondo

Introduction

Pharmaceutical industry research and its dissemination are done at least partly for marketing purposes, to establish influence or dominance over areas of medical knowledge. Indeed, medical research, education, and marketing are often fused, with the result that doctors seeking knowledge about conditions and treatments typically turn to agents of pharmaceutical companies. These agents include not only local sales representatives but also industry-sponsored researchers and educators, conduits in a knowledge system that enables companies to "sell without selling," a marketing ideal (Oldani 2004).

This chapter describes some important vectors of pharmaceutical companies' efforts toward hegemony of medical knowledge: the ghost management of research and publication and the shepherding of knowledge dissemination. Rather than focusing on individual knowledge claims and their justifications, the chapter highlights forces available to shape terrains on which claims are produced, distributed, and consumed. So doing underscores the significance of the political economy of knowledge, as a counterpart to traditional epistemology, in philosophical studies of medical knowledge. The now quaint-sounding discipline of political economy is the study of the circumstances of production, distribution, and consumption of goods; here the framework is applied to knowledge.

In his *Prison Notebooks* (1971), the Italian Marxist theorist Antonio Gramsci discusses how, when dominant interests have *hegemony* over key institutions, power can routinely operate without coercion. A major challenge in discussions of hegemony is to identify and describe concretely the mechanisms that shape cultural

S. Sismondo (✉)
Department of Philosophy, Queen's University, Kingston, ON, Canada
e-mail: sismondo@queensu.ca

© Springer Science+Business Media B.V. 2017

D. Ho (ed.), *Philosophical Issues in Pharmaceutics*, Philosophy and Medicine 122,
DOI 10.1007/978-94-024-0979-6_4

institutions and the views that they make and embody. Perhaps emblematic of the challenge, Gramsci states: "Every social group…creates together with itself, *organically*, one or more strata of *intellectuals*, which give it homogeneity and an awareness of its function, not only in the economic but also in the social and political fields" (Gramsci 1971: 5, emphasis in original). If Gramsci is right, then that organic creation will tend to be hidden from view, naturalized within cultures.

Transposing Gramsci's idea of hegemony from culture at large into the more defined sphere of medical knowledge allows for a close specification of mechanisms of cultural control, even given the secrecy of the pharmaceutical industry. At least in this area, the creation of intellectuals and the domination of institutions are not at all organic, but are instead the result of deliberate and careful actions. This chapter looks at attempts at hegemony over medical knowledge through: contract research, publication in medical journals, the creation of medical culture via sponsorship of key opinion leaders (KOLs) and continuing medical education (CME), and the continued dissemination of that culture via sales forces. There are rough parallels between these institutions and those on which Gramsci focused: the press, schools, religion, and popular arts. Gramsci's central concern was strategic: to displace conservative (in particular fascist) hegemony over key institutions with socialist hegemony. Perhaps such strategic thinking would be applicable here, but the goal of this chapter is primarily description.

In addition to secondary sources, the material presented in this chapter is derived from a wide variety of sources within the pharmaceutical industry and its closely connected subindustries and professions. The author and research assistants have attended and taken detailed notes at a number of conferences, meetings, and workshops on such topics as publication planning, sponsored clinical trials, KOL management, and the running of speakers bureaus. The author has also collected brochures from firms that provide services to the pharmaceutical industry, articles in industry publications, and reports. The chapter draws lightly on more than a dozen interviews with KOLs in the United States, all of whom earned more than $100,000 from their work with the industry in 2009. The result is a high-level overview, but built on the perspectives of people working on the ground within or closely connected to the pharmaceutical industry.

The Ghost Management of Pharmaceutical Research and Publication

A great number of studies and several meta-analyses have shown that journal articles on industry-sponsored trials report results on efficacy and harm, and draw overall conclusions, that are all favorable to the sponsors (Lundh et al. 2012). For many areas of medicine, articles reporting industry-sponsored research are nearly certain to come to positive conclusions. A standard explanation of industry's favorable position focuses on the conflicts of interest that academics involved in industry-sponsored research have. The concern is that more-or-less independent researchers act to align their studies with funders' interests, through those studies' design, implementation, interpretation, and/or publication.

Let us consider the common image of industry-sponsored research. First, an independent researcher designs and proposes a study. Then, looking for support, that researcher approaches one or more drug companies. A company may choose to fund the research, either out of interest in the results or to buy goodwill, or both. Then, the researcher performs the study, writes up a few articles on it, and submits them to journals. Along the way, the researcher might bias the study to support the sponsor's interests by, for example, testing the experimental drug against an inappropriate comparator or in an unrepresentative population or choosing statistical tests with success in mind.

However much one believes in the power of conflict of interest, it seems questionable that conflicts of interest stemming merely from research funding could produce fairly consistent positive results by independent researchers. Luckily, we do not have to believe that researchers are so easily influenced, because the common image of sponsored research is wrong on key details of the actual processes of pharmaceutical research production. One is that most clinical trials are not in any interesting sense performed by independent researchers. A second is that the independent researchers who act as named authors on journal articles often neither design the studies nor write the articles. Instead, the pharmaceutical industry manages the process of producing and publishing research about drugs, from start to finish, from behind the scenes. Two main sets of actors in the process are contract research organizations (CROs) and publication planners. Together, they intervene substantially in the production and dissemination of pharmaceutical knowledge, in ways superficially similar to but substantively different from typical scientific and academic research and publishing activities. This whole process is the "ghost management" (Sismondo 2007, 2009) of pharmaceutical knowledge, and it is partly through this process that companies establish positive scientific profiles for their drugs.

Contract Research Organizations

> Companies can now pick and choose populations…in order to get a most pronounced drug benefit signal as well as a 'no-harm' signal.—Former CRO chief executive (quoted in Petryna 2009: 17)

The pharmaceutical industry sponsors a majority of the 40,000 new clinical trials initiated each year and 60,000 trials ongoing at any one time and provides roughly half of all funding for clinical trial research (CenterWatch 2009). Of this research, 70–75% is done by CROs, for-profit companies that perform research and development (Fisher 2009; Mirowski and Van Horn 2005). An indicator of their rising significance is that CRO revenues increased from US$1 billion in 1992 to an estimated $27 billion in 2010.

CROs are involved in all different aspects of pharmaceutical industry research, from preclinical work through all phases of clinical research, and they perform 95% of laboratory services related to trials. According to a 2005 survey of pharmaceutical company departments that hired CROs, outsourcing serves a number of purposes. It saves costs (74%), protects the sponsoring company from risk (70%), enhances quality (67%), saves time (59%), and increases productivity (52%) (CenterWatch 2009). In addition, though not arising in the survey, CROs provide something that academic researchers leading clinical trials will not: data to pharmaceutical companies with no strings attached. The data from a CRO study is wholly owned and controlled by the sponsoring company.

CROs conduct trials with one eye to the drug approval process and the other to the marketing of products. For both purposes, the pharmaceutical companies sponsoring the research prefer, and often need, trials that meet high formal standards within medicine—often randomized, controlled trials, considered the "gold standard" of medical science. This requires access to large populations and often populations in multiple countries. Available populations are the result of a number of factors, and recently these have been pushing more trials to the developing world and Eastern Europe. Costs per patient are considerably lower in the developing world than in North America or Western Europe. Many people in the USA and Western Europe are unwilling to participate in trials, and many are ineligible because they are already being treated. Wealthier countries can be "treatment saturated," whereas elsewhere there are many potential subjects who are "treatment naïve." Thus countries like India, for example, are well positioned to provide subjects; India's *Economic Times* wrote in 2004: "The opportunities are huge, the multinationals are eager, and Indian companies are willing. We have the skills, we have the people …" (Shah 2006: 17).

Most important for the argument of this chapter, though, is a point mentioned above: CROs have no interest in publishing the results under their own names. They simply produce data for their clients, and clients can use it to best advantage.

Publication Planning

> Adis Communications works in partnership with clients to position their products at the right place, at the right time through: Hundreds of well-respected, and high-impact factor journals …. (Adis Communications 2006)

Although industry-sponsored articles about drugs in medical journals are often "authored" by academics, authorship does not necessarily represent significant contributions to the research, design, or writing of the articles. If a pharmaceutical company is behind it, an article on a drug is almost always part of a dedicated "publication plan," designed and implemented by "publication planners." In this process, authors typically make only modest contributions to articles. Here's one email from a pharmaceutical company to an eventual author: "It will not be you personally who will have to write those articles but a ghost writer will do this for us/you and you kind of give your good name for this publication!" This email was accidentally included in correspondence forwarded to a journal editor in 2010, who quoted it at a publication planning conference.

From the few cases where there is hard data, it appears that roughly 40% of medical journal articles mentioning major new in-patent drugs are parts of publication plans (Healy and Cattell 2003; Ross et al. 2008). This includes not only articles reporting on company-led trials but also some reviews, meta-analyses, and commentaries. Some planners work within pharmaceutical companies, but more work in independent agencies: more than 50 agencies advertise publication planning on the Internet. Some of these boast that they have hundreds of employees and handle many hundreds of manuscripts per year (individual planners handle dozens of manuscripts per year). There are two competing international associations of publication planners that organize meetings and seminars. One of those associations, the International Society of Medical Planning Professionals, has more than 1000 members. Both it and its competing association, The International Publication Planning Association, hold annual conferences, and the latter hosts regional conferences. Publication planning, then, is a sizeable activity.

Publication plans aim to maximize the value of research, in terms that are simultaneously commercial and scientific. Thus, plans lay out the terms for constructing articles that establish consistent profiles for drugs. The publication planning team should be put in place early, said one presenter at a 2007 workshop for new planners, "before too much data has gone unpublished." Ideally, it would be in place for research design, and this is especially important when there is "need to create [a] market" or create an "understanding of unmet need."

It is worth quoting at length one publication plan's description of planning itself:

Strategic publication planning provides the tactical recommendations necessary to develop a scientific platform within the biomedical literature to support the market positioning of an established product or the launch of a new product. The process of publication planning includes:

- An analysis of the characteristics of the market into which the product will be launched
- An analysis of competitive issues
- The expected product profile
- Identification of issues relevant to the disease state or primary indication for the product
- Development of a series of key communication messages addressing the major issues
- The availability of clinical and preclinical data to support the key communication messages
- Recognition of appropriate target audiences for each of the recommended publication tactics
- Recommendations for publication vehicles (e.g., journals, meetings, congresses, etc.) for each publication activity (Wyeth 2002)

On its standard self-presentation, publication planning should be, though isn't always, independent of marketing. It should be in the service of scientific knowledge about results. However, in their interactions with each other and with the pharmaceutical industry, planners recognize that their work has marketing value. "How are we going to create publications that have the right message, and a memorable message, for prescribers?" asked a planner, addressing attendees at the introductory workshop mentioned above. The websites of publication planning firms promote their work on that basis. Watermeadow Medical says that its "mission is to ensure that…messages reach and energize…target customers," including through "hard-core scientific writing" (Watermeadow Medical 2007). Envision Pharma's site says that "data generated from clinical trials programs are the most powerful marketing tools available to a pharmaceutical company" (Envision Pharma 2006). Ultimately, publication planning needs to generate revenue by providing information that increases sales.

Publication plans exist to serve the marketers, and yet, to some extent planners see part of their job as constraining marketers' influence. It is only by stifling the marketing department's efforts to hype the product that publication planners can do effective marketing to scientific audiences. Thus planners have to convince the marketers that their more subtle approach, with its limited range of tools, is the right one. This is a version of a common tension occurring when the most persuasive rhetoric is not marked as an explicit attempt to persuade. As already mentioned, to "sell without selling" is a sales and marketing ideal. At least some of the time, marketing is best done if it is invisible.

Science can indeed mesh smoothly with marketing. The director of one publication planning agency writes in an article that science and marketing are equal partners (Bohdanowicz 2005). Therefore, "Where shall we publish this study?" is paired with "Who are our customers?" And "What can we claim from the results?" is paired with "What are our customers' needs?" In this image, science and marketing together determine what the research says and how the products can be sold.

Scientific standards are important, though. Meeting them constitutes part of what is considered ethical behavior and underpins the distinction between doing publication planning and doing public relations. After planners persuade their sponsors that their work will provide a good return on investment, they want to obey ethical guidelines in the hands-on work they do and to adopt high scientific standards for the writing of each paper. In addition, publication planners can only succeed if their work displays high standards, so that their papers will be published to best advantage. Medical journals have high rejection rates, as high as 95% in the case of journals such as the *Journal of the American Medical Association* and the *British Medical Journal*. Meanwhile, publication planners claim to have very high success rates, for example, an "acceptance rate on first submission of 94% for abstracts and 78% for manuscripts" (Gardiner-Caldwell Group 2007).

Authorship problems represent a major ethical concern for publication planners, in particular because of difficulties meeting the International Committee of Medical Journal Editors' (ICMJE) criteria for authorship (ICMJE 2014). Academic authors, often referred to as KOLs, are valuable to the credibility of the manuscript and therefore essential to the whole project of publication planning. To leave analysis and writing to authors, though, will often result in a manuscript that does not serve the company's interests. One experienced planner addressing colleagues at a 2011 conference jokingly illustrates the point, waving an imaginary manuscript in the air: "What is this? They're promoting the competitor!" Another calmly affirms that "the approach of having an industry-authored [industry-written] first draft is a good one." Moreover, as depicted by planners, authors typically make few substantial contributions to the manuscripts they author, are slow to respond, and miss deadlines. Some expect prominence in authorship order and even try to violate ethical practices, for example, by trying to remove acknowledgment of medical writers.

While the planners complain about deadbeat authors, they create the conditions for those deadbeats. According to one speaker's estimate, 50% of companies show only the penultimate manuscript to authors, to solicit their input. It is likely that authors will have little to add to a well-crafted manuscript essentially ready for publication. That becomes especially likely if authors are given tight deadlines. According to one whistle-blowing medical researcher, part of the problem he faced was that he received abstracts only after they were submitted (and accepted) for meetings and received manuscripts only days before the planners' deadlines for journal submission. The orderly and efficient rollout of presentations and papers means that authors are likely to contribute little.

To produce a manuscript, planners must coordinate work by multiple parties they do not want to become authors, such as company statisticians, company and agency researchers, and medical writers. Perhaps some of these would meet authorship criteria, or perhaps no single person would, because the ICMJE maintains a concept of authorship that does not easily apply to cases of corporate production of manuscripts. On the whole, research as managed by publication planners is hard to fit into the ICMJE's framework, although it directly opposes the implicit ethical stance adopted by those criteria.

In this largely unseen process, pharmaceutical companies initiate and fund the planning, research, analysis, writing, and the placing of papers, and typically maintain control of data throughout. In the corporate production of knowledge, medical writers, authors, planners, company scientists and statisticians all play roles, with the end goal of supporting the marketing of products.

Industry-Sponsored Dissemination of Medical Knowledge

Sales Representatives

> [T]hese tremendous R and D budgets and the entire flow of knowledge and information used to discover new products rests on the ability of the industry to convince those who can write a prescription, or a script (doctors, nurse practitioners, physician's assistants, etc.), to write that script for their particular product.—Michael Oldani (2004: 329)

Publications address many audiences, but to pharmaceutical companies the most important audiences are generally made up of prescribing physicians. Simply by being published, articles may reach some physicians and researchers who influence physicians. But companies do not rely on this route and distribute their preferred knowledge directly to physicians by sales representatives, key opinion leaders (described below), and occasionally by direct mailings—the pharmaceutical company Merck is reported to have bought 900,000 copies of a *New England Journal of Medicine* article reporting a large trial of Vioxx (Smith 2006), an amount that suggests at least some mass mailings.

At the ground level, pharmaceutical sales representatives use reprints of publications for promotional purposes. Addressing an audience of publication planners at a large conference, a former sales representative, now an industry consultant, gives a bit of a pep talk: "Folks, they're dying for your work, by the way. Field reps are dying every day for more of your work. You know that, right? Because that's what doctors are going to see." In the course of discussing legal developments around interactions between companies and physicians, she indicates how articles create opportunities for conversations. She imagines a situation in which she is not permitted to discuss science suggesting the off-label use (i.e., use not sanctioned by regulators) of drugs:

> Those of you who've been a sales rep know how difficult that would be. First reprint I've gotten now in three years, and I've got it gripped, I've got it in my hand .... So now I'm now in an office and I've got this reprint and I think 'Hi doc, good to see you today. By the way, and one more thing, here's this reprint. Goodbye.'

This, she indicates, would be uncomfortable, unusual, and unproductive. For sales representatives, distributing reprint of scientific article is an opportunity to initiate discussions. Those discussions will invariably be of the article, but also about how to use its information—setting up possible prescriptions and sales.

Of course, pharmaceutical sales representatives do much more than transmit knowledge. Providing information about drugs is only one component of their jobs. Sales reps also provide samples, as gifts and to get physicians used to prescribing their products. Where these haven't been made illegal, they provide other, larger, gifts to physicians (see Oldani 2004). Perhaps most commonly, they provide friendship (e.g., Fugh-Berman and Ahari 2007). Ultimately, though, it is the fact that sales reps provide information, whether in the form of scientific reprints or product information sheets, that legitimizes their presence in physicians' offices. The transmission of medical knowledge is what allows sales reps to make their pitches, offer their friendship, and convince physicians to prescribe specific drugs.

But while sales representatives might be prominent vectors for the distribution of knowledge, their sales role makes them suspect. According to another former pharmaceutical sales representative, "[t]here are a lot of physicians who don't believe what we as drug representatives say. If we have a KOL stand in front of them and say the same thing, they believe it" (Moynihan 2008). For this reason it is worth taking a detailed look at key opinion leaders.

Key Opinion Leaders

> I would give them all the information that I wanted them to talk about. I would give them the slides. They would go through specific training programs on what to say, what not to say, how to answer to specific questions, so that it would be beneficial to my company.—Former sales representative (quoted in Moynihan 2008: 1402)

The term "key opinion leader" descends from work on structures of influence in such domains as politics, fashion, culture, and medicine, by the sociologist Paul Lazarsfeld and his students (e.g., Katz and Lazarsfeld 1955). As the term itself suggests, pharmaceutical companies engage key opinion leaders (normally "KOLs") primarily to influence others, to lead opinions in the directions that the companies prefer: KOLs are key mediators between pharmaceutical companies and physicians. For this reason, relations between the companies and KOLs are ideally, from the point of view of the companies, part of general "KOL management" (a standard phrase) plans.

There are different kinds of KOLs. High up in the ranks of prestige, and paid accordingly, are researcher KOLs. These are usually academics with significant research profiles, whose value to pharmaceutical companies might stem from any number of activities. As discussed already, they might be engaged by publication planners to serve as authors on medical journal articles stemming from company-led research. In addition, they might contribute to research by recruiting patients for trials, by responding to invitations to conduct trials on specific topics, or by initiating their own trials. They might be consulted on any of medical, marketing, or research issues. And finally, researcher KOLs are often paid to speak to other researchers, patient groups, or to physicians at continuing medical education sessions, with the eventual goal of increasing sales. One speaker at a 2012 meeting on KOLs, enthusing about a new approach to network analysis, says, "So it's really very, very interesting and starts to give us the tool and the power to be able to actually look at these network maps and start to think about the implication in terms of the things that we are doing commercially." A marketing firm writes in overview: "Interacting with qualified investigators, physicians experienced in regulatory reviews, well-known and respected speakers, and highly published authors will help to efficiently manage tasks within the critical path of the product and disseminate the message of the product to the end prescribing audience" (InsiteResearch 2008).

These well-known physicians and researchers are not usually hired simply to present a company's script but are chosen and/or brought on board with care. Independent agencies identify KOLs who could serve the pharmaceutical companies' needs and may design communication plans for companies to build relationships and knowledge with their prospective collaborators. The KOLs are nurtured through seminars, close contact, advisory boards, and publications.

At lower levels of prestige, but equally valuable to pharmaceutical companies, are ordinary physicians, either general practitioners or specialists, who are paid to speak to other physicians. They become part of the "speakers bureaus" for particular drugs and give talks in clinics over lunchtime or after dinner, with other physicians assembled by sales representative around a buffet lunch or a pleasant dinner (Reid and Herder 2013). More occasionally, they may speak at community events, again organized by sales representatives. In this context, KOLs effectively become salespeople, as well as conveyors of scientific information: according to a Merck study, the return on investment from KOL-led meetings with physicians was almost double the return on meetings led by sales reps (Hensley and Martinez 2005). It is common for drug companies to measure prescriptions before and after KOLs' talks—companies buy prescription data from health information services companies, which buy it from pharmacies—and they expect an increase (Moynihan 2008). One marketer speaking at a 2012 conference on KOLs essentially defined promotion in these terms: "you have a key opinion leader engagement with a group of doctors, and you measure sales before and after the engagement."

KOLs do more than sell, though. Both physician and researcher KOLs, but especially the latter, may be engaged for larger marketing activities, paving the way for later sales efforts. For example, they may be engaged first to learn about and then speak on diseases, rather than drugs:

> Another common objective…is to educate the marketplace and drive awareness of a particular disease state, mechanism of action, or existing treatment alternatives. A goal within this objective may be to successfully engage with key opinion leaders by completing a set number of advisory boards. (CampbellAlliance 2011)

Marketing involves a full range of activities with the goal of coordinating products, distribution networks, and demand (e.g., Applbaum 2004). With this in mind, pharmaceutical companies might not only want to communicate with physicians and others about their products but also to create awareness of new opportunities and approaches and interest in and concern about particular conditions and to introduce fears about alternatives. For all of such goals, KOLs are excellent conduits.

The marketing company InsiteResearch (2008) argues that the term "KOL management" is the right one for interactions with KOLs. Drawing on a dictionary definition, the company argues that, in general, management should involve "handling, direction, and control," precisely what is needed to make KOLs effective. Thus a speakers bureau program begins with a training session, to ensure that speakers are well versed in the positive aspects of the product and able to speak effectively about them. For example, Wave Healthcare claims on its website:

> It's vital that advocates are able to communicate and influence colleagues with clarity and conviction. To ensure speakers are at the top of their game, we have developed a communication skills programme for clinicians. (Wave Healthcare 2011)

Another such firm, KnowledgePoint360, which owns Physicians World Speakers Bureau, offers programs for training speakers, and its promotional material appears to treat KOLs and employees in the same terms: "Whether it is for external resources, such as speakers, or internal staff, including sales representatives and medical science liaisons, a robust training program is critical to the long-term success of any pharmaceutical, biotech, or medical device company" (KnowledgePoint360 2010; see also Carlat 2007). Pharmaceutical companies work with physicians to make them "product champions" and pay them generously for their lectures (Moynihan 2008).

Typically, physician KOLs are nominated by sales representatives, who have a sense of their abilities. Sales reps will know "what's their stage presence?" argued one company employee making a presentation at a 2014 conference on speakers bureaus. They will also know if "he looks good in a tie," suggested another pharmaceutical company manager at the same conference—who quickly disavowed that kind of recommendation, because it doesn't relate to the KOL's knowledge or communication skill and so might be looked at askance by regulators.

Though physician KOLs' main role is to communicate facts and views to other physicians, with the goal of increasing sales, pharmaceutical companies might sometimes recognize the value of having speakers who are also customers. One company manager, addressing the audience at a 2011 KOL management conference, raises the specter of an investigation of a speakers bureau program: "When you say 'I need 700 to 1000 speakers in this activity', the questions [that are] going to get pushed back to you in investigations are, 'Why do you need so many? How many is each speaker going to do? Why did you need a thousand?'" His concern here is that investigators will conclude that speakers' fees are inducements to prescribe or payments to receive (rather than transmit) advertising messages. That would constitute illegal marketing.

After physician KOLs have been trained, they become part of a speakers bureau for a company and wait to be offered engagements. As already mentioned, sales representatives handle all of the details: transportation is arranged, the time and place are set, invitations are sent and resent, and the equipment is set up and the food laid out. All that the KOL has to do is to make the presentation. Most of the time, that is straightforward. In the USA, speakers are not permitted to adjust the prepackaged PowerPoint slides or to deviate from their scripts when doing what are known as promotional talks, the bulk of the engagements. Says one psychiatrist, who earns large sums in speaking fees:

> So if I am doing a promotional program for a company, I have to use the slide deck that they provide me—I am not allowed to alter it in any way and every word in that slide deck is basically reviewed by their own internal counsel... .

In addition to the slides and the scripts, answers to standard questions also are scripted, and speakers are trained not to answer any questions in ways that might either be illegal or run against company interests.

The KOLs are sometimes aware that they are being used by pharmaceutical companies, though they find ways to defend their actions. One endocrinologist interviewed is quite critical of the industry and especially of its role in promotional talks, though he gives them regularly. "The reason for giving the promotional talks is to help the company sell its drug—I mean that's basically—that's what a promotional talk is." A hospital-based hematologist echoes this point but manages to find educational value despite the problems. "The honest answer is that promotional talks are not really for educating so—and I give plenty of promotional talks—... but some speakers are better than others at bending it into an educational talk." The value of their work in terms of education is echoed by others. Every single one of the KOLs interviewed for this project invoke education as a reason for speaking on behalf of companies, even when they are doing purely promotional speaking. In interviews, KOLs take pride in their teaching, and teaching is how they frame even promotional talks. "I am educating fellow physicians. I spend my day educating patients, I spend some of my evenings educating fellow physicians," explains one.

But in this context, it is the pharmaceutical companies' preferred messages that are being communicated. As we have already seen, pharmaceutical companies produce large quantities of data, shaped and arranged to support their interests. When KOLs serve as authors on company manuscripts, they give their weight to the establishment of that knowledge in the medical community as a whole. But pharmaceutical companies are not content to let that knowledge sit in medical journals, where it does little good. While there are senses in which KOLs are providing something like education, often involving scientific information, the education they are providing has been shaped and created by the companies for which they are working. KOLs do distribute knowledge.

If done correctly, then, KOL management will spread knowledge, change opinions, and change prescribing habits—which will generate a good return on companies' investment.

Orchestration of Continuing Medical Education

> Based on my very direct experience, quite frankly, the CME lectures which everybody espouses as being appropriate interaction for the best, can be the most biased presentations of any you'll ever see given—and you don't ever trace back the funding for the CME group to the couple of companies giving the vast majority of the money to one of those speakers

bureaus. So while CMEs are given a veneer of legitimacy they actually can be very dangerous to the public educational experience.—Interviewed clinical researcher (oncologist) KOL

In many jurisdictions, physicians must take continuing medical education (CME) courses to maintain accreditation. CME is supposed to be independent of corporate interests—sponsors are not allowed to control the content of courses. For pharmaceutical companies, this is the best kind of marketing: directed at receptive audiences that need to educate themselves and provided by sources that the audiences have reasons to trust.

The independent agencies that organize most of these courses are allowed to provide organization, pay for speakers, help speakers prepare their talks, and provide entertainment for participants. In 2012, commercial support for CME (including advertising and related income) in the USA accounted for roughly 40% of income for accredited CME providers, a considerable reduction from a few years earlier (Accreditation Council for Continuing Medical Education 2012).

Accredited CME providers are subject to regulation, the most important aspect of which is that sponsors like pharmaceutical companies may not control the content of courses. In the USA, pharmaceutical companies may provide funding for CME, may provide organization, pay for KOL speakers, help them prepare their talks, and provide entertainment for participants. In some cases, even fully independent organizations may invite them to influence content: for example, one letter by a Canadian medical organization soliciting funds for a CME conference stated that "major sponsors will be given the opportunity to nominate participants to represent industry's interest and to participate actively in the conference" (Brody 2007: 208). But in theory, the company must allow speakers complete freedom when it comes to the actual content.

Yet, for pharmaceutical companies, it is only a modest challenge to align KOLs with their own interests when it comes to CME. If the above measures of support are not enough, companies have further systems and methods of orchestrating CMEs indirectly. If sponsors have chosen their speakers well, supported the research of these speakers, and given them templates and slides for their talks, the courses will convey preferred messages.

The companies attempt to carefully manage their KOLs, their promotional talks, and their contributions to CME. At the very least, those talks tend to strongly endorse the sponsor's products. As one medical education and communication company advertised: "Medical education is a powerful tool that can deliver your message to key audiences, and get those audiences to take action that benefits your product" (quoted in Angell 2004: 139). Both promotional and CME talks, then, are part of pharmaceutical companies' promotional campaigns. Any education their talks provide and any health benefits that result from it have to be understood as shaped by the sponsoring companies' interests. According to an industry education specialist, the ideal for CME is "control—leaving nothing to chance" (Bohdanowicz 2009).

Summary and Conclusion

> So I do agree that in the past it was a little excessive and there was probably too much influence in a negative way but now I think it's the other way around—[distrust and regulation of pharmaceutical marketing is] stifling innovations and when I meet, you know I'll be honest, when I meet doctors who refuse to attend any promotional events they honestly are usually the ones that are the least educated about products in our field.—Interviewed physician (psychiatrist) KOL

Pharmaceutical marketing is centered on the production and distribution of information, mostly forms of medical science. The industry provides roughly half of all funding for clinical trials—often randomized, controlled trials, the most valued form of medical knowledge—and sponsors a majority of the new trials initiated each year. The bulk of that funding goes to CROs, which make no claim on the data they produce, simply handing it over to the companies that hire them to use as they want. Using this and any other available data, the pharmaceutical industry produces a significant portion of the scientific literature on in-patent prescription drugs, with as many as 40% of the articles on recently approved drugs in the more prestigious medical journals having been ghost managed for companies. Publication planners create teams of professionals to create those articles and choose KOLs to serve as authors and journals to which to submit them.

Companies' interests can thus be expected to influence a myriad of legitimate choices in the design, implementation, analysis, description, and publication of clinical trials. We can reasonably expect, and there is abundant evidence, that companies make those choices to support their commercial interests. Even if companies are not completely coherent actors, they are coherent enough in their goals that choices in all the different stages of research and communication can point in the same direction. The result is still recognizably medical science, and may even be high quality science, but it is science serving particular and clear interests.

This continues with the communication of medical science in the field. The science that sales representatives give to physicians is invariably science that serves their companies' interests. When they give talks, KOLs contribute to the enormous influence that the pharmaceutical industry has on medical knowledge. The form of CME in which KOLs participate is one thoroughly shaped by the interests of the companies that sponsor it. What is communicated will often be sound medical science, which is why KOLs are willing to communicate it; nevertheless, it will be science chosen to help sell a product.

This chapter is broadly about knowledge, but the specific issues at stake are not especially salient within the traditional philosophical project of epistemology. Epistemology is centrally the study of justification, and in particular the study of the justification of individualized beliefs. Undoubtedly, some of the claims that pharmaceutical companies make and promote are poorly justified, and undoubtedly some are false in egregious ways. On occasion, there are significant scandals involving errors, falsehoods, and gross manipulations circulated by pharmaceutical companies. But by and large, these companies are working within the medical mainstream, producing data of reasonably high quality using the most valued of research tools; they go on to analyze it using standard statistical means and construct articles that pass the scrutiny of peer reviewers at many of the best medical journals. The problems of knowledge in the pharmaceutical industry discussed in this chapter are not primarily problems of justification.

However, seen in terms of political economies of knowledge, there are very large concerns. As this chapter has illustrated, largely unnoticed issues of influence and control permeate important areas of medical knowledge, whether the claims that constitute that knowledge involve issues of justification or not. Individual companies with stakes in specific medical topics have the capability to influence knowledge so significantly that their preferred science becomes dominant. They have the capability to achieve hegemony over understandings of particular diseases, symptoms, treatment options, trajectories and side effects. By virtue of the enormous resources at their disposal, they have dominant positions within the political economy of medical knowledge. This flood of pharmaceutical knowledge is not created for broad human benefit, but to increase profits; at least some of the time those two goals will be very opposed. We might ask not whether this or that piece of pharmaceutical knowledge is justified or true, but note instead that the structures of knowledge that create it concentrate power in very few actors, who in turn have very narrow interests.

We can see an aspect of pharmaceutical companies' influence in the statement at the head of this section. For the KOL quoted, being an educated physician is closely linked with knowing the material that pharmaceutical companies are communicating. In fact, most of the KOLs interviewed for this project defend their giving promotional talks in related terms: if physicians are "not educated enough, the public will suffer," says one; "oh, it helps other patients elsewhere, it's spreading the word—it's spreading the gospel," enthuses another. For them, pharmaceutical industry influence has become normalized, to the extent that they see the industry as the source of the most valuable medical information. The industry has achieved a level of hegemony over parts of medical education and thus over what physicians see as treatable diseases and how they should be treated.

Pharmaceutical companies ghost manage the production of the research, they shepherd the KOLs who disseminate the research as both authors and speakers, and finally they orchestrate the delivery of CME courses. In so doing, they attempt to be the ultimate sources of the information physicians rely on to make rational decisions about patient care. In this we can see the importance of the hegemony of knowledge gained through pharmaceutical industry strategy.

Acknowledgment This chapter presents material from a large project on the political economy of pharmaceutical knowledge. That project was funded by a grant from the Social Sciences and Humanities Research Council of Canada (#410-2010-1033). Some additional research was funded by a grant from the Canadian Institutes of Health Research (#2009-11-02). Zdenka Chloubova performed many of the interviews quoted here. Khadija Coxon assisted with other research that made its way into this chapter and helped considerably with the organization of a draft. Apologies are owed to Ernesto Laclau and Chantal Mouffe for the chapter's title.

References

Accreditation Council for Continuing Medical Education. 2012. *Annual Report*. http://www.accme.org/sites/default/files/630_2012_Annual_Report_20130724_1.pdf. Accessed 3 Oct 2013.

Adis Communications. 2006. http://www.pharmalive.com/content/supplements/gpms/2004/adis.cfm. Accessed 20 Dec 2006.

Angell, M. 2004. *The Rruth about the Drug companies: How They Deceive Us and What to Do about It*. New York: Random House.

Applbaum, K. 2004. *The Marketing Era: From Professional Practice to Global Provisioning*. New York: Routledge.

Bohdanowicz, H. 2005, September 21. A Guide to Strategic Communication Planning. *Pharmaceutical Executive Europe*.

———. 2009. The Synergy of Public Relations and Medical Education. *Communiqué* 24: 14–16.

Brody, H. 2007. *Hooked: Ethics, the Medical Profession, and the Pharmaceutical Industry*. Lanham: Rowman & Littlefield Publishers.

CampbellAlliance. 2011 *Communicating the Value of Medical Affairs*. Brochure for a Whitepaper.

Carlat, D. 2007, Nov. 25. Dr. Drug Rep. *The New York Times*.

CenterWatch. 2009. *State of the Clinical Trials Industry: A Sourcebook of Charts and Statistics*. Boston: CenterWatch.

Envision Pharma. 2006. http://www.envisionpharma.com/publicationsPlanning/. Accessed 20 Dec 2006.

Fisher, J. 2009. *Medical Research for Hire: The Political Economy of Pharmaceutical Clinical Trials*. New Brunswick: Rutgers University Press.

Fugh-Berman, A., and S. Ahari. 2007. Following the Script: How Drug Reps Make Friends and Influence Doctors. *PLoS Medicine* 4(4): e150.

Gardiner-Caldwell Group. 2007. http://www.thgc-group.com/. Accessed Nov 2007.

Gramsci, A. 1971 *Selections from the Prison Notebooks of Antonio Gramsci*, Eds. and trans. Q. Hoare and G.N. Smith. New York: International.

Healy, D., and D. Cattell. 2003. Interface between Authorship, Industry and Science in the Domain of Therapeutics. *The British Journal of Psychiatry* 183: 22–27.

Hensley, S., and B. Martinez. 2005, July 15. New Treatment: To Sell Their Drugs, Companies Increasingly Rely on Doctors. *Wall Street Journal*.

ICMJE. 2014. *Defining the Role of Authors and Contributors*. http://www.icmje.org/recommendations/browse/roles-and-responsibilities/defining-the-role-of-authors-and-contributors.html. Accessed 30 Jan 2015.

InsiteResearch. 2008. The Prescription for KOL Management. *Next Generation Pharmaceutical* 12. http://www.ngpharma.com/. Accessed 28 Mar 2011.

Katz, E., and P. Lazarsfeld. 1955. *Personal Influence: The Part Played by People in the Flow of Mass Communications*. Glencoe: The Free Press.

KnowledgePoint360. 2010. *Promotional Brochure*.

Lundh, A., S. Sismondo, J. Lexchin, O. Busuioc, and L. Bero. 2012. Industry Sponsorship and Research Outcome. *Cochrane Database of Systematic Reviews* 12: MR000033.

Mirowski, P., and R. Van Horn. 2005. The Contract Research Organization and the Commercialization of Scientific Research. *Social Studies of Science* 35: 503–534.

Moynihan, R. 2008. Key Opinion Leaders: Independent Experts or Drug Representatives in Disguise. *British Medical Journal* 336: 1402–1403.

Oldani, M. 2004. Thick Prescriptions: Toward an Interpretation of Pharmaceutical Sales Practices. *Medical Anthropology Quarterly* 18(3): 325–356.

Petryna, A. 2009. *When Experiments Travel: Clinical Trials and the Global Search for Human Subjects*. Princeton: Princeton University Press.

Reid, L., and M. Herder. 2013. The Speakers' Bureau System: A Form of Peer Selling. *Open Medicine* 7(2): e31.

Ross, J.S., K.P. Hill, D.S. Egilman, and H.M. Krumholz. 2008. Guest Authorship and Ghostwriting in Publications Related to Rofecoxib: A Case Study of Industry Documents from Rofecoxib Litigation. *Journal of the American Medical Association* 299(15): 1800–1812.

Shah, S. 2006. *The Body Hunters: Testing New Drugs on the World's Poorest Patients*. New York: New Press.

Sismondo, S. 2007. Ghost Management: How Much of the Medical Literature is Shaped Behind the Scenes by the Pharmaceutical Industry? *PLoS Medicine* 4(9): e286.

———. 2009. Ghosts in the Machine: Publication Planning in the Medical Sciences. *Social Studies of Science* 39(2): 171–198.

Smith, R. 2006. Lapses at the New England Journal of Medicine. *Journal of the Royal Society of Medicine* 99(8): 380–382.

Watermeadow Medical. 2007. http://www.watermeadowmedical.com/. Accessed 29 Nov 2007.

Wave Healthcare. 2011. *KOL Training*. http://www.wavehealthcare.co.uk/en/1/koltraining.html. Accessed 25 Mar 2011.

Wyeth. 2002. *Publication Plan 2002—Premarin/Trimegestone HRT Working Draft*. http://dida.library.ucsf.edu/tid/awb37b10. Accessed 19 Aug 2013.

A Call to Stop Treating Doctors Like Delinquent Adolescents and Medical Product Companies Like Criminal Enterprises

Lance K. Stell

> The fear that people would make bad decisions if given truthful information cannot justify content-based burdens on speech. That precept applies with full force when the audience—here, prescribing physicians—consists of 'sophisticated and experienced' consumers. (Sorrell v IMS Health 2011)

For the past 20 years or so, we have been deluged with moralistic attacks on physicians' multifaceted relationships with makers of branded medical products. *Availability entrepreneurs*, (Kuran and Sunstein 1999) using "conflict of interest" (COI) tendentiously to frame these relationships, have created a corruption-tinged moral panic (Rohloff and Wright 2010). A commentator warned early on that by turning away from evaluating research quality toward relationship risk framing, COI might morph into "a new McCarthyism" (Rothman 1993). The warning was brushed aside. COI-hunting articles swelled from a cascade to a tsunami. A few offered resistance (Stell 2002, 2005, 2009; Stossel 2005, 2008; Huddle 2008; Hirsch 2009; Lesko et al. 2012; Barton et al. 2014), but thousands of opposing commentaries and research articles poured over and around them. Not surprisingly, conformity trumped dissent (Sunstein 2002), and the response was tepid from those with the most at stake—drug/device innovators, translational scientists, the medical product industry itself, and ultimately patient groups (Stossel and Stell 2011).

COI entrepreneurs have swamped high-impact professional journals with one-sided opinions and studies, and editors have been complicit. These entrepreneurs have repeatedly marketed anecdotes about corruption to the national media, thereby creating a social-signaling echo-chamber, while mostly insulating themselves from dissent. With the support of Congressional converts, they have put in place *social-distancing* policies designed to contain the supposed-contagion of commercial

L.K. Stell (✉)
Professor of Philosophy (Emeritus), Davidson College, Davidson, NC 28036, USA

Emeritus Professor of Surgery and Internal Medicine, Carolinas Medical Center, Charlotte, NC 28232, USA
e-mail: Lastell@Davidson.edu

© Springer Science+Business Media B.V. 2017

D. Ho (ed.), *Philosophical Issues in Pharmaceutics*, Philosophy and Medicine 122,
DOI 10.1007/978-94-024-0979-6_5

influence. They have encouraged a vast, costly shaming apparatus to stigmatize physicians who accept payments (down to the $10 level) for working with their industry partners to develop new drugs and devices and to disseminate clinically relevant information to their colleagues over dinner at nice places.

AMSA's PharmaFree Program, guided by remarkable ethical certitude and underwritten with funds from the Macy Foundation and the 2004 Neurontin® settlement, issues annual report cards grading more than 400 medical schools and medical centers on the aggressiveness of their social-distancing policies. Many major medical centers now cloister their academic physicians, forbidding them to visit with drug reps on or off campus. Medical scientists and experienced clinicians are prohibited from giving paid product-update lectures to colleagues over dinner. Medical students treat themselves, residents, and fellows like hothouse plants—sheltered from and taught to fear or even to despise the supposedly professionalism-destroying undue influence of an industry whose products they will be or actually are prescribing to their patients. Commercial sponsorship for accredited continuing medical education has declined 50% since 2006, because the COI entrepreneurs believe that sponsorship biases education, no matter how long the arms-length requirements imposed by the ACCME.

To rationalize these policies, COI entrepreneurs imported gift-exchange theory from social science to portray academic physicians and other life-scientists as "cheap dates," eager to sell their professional integrity for a logo-marked pen, a coffee mug, a pizza, or a bag of bagels. When amended industry codes (PhRMA 2008 and AvaMed 2009) prohibited gifting (effective, January 2009), COI entrepreneurs redefined the vexed category "gifts" to "gifts or payments," whether direct or indirect. Federal law now requires that any "transfer of value" of $10 or more a year received from industry must be reported by each doctor and medical center on a searchable national website. The law has no provision for this costly, error-prone enterprise to sunset if measurable improvements in medical professionalism or in public confidence, proven by data, fail to materialize. Nor does the provision's survival rest on proving with data that patients (rather than media interests) regularly use the website for scrutiny of listed physicians and medical centers.

The moral panic has characterized such once-respected medical product companies as RICO enterprises for exercising its constitutional privilege of commercial speech to promote off-label product uses. FDA admits the value of off-label use and has opined that off-label promotion is not per se illegal (Caronia v U.S. 2012, decision not appealed). The core argument is that industry marketing *causes* doctors to prescribe off-label and to make "false claims" for government reimbursement—in violation of a Civil War-era law that allows triple damages for each count. The national media portray False Claims Act (FCA) whistle-blowers who get 25–30% of a lawsuit's settlement (the record award to a whistleblower stands at $700 million) as altruistic guardians of the public good. Under threat of being "debarred" from doing any further business with the government, makers of branded medical products have accepted settlements in the billions of dollars (current record: GSK $3 billion) and submit to "corporate integrity agreements," requiring their expending hundreds of millions of dollars more to clean up their business practices, while being monitored by compliance overseers.

The COI entrepreneurs are correct: *No Free Lunch*! The costs they have imposed must come from somewhere, at least some in product price increases and lost research and development resources. The drug rep workforce has decreased 25%. Industry funding for continuing medical education has declined 50% since 2006. Commercial support for undergraduate medical education has declined 34.7% since 2007 (ACCME 2011 Report). Collaboration between medical product companies and physicians has been made more difficult. FDA has increasing difficulty obtaining the very best expert advice, because the most knowledgeable life-science experts are "conflicted" in every imaginable dimension. Information flows have been impeded, knowledge costs have been increased, the value of dissent in error discovery and correction has been short-circuited, all because, at bottom, medical professionals as a whole supposedly cannot be trusted to behave like "sophisticated consumers" of medical product information—which the Supreme Court expects them to be. An already painfully slow turning engine of bench-to-bedside "translational science" turns slower. Its productivity has fallen 80-fold over the past 50 years (Scannell et al. 2012). Further distancing health scientists and physicians from medical product development will not stem the decline. Neither the COI entrepreneurs nor anyone else has been held accountable to prove with data that the aggregate costs of this moral panic have made doctors or companies more ethical and the public more trusting and safer or that the whole exercise has been "worth it," all things considered.

The COI Entrepreneurs' Argument in a Nutshell

1. All relationships between makers of branded products and physicians create intolerable *conflicts of interest* for physicians, especially financial conflicts of interest, the kind most ethically toxic to professional judgment.
2. Myriad studies have concluded that marketing unduly influences physicians' prescribing decisions.
3. A widely publicized systematic review of studies examining medical product company marketing has been cited by professional thought leaders for finding overwhelmingly "negative results on clinical care" or patient harm.
4. A commitment to professionalism implies that conflicts of interest should be minimized—eliminated, reduced, or otherwise managed—and, to enable appropriate credibility discounting, always disclosed.
5. Since physicians have proven themselves *susceptible* to Medical Pharmaceutical Industry's (MPI's) *undue influence*, *social-distancing policies* should prevent/deter their meeting with drug/device reps anywhere. Taking gifts, monetary payments (direct or indirect), or food from MPI should be *stigmatized* (AMSA, PharmedOut, No Free Lunch: Take the Pledge).
6. All MPI transfers of value, direct or indirect, no matter how small should be publicly disclosed—by every physician, medical center, and medical school on an annually updated, searchable website.

Conflict of Interest

> Every physician-patient encounter is a conflict of interest. Every physician-payer encounter is also a conflict of interest. (Todd 1991)

Incentive misalignments are ubiquitous (Langbein 2005). In every relationship, the parties face incentives, varying in strength, to act contrary to and without regard for the interests of the other(s). Relationships are always "conflicted" to some degree: husbands with wives, parents with children, siblings with each other, friend with friend, partner with partner, lawyers with clients, research subjects with investigators, and physicians with patients. Indeed, the relationship we each have with ourselves is "conflicted." Self-help books are premised on the conflicts between the motivating effects of what we take an interest in and what is in our own best interests. Each of us makes most of our own trouble. Each is his own worst enemy. But the COI frame, by emphasizing relationship risk, ignores convergent interests which may be predominant, deeper, more persistent, and stronger. The COI frame creates a biased picture, inviting the false belief that non-conflicted relationships exist. Hence it has been observed that conflict of interest is "an epithet" (Langbein 2005).

There are several conceptions of COI but two are most common in the medical literature. One says that a physician is conflicted when a reasonable person/observer would [should?] perceive the physician's relationships or circumstances as role compromising (Erde 1996; Brody 2007; Brennan et al. 2006).

Conflicts of interest occur when "…motives that caregivers have and/or situations in which we could reasonably think caregivers' responsibilities to observe, judge, and act according to the moral requirements of their role are or will be compromised to an unacceptable degree" (Shimm and Spece 1996).

The other conception says that a physician is conflicted when a second party or third party or a relationship or a situation [to a high degree of medical certainty?] has exerted, is exerting, or will exert "undue influence" on the physician's actions (e.g., prescribing) adverse to the interests of a second party (e.g., a patient) or third party (e.g., a payer or society) (Thompson 1993; IOM 2009). Each conception sometimes adds a proviso, "potentially" conflicted, a nuance I ignore.

The Perceptions of a Reasonable Person

The "reasonable person" theory of COI should be rejected. Diagnosing others' motives, even when performed by a reasonable person, is dubious. Appearances on which the diagnosis will be based and the attribution made are susceptible to manipulation. Under experimental conditions, a group of otherwise reasonable persons can be made to perceive objects as moving when they are stationary, even to the point of offering varying estimates of how far a motionless item moved (Sherif 1937).

An individual who expresses conformity with others' reported perceptions (e.g., that the object moved) serves his own interest because reporting convergence (despite that he may actually perceive and believe otherwise) better secures his interest group approval. A dissenter, by expressing nonconformity, better serves the group's interest (in error discovery/reduction) but at personal cost (isolation, group disapproval).

Even when firsthand (not hearsay) perceptions offered in evidence are subjected to adversarial examination, juries have convicted witches; individuals have been condemned and their careers ruined for having engaged in "un-American activities"; and courts of law have awarded damages when an obstetrician's substandard care can be made to appear as the proximate cause for a child's cerebral palsy (the cause(s) of cerebral palsy remain unknown). The reasonable conception of COI makes no provision for reasonable disagreement nor for adversarial examination of appearances of impropriety. Conformity pressures toward groupthink and error are ignored.

Undue Influence

The undue influence theory of COI aims to protect freedom of choice by nullifying those made under *undue influence*. The concept "undue influence" is problematic. One recent legal treatise characterized it as "one of the most bothersome concepts in American law" (Dukeminier et al. 2005; Scalise 2008; Madoff 1997). The concept is bothersome because the value it protects, decision-maker autonomy, is at least as bothersome. Analyzing a vexed concept by reference to another will be, well, vexing.

The "undue influence" theory of conflict of interest entered the medical COI literature in 1993 as follows:

> A conflict of interest is a set of conditions in which professional judgment concerning a primary interest (such as a patient's welfare or the validity of research) tends to be unduly influenced by a secondary interest (such as financial gain). (Thompson 1993, emphasis mine)

The IOM (2009) adapted but possibly strengthened Thompson's definition in its book-length report on conflicts of interest.

> A conflict of interest is a set of circumstances that creates a risk that professional judgment or actions regarding a primary interest will be unduly influenced by a secondary interest. (Emphases mine)

What does it mean to say that there is "a risk" that one interest "will be" unduly influenced by another? Risk means "probability of harm." Is "probability of harm" objective, based on statistical, numerical data? Or is it dependent upon readily recalled anecdotes and perceptions made available to a reasonable but suggestible observer?

An undue influence inquiry determines whether an agent (a physician) was "susceptible" to influence in the circumstances and whether the influence complained of as "undue," "improper" influence which, by some mechanism, substituted the influencer's volition for that of the agent, overpowering his will without convincing his judgment, to the patient's detriment. An undue influence inquiry does not exclude that the influencer and patient are the same person.

Evidence that Patients Exert Comparatively Irresistible Influence on Physicians Prescribing

Using a survey of 211 patients, Keitz et al. (2007) determined patient pre-visit expectations for meds, tests, and specialist referrals, post-visit fulfillment of expectations, and the effects on satisfaction and trust. They found:

- Overall: 66% of patient expectations were met.
- Physicians suggested alternatives 21.6% of the time.
- Expectations met for meds, 75%.
- Expectations met for tests, 71%.
- Expectations met for referrals, 40%.

Of 138 physicians, 62 (44.9%) reported that they would not have ordered for meds, referrals, and tests had the patients not directly asked and acquiescing to eight requests (12.9%) made them uncomfortable. The study concluded that patients generally received what they asked for and altered physician behavior nearly half of the time. Patient satisfaction and trust were unchanged regardless of whether or not expectations were met.

It will be objected that direct-to-consumer (DTC) advertising prompts patients' requests for meds, tests, and referrals, which physicians, who are increasingly subject to patient satisfaction surveys, find irresistible (at least when compared with drug rep detailing). Were patients more ignorant of their options, *ex ante*, they might make fewer focused requests for their physicians. Thus, banning DTC has been proposed. I leave that proposal aside.

Legal Analysis of Undue Influence

> A classic English case in the undue-influence literature analyzed the concept as follows: The test for undue influence is the presence of agency and whether or not the individual will had been overpowered—the inequity of the act consists in compelling a person of ordinary firmness to do what he does not want to do...To be undue influence in the eye of the law there must be—to sum it up in a word—coercion. (Wingrove v Wingrove 1885: 82–83)

A more modern American source expands the analysis as follows:

> Any action that subverts the will by solicitation, flattery, putting in fear or some other manner...kindness or attention is not enough...[Its effect must be] Psychological domination resulting in an inability to resist the will of another. (Estate of Auen 1994: 565)

Susceptibility: Physician Narcissism

Members of the medical profession (incidence unspecified) have been diagnosed with "entitlement syndrome," a symptom associated with narcissistic personality disorder. Supposedly, the selective educational process, disciplined delayed gratification, initial commitment to service, and long years and hours of training eventually confer high status. Supposedly this makes it appear to some physicians in training that gift offers of pens, mugs, and nice dinners are deserved tribute, marks of appropriate respect shown by drug rep supplicants. But if so, it is argued, their syndrome makes them susceptible to gift-exchange bribery by the reps. The vain, affirmation-seeking physician, hungry to satisfy an acquired taste eagerly dispensed rep flattery, reciprocates by irrationally prescribing the costly products from the drug rep's bag (rather than cheaper generics) (Brody 2007: 192).

This explanatory mechanism doesn't make sense. The formerly committed, but now entitled, narcissistic physician, who accepts tribute from inferiors as her due, would not feel any obligation of reciprocity.

It will be objected that drug reps successfully use "key opinion leader" (KOL) designation as flattery to entice physicians to lead astray their colleagues at product promotion events. Physicians who attend such events are much more likely to prescribe and make formulary addition requests for the sponsor's products than non-attendees. However, the physician samples in these studies (attendees vs. non-attendees) are self-selected. The marginal effect of the KOL on attendee motivation to prescribe and to make formulary addition requests must be confounded by selection bias. Besides, drug-talk attendees voluntarily take a weeknight or weekend night to hear what the KOL has to say and to visit with their colleagues. Their opportunity cost—time away from home/family—must be judged "worth it" all things considered. Prohibiting attendance at such gatherings is oppressive.

The standard of "firmness" for measuring physicians' susceptibility to influence should be higher than for a person afflicted with narcissistic personality, higher than for "the man of ordinary firmness," and, indeed, higher than for a "medical student of average firmness" or for other trainees such as residents and fellows.

We do not allow trainees to set the standard for measuring the adequacy of a licensed physician's clinical decisions. How does their naiveté qualify them to set the professionalism standard for physicians when they have yet to attain that status? Neuroscientists claim to have discovered that prefrontal lobe development, where brain functions for impulse control and judgment have been localized, is not complete until age 25 or thereabouts (Lebel and Beaulieu 2011). It is not obvious why medical students whose frontal lobe connections have only recently matured and who have little or no experience in dealing with commercial product information should set the standard for physician exposure to marketplace influence. Howsoever, the neuroscience of judgment may turn out to be that the COI entrepreneurs have not proved with data that physicians are so susceptible to having their wills overpowered and without having convinced their judgment that they must be protected for their own and their patients' sake. This will be further discussed below.

COI Management

No one knows what "management" means in practical terms. The Institute of Medicine report on conflict of interest says that managing COIs should involve "… judicious practical judgment sensitive to the institutional context." In other words, managing COIs should be a "slow-thinking" (Kahnemann 2011) process—balancing costs and benefits, considering relevant differences. But real-world COI management has tended to favor "fast-thinking," zero-tolerance, categorical, heavy-handed administration, i.e., "banning and prohibiting." That's what "taking COI seriously" typically means.

When "slow-thinking" nuanced management review of an actual or proposed commercial relationship is undertaken, it would tend to be well, slow, and responsive to contextualized relevant differences. But by avoiding blanket judgments in favor of exceptions, the process risks perceptions of favoritism and arbitrary indulgences. Delayed, slow arrival at assessments of COI imposes a time-tax on relationship proposals that discourage collaboration. Slow thinking does not weigh risks symmetrically. Scandal risk (headline liability) is weighed more heavily than the risks of foregone benefit. The latter's costs are uncalculated and invisible and will not make news in the morning paper. On the other hand, no administrator will be fired for nixing a proposed collaboration.

A policy of avoidance of "relevant" COIs creates a presumption, a default against relationships with medical product companies. This on the assumption that such relationships are of no value, or are so shot through with corruption that the outcome of the relationships must be tainted. This thinking is exemplified in the DeAngelis-Fontanarosa Rule (J Am Med Assoc DeAngelis and Fontanarosa 2008) that imposed additional "independent" statistical review-costs on author submissions reporting research sponsored by medical product companies. This COI management policy was anecdote based, not evidence based.

A study (Woolley et al. 2011) of manuscripts withdrawn for research misconduct (e.g., falsification, fabrication, and plagiarism) certainly did not confirm *JAMA*'s discriminatory policy. On the contrary, "of the 463 retracted publications retrieved, 213 (46%) were retracted because of misconduct. Publications retracted because of misconduct rarely involved declared medical writers (3/213; 1.4%) or declared pharmaceutical industry support (8/213; 3.8%); no misconduct retractions involved both declared medical writers and the industry." Vera-Badillo et al. (2013) have observed: "The pharmaceutical industry is increasingly influential in clinical trial sponsorship with data showing an increase industry sponsorship of phase III RCTs from 24 to 72% over a 30 year period. In our cohort of trials, 67% were industry sponsored, but we found no association between industry sponsorship and biased reporting of either efficacy or toxicity, and no association of for-profit sponsorship with change of the PE [primary endpoint] between that listed in trial registries and the final publication" (1241).

Full disclosure, another nostrum, is problematic. Cain et al. (2005), in a laboratory guessing-game experiment played by a sample of undergraduate students, found that disclosure of misaligned incentives, *where there is no ongoing relationship to encourage reliance*, was associated with somewhat reduced reliance on conveyed information but also even more reduced reliability of that information. Koch and Schmidt (2010), however, found that established relationships encouraged reliance, also greater reliability and more truthful information. Ben-Shahar and Schneider (2011) found little empirical support for disclosure efficacy across a wide range of commercial and healthcare settings. No one reads boilerplate disclosures nor knows what to make of the information. Is physician's full disclosure of his commercial/noncommercial interests a "trigger warning" to patients that they should take precautions, or is it a boast of his having noteworthy relationships? Which relationships really matter? Won't the discloser's characterizing the importance of his own relationship risks make him a "judge in his own case"? Isn't the whole point of asking for the disclosure some evidence that we can't trust the discloser to tell the whole truth or to give an unbiased interpretation? Are the most current relationships most important? What about favorably remembered ones from long ago? What about deeply felt indebtedness to mentors and colleagues from residency or a fellowship? What about distinguishing hated, hectoring, unreasonable but well-paying commercial sponsors from those lower-paying but fondly remembered? Is a current high-dollar grant unduly influential? Suppose the doctor sincerely regrets having gotten himself into it. Disclosures and omitted disclosures tell us none of these things.

"Nothing to declare" is also non-illuminating. Ideological, religious, ethnic, and personal biases are not disclosed. A doctor's commitments to her various value-causes and her tolerance for or distaste of commercial behavior, matters to her judgment. Disclosure of "significant" or "relevant" COIs in the recent past is also wanting. Examples of "significant" are unhelpful because no one knows what they mean in an individual case. The IOM insists that even "small gifts" of trivial value may (may!) influence a physician's prescribing behavior. By implication, *any* commercial influence counts as "undue influence."

What Do Patients Want to Know About Relationship Risk?

Kao et al. (2001) found that 54% of patients wanted to be informed about their physicians' payment method. Seventy-six percent said that an incentive to reduce utilization would reduce quality of care. This suggests that a majority of patients want information to enable their discounting the credibility of physicians' medical recommendations, giving them a reason to seek a second opinion. However, respondents answered without considering or incurring any costs, even hypothetical costs. For example, shouldn't the survey question ask: "*In an 8–10 min office visit with your* physician, *would you prefer that she/he* spend more time (how much more?) providing updates on *his/her* reimbursement *and consulting relationships* and less time (how much less?) discussing *your* medical condition, interpretation of tests, and treatment options/*recommendations?*" Survey questions that suppress consideration of trade-offs are worthless.

McDaniel et al. (2007) found that patients *preferred* that they be the focus of the clinical interview and that physicians should limit their self-disclosures. Pearson and Hymans (2002) found that physicians dislike discussing with patients how they are incentivized. “Nearly every physician avoided discussing the role of incentives and stressed instead that he or she could be trusted under any circumstances.”

Disclosure enables “outing,” stigmatizing, hectoring, and exclusion of “conflicted” doctors. To what purpose? As a theoretical exercise, the Public Citizen Group (Lurie et al. 2006) examined whether FDA expert panel voting would have changed if conflicted experts had been excluded. They found no difference. Prompted by the Lurie paper, an FDA research group (Gieser 2009) used an expanded dataset finding similar to Lurie that there was no statistically significant relationship between conflict rates and voting outcomes ($\alpha=0.05$) when considering financial ties with the product sponsor (“index conflict”), a competitor of the sponsor (“competitor conflict”), or the product sponsor or a competitor or both (“any conflict”). Overall, ERG found no evidence to suggest that having a financial conflict of interest tends to increase votes in favor of that interest.

A similar exercise was run for drafters of clinical practice guidelines (Jones et al. 2012). The authors found that excluding conflicted experts would not have made any difference in voting. What about excluding them from discussion? Remaining panelists complained that excluding conflicted experts robbed the discussion of intellectual depth and experiential enrichment.

After all these years of COI hectoring, disclosing, and managing, we have yet to receive any justified, useful guidance about who should be excluded absolutely, who should be excluded from guideline discussion, and who should not vote.

The COI Cascade

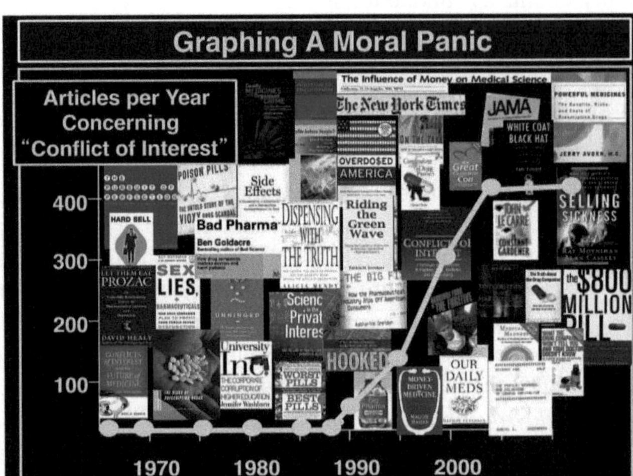

A PubMed search using “conflict of interest” as a search term returned one article from 1974. Three appeared in 1975. The number reached 98 in 1991. In 1995, 252 appeared. In 1997, the number reached 236. In 2009, Lanier (editor, Mayo Clinic

Proceedings) returned more than 7300 articles, 70% published over the previous decade. *British Medical Journal*, *Journal of the American Medical Association*, *The New England Journal of Medicine*, and *The Lancet* registered more than 195, 170, 160, and 125 COI items each. The evidence is that COI was made increasingly "available" in the professional and public media (PubMed Spread Sheet 1974–2013).

PubMed Spread Sheet

Pubmed – conflict of interest [MeSH Terms]			
Year	Count	Year	Count
2014	131	1993	147
2013	422	1992	147
2012	414	1991	98
2011	449	1990	48
2010	566	1989	24
2009	513	1988	11
2008	502	1987	17
2007	525	1986	12
2006	500	1985	16
2005	510	1984	7
2004	489	1983	10
2003	495	1982	29
2002	482	1981	16
2001	366	1980	8
2000	297	1979	4
1999	231	1978	4
1998	281	1977	7
1997	236	1976	2
1996	192	1975	3
1995	252	1974	1
1994	174 search date		10/22/14

An availability cascade has been described as a self-reinforcing process of collective belief formation, by which an expressed perception triggers a chain reaction that gives the perception ever increasing plausibility through its apparent availability in public discourse. The driving mechanism involves a combination of informational and reputational motives: individuals endorse the perception partly by learning from the apparent beliefs of others and partly by conforming, even distorting their public responses in the interest of maintaining social acceptance (Kuran and Sunstein 1999).

Availability entrepreneurs are activists who manipulate the content of public discourse, who strive to trigger and sustain availability cascades likely to advance their agendas. Availability campaigns sometimes yield social benefits, but they also bring harm, by hijacking social resources for precaution-taking that would better promote the general welfare when used elsewhere. Most cascades tend to die out. But their institutionalized effects linger and are very difficult to undo and often become permanent.

Fair-Balanced Treatment of a Controversial Issue?

Did high-impact journals afford a fair-balanced discussion of a COI? No. A comprehensive review (Lesko et al. 2012) examined research and opinion articles published between the 1980s and 2008 in three high-impact medical journals regarding relationships between companies in the medical product industry and academic physicians. Journals included were *Journal of the American Medical Association*, *The Lancet*, and *The New England Journal of Medicine*.

Of the 108 articles that met the reviewers' inclusion criteria, only 12 (11%) were neutral or emphasized relationship benefits. All of those mentioned or discussed an opposing point of view. Six attempted to refute the benefit-denying viewpoint. None of the relationship-benefit emphasizing articles drew inferences about patient outcomes. In contrast, 16 of the research papers (15%) emphasized risks associated with industry relationships. None of these mentioned or attempted to refute an alternative viewpoint.

A majority of the articles (80 or 74%) were commentaries emphasizing risks of relationships with companies. 76% drew inferences about patient outcomes. Only 7 (6%) discussed critically and attempted to refute that relationships with industry were beneficial. The study authors concluded that these journals evidenced an anti-industry publication bias and had participated in a conformity cascade—where normative positions do not emerge from objective, balanced weighing of evidence but from social acceptance pressure to conform with the opinions of fellow editors all of whom in effect practice discrimination preferring to accept manuscripts from "COI entrepreneurs" (not the term used by the study authors).

COI "SCHOLARSHIP:"
108 *JAMA, LANCET* &
NEW ENGLAND JOURNAL OF MEDICINE PAPERS
CONCERNING PHYSICIAN-INDUSTRY RELATIONSHIPS
(*Nature Biotechnology 30*: 320, 2012)

Emphasis:	Relationship risks	Relationship benefits
	89%	11%
Provided evidence	44%	100%
Drew inferences about patient outcomes	76%	0%
Acknowledged alternative viewpoint	43%	100%
Critiqued alternative viewpoint	3%	100%

Since the mid-1970s, social psychologists and behavioral economists have shown experimentally the mind's susceptibility to availability heuristics—they guide the mind, enable substitution of easier for harder problems, and give salience to some features of experience while ignoring others. They over-invest with significance vivid, emotion-laden instances with "representativeness" of social reality.

When a social environment is seeded with a risk heuristic, like COI, and is regularly re-seeded by availability entrepreneurs and media-driven corruption stories, a cascade can become a "moral panic." A white-glove inspection of the environment finds dirt everywhere (Stossel and Stell 2011): "one sees and hears about it all the time." Precautionary, preventive measures are enacted. When found wanting, more costly, more stringent measures are proposed. Yet, the social environment is never purged, never is found more ethical, nor are desultory regulations ever repealed, never pronounced an overreaction, a failure.

Medical Product Marketing Unduly Influences Physicians' Prescribing with Patient Harm Resulting

- "The systematic review of the medical literature on gifting by Wazana found that an overwhelming majority of interactions had <u>negative results on clinical care</u>" (Brennan et al. 2006).
- "There is a fairly extensive accumulated database tending to show that the relationships that have grown up between physicians and drug reps in the past sixty years have <u>deleterious consequences</u>" (Brody 2007: 195).
- "But conflicts of interest (COIs) at both the individual and institutional levels can have a negative effect beyond higher drug prices. This includes influences on physicians that can lead to <u>negative patient implications</u>, including (1) incorrect drug claims and uses; (2) formulary requests for medications with no significant advantage over existing medication; (3) irrational prescribing; (4) fewer prescribed generics; and (5) compromising integrity in clinical decision making that <u>can lead to patient safety issues</u>" (Mackey and Liang 2013) (my emphases).

What Evidence Supports These Remarkable Claims?

The Brennan paper aspired to be a professionalism manifesto, a social-signaling mission document, for academic medical centers. The authors claimed that physicians' relationships with medical product companies created financial conflicts of interest which "...occur when physicians are tempted to deviate or do deviate from their professional obligations for economic or other personal gain." Especially toxic was the drug rep practice of reminder-item gifting because it insidiously triggered a powerful but underappreciated psychological impulse to reweigh information and choices in light of the gift.

Gift-exchange theorists make an empirical argument for unconscious reciprocity. They say that getting a recipient to accept small gifts (a pen) is easy because small gift offers trigger nearly hard-wired, reflexive gift acceptance. The icebreaker makes the recipient more willing to accept gifts of increasing value—free travel and expense-paid reimbursement to a product promotion seminar, tuition-free accredited CME, and, ultimately, accepting honoraria for peer-to-peer speaking and well-paid consulting gigs. Imperceptibly, *liking*, and ever stronger *dependency* on the gift-giver results. Ultimately, the recipient is "hooked," and her *independent medical judgment* is destroyed. Brainwashed by kindness/flattery, the once evidence-demanding, skeptical physician becomes a Manchurian Candidate, not merely for the interests of one pharmaceutical company, but for the industry as a whole ("industry favorable"). Brennan et al. claimed that gifting "often" caused physicians to have positive attitudes toward the reps. They further claimed on Wazana's authority for having found, much more ominously "…an overwhelming majority of interactions had negative results on clinical care."

What Wazana reported was: "Although some positive outcomes were identified (improved ability to identify the treatment for complicated illnesses), most studies found negative outcomes associated with the interaction. These included an impact on knowledge (inability to identify wrong claims about medication), attitude (positive attitude toward pharmaceutical representatives; awareness, preference, and rapid prescription of a new drug), and behavior (making formulary requests for medications that rarely held important advantages over existing ones; non-rational prescribing behavior; increasing prescription rate; prescribing fewer generic but more expensive, newer medications at no demonstrated advantage.)…No study used patient outcome measures" (Wazana 2000: 378, my emphases).

Inability to identify wrong claims about medication reasonably counts as "a negative outcome." Finding an association between that troubling inability and drug rep exposure cannot rule out that the inability may have been present prior to drug rep exposure. Without more, it cannot be attributed to the drug rep's influence (e.g., gifting reminder items, handing out reprints), even if the inability overlaps with the products in the rep's bag. The inability might (or might not) extend to discern many incorrect medication claims. For example, an early study (Haayer 1982) found that rational prescribing by physicians overall (as judged by an expert panel composed mostly of pharmacologists and pharmacists) occurred less than 50% of the time, irrespective of where their information came from. Indeed, none of the standard professional sources of information seemed to have a great impact on physician prescribing quality.

Developing a positive attitude toward ("liking") drug reps is not in itself a negative outcome. Here's a plausible justification for it. Suppose physicians appreciated having an increased ability to identify the (correct) treatment of complicated diseases—a positive outcome Wazana found associated with rep exposure but not mentioned by Brennan et al. Suppose those physicians attributed (at least some of) that ability to drug rep exposure and were grateful. Their attribution might (or might not) be correct. Their reactive attitude, *liking*, may (or may not) be well-placed. A study suggesting additional good reasons for positive attitudes toward drug reps will be outlined below.

Awareness and rapid prescribing of a new drug is not necessarily a negative outcome and, to the contrary, is often a positive outcome. Intentionally imposing delay, as has been proposed (Psaty and Burke 2006), denies patients the promised benefits associated with FDA marketing approval ("safe and effective"), which was based on the best-available evidence. How is it rational to withhold use of a newly approved, first-in-class drug that has no alternative (besides doing nothing)? There are risks to doing nothing, also to delay and to too-rapid prescribing of new products. Standing pat with familiar therapeutic options indulges status quo bias (Sunstein 2005) and risks "clinical inertia"—knowing better but not doing better (Phillips et al. 2002).

Status quo deserves no privilege per se. A recent study reviewed original articles published in a high-impact journal over a decade. It found 146 medical practice reversals. The authors concluded that reversal of medical practice is common and occurs across all classes of medical practice (Prasad et al. 2013). A "survival analysis" of clinical practice guidelines found that more than ¾ of the AHRQ guidelines needed updating. It recommended that guidelines should be reassessed for validity every 3 years (Shekelle et al. 2001). A survival analysis of systematic reviews directly relevant to clinical practice found that signals for updating occurred frequently and within a very short time (Shojania et al. 2007).

Newer medications may be better overall than older ones—safer and more effective with more convenient dosing and fewer, less severe side effects. Relationships with industry facilitate the development of these products and the dissemination of information regarding their use (Lichtenberg 2006). A rule of thumb "never be first to adopt nor last" may not pass muster-review with medical "mindfulness"(Epstein 1999) and ill-serve one's patient. Therapeutic progress requires that somebody go first.

Increased prescription rate may also be questioned as a negative outcome. A review of harm caused by omissions concluded that more harm likely results from omission than commission and our inattentiveness to nonevents blinds us to the risk (Hayward et al. 2005).

A prescribing bias for generic medications may (or may not) be justified. According to FDA, more than 80% of scripts are filled by generics currently. They tend to be less costly than branded "equivalents." Seventy-eight percent of prescriptions cost patients $10 or less (IMS Health 2014). However, generic sales generate profits for the makers of generics. Generic sales provide larger profit margins for pharmacy companies and pharmacy benefit management companies (PBMs) than branded products. Both pharmacies and PBMs profit from a default rule that presumes equivalence between "in-class" products (generic and branded alike) until proof shows otherwise. Having an interest in setting and applying the standard of proof may have incentive effects. Thus promoters of generics, including "counter-detailers," promote the interests of companies that make generics, the interests of pharmacies whose profits are larger on generic products and PBMs that negotiate contracts on behalf of employers and patient groups. A Federal Trade Commission analysis of 95 drugs reported that a pharmacy's gross profits per prescription increase when a first-to-file generic wins 180 days of market exclusivity under an Abbreviated New Drug Application (ANDA). But, a pharmacy's gross profits per prescription are even higher when an authorized generic (AG) competes with the successful, first-to-file ANDA during the 180 day period. In 2006, Merck decided to sell Zocor to major managed care companies as an AG but to be priced lower that the versions marketed by Teva and Ranbaxy under their FDA-awarded 180 days of exclusivity under first-to-file ANDA. Market price of Simvastatin declined more quickly than it would have without AG. AG competition cost these generic makers millions in revenue, prompting the Generic Pharmaceutical Association to lobby for prohibiting the marketing of AGs.

Prior-authorization and automatic generic substitution rules that defeat "dispense as written" prescriptions also promote the interests of big companies such as CVS Caremark. But those rules also tend to delay discovery of marginal benefit and marginal risk associated with newly approved, branded products. No data say that foregone marginal benefit and delayed risk discovery are opportunity-costs worth paying. Demanding high-quality cost-effectiveness studies, comparing "presumed equivalent" generics, with newly approved drugs and devices serves the financial interests of all those who benefit from status quo bias.

What Are "Negative Results on Clinical Care"?

Brennan et al. do not define them. It was not among the outcome measures cited in Wazana's paper. Brennan and colleagues do not themselves establish any connection, associational or causal, between the "negative outcome" measures Wazana actually reports and the outcome "clinical care," which insinuates without actually saying "patient outcome." Brennan et al. have never admitted nor corrected their inaccurate citation. Scholars (e.g., Mayes and Laing) continue to rely on Brennan's citation of Wazana to this day.

Howard Brody, one of the most trenchant critics of MPI relationships with physicians did not rely on Wazana's review. Indeed, he found (Brody 2010) that the Wazana paper is "...now nearly worthless for two reasons. First, it's woefully out of date; second, if you read it carefully, it "proves" almost nothing—perhaps that more contact with drug reps makes docs who serve on formulary committees more likely to recommended company drugs."

Wazana's 2000 paper was "a systematic review." It assembled, organized, and assessed the quality of 29 studies already in the literature. Is it fair to trash Wazana's review for failing to produce proof stronger than the reviewed articles? Assembling evidence from studies showing an association between marketing and changed prescribing behavior, some positive, some negative, cannot add up to a strong negative proof.

Wazana did include a fairly old study (Haayer 1982) done in the Netherlands, which found a significant association between reliance on industry-provided information, older-age physicians, and "non-rational prescribing." But overall, less than half (48%) of physicians' prescribing decisions were judged "rational" by an expert panel composed of a clinical pharmacologist, a pharmacologist, a general practitioner physician, two pharmacists, and a physician who advised on pharmacology for a health insurance fund.

Haayer does not suggest it, but his study provides some evidence that the prescribing privilege, currently monopolized by physicians, should be at least shared with pharmacists or perhaps wrested from physicians entirely and given over to pharmacists, who would more rationally dispense medications for conditions diagnosed by physicians. Pharmacists dominated the panel that set the "gold standard" for rational prescribing. Despite the age on the Haayer study, Brody (2007) relied on that 1982 Dutch paper as well.

What about Mackey and Liang? Neither do they produce data-evidence of their own. For authority, they rely on the Brennan paper (which cites Wazana), on the Wazana paper, and on a Pew study discussing conflict of interest. The Pew paper provides no data-evidence that branded-product marketing *is significantly associated with* patient harm, let alone that marketing *tends to cause* patient harm.

A Better Systematic Review of Industry Influence?

Spurling et al. (2010) published a systematic review of 58 studies that addressed whether drug company information *has an impact* (influences, tends to cause changes) on doctors' prescribing. Indeed, Brody praises the Spurling group for producing just the scholarship that Wazana supposedly failed to produce. What did they find? On one important point, they found what Wazana found: "...none of the [reviewed] studies that we found examined clinical outcomes."

The Spurling group said that "the limitations of studies reported in the literature...mean that we are unable to reach any definitive conclusion about the degree to which information from pharmaceutical companies increases, decreases, or has no effect on the frequency, quality or cost of prescribing" (2010: 19). Only two studies among the 58 reviewed were randomized. These failed to test the marketing strategies that companies actually use in the field. The remainder of the studies were observational, which enabled measuring associations, not causes.

Spurling and his colleagues say explicitly that they could not rule out that even where associations were established, they might have arisen from confounding, bias, or chance. These authors did opine that the associations found by the studies *reasonably* could be attributed to marketing *influence*—reasonably, because companies are trying to change physicians' behavior after all, but any causal relationship thus discovered could be bidirectional. They further state that they could not rule out that the authors of the studies they reviewed merely produced results consistent with their own ideological bias. Indeed, sponsors' funding biases may have attracted study proposals from biased authors. If so, bias-driven findings may have been produced without authors' and sponsors' awareness.

In conclusion, Spurling et al. raise the possibility, *in theory*, that MPI advertising actually may be beneficial in various ways, such *as* "increased prescribing of drugs that provide better [not to mention "good"] outcomes or improved cost-effective use of health care resources." But unlike Wazana's review, which did find some evidence of marketing's benefit (an associated increased ability to identify and treat complicated diseases), none of the studies Spurling et al. reviewed found *any* associated benefit. Dissemination of comparative effectiveness research (CER) has been touted as beneficial marketing—by reducing uncertainty and reducing cost. The Affordable Care Act allocates $3.5 billion through 2019 for these purposes. However, the evidence for CER's power to change physicians' behavior is not encouraging (Timbie et al. 2010). Indeed, asking research subjects to participate in studies that have a vanishingly small chance of changing physicians' behavior, a major rationale for such studies, is dubious ethically.

Arguably, the Spurling group's review exposed a strong "omission bias" in the studies they reviewed. Apparently none even looked for beneficial associations between product marketing and prescribing. So, none was found, therefore, none could be reported. That's omission bias—the marketing-friendly dog was not invited to bark.

Does It Strain Credibility that Industry Marketing May Be associated with Some Positive Patient Outcomes?

To win FDA marketing approval for an investigational new drug, companies must submit clinical data, measuring relevant endpoints that adequately support a marketing application. These data typically come from randomized clinical trials. Companies must submit all data collected. It is fraud to omit any. The data must be convincing to expert panels and ultimately to the FDA, so that the study results are reproducible—making the product safe and effective when used according to the label approved for it. The label's content, negotiated between FDA and company representatives, must by law provide information to patients for proper use. It is a serious criminal offense to "misbrand" a product for sale in interstate commerce. Before receiving a license, each individual production facility must prove to FDA that it can reliably manufacture the approved product within tolerances set by the agency.

In the final sentence of their paper, Spurling and colleagues speculate that the all-things-considered indeterminacy of the evidence, when considered from the precautionary principle's perspective, suggests that prudent physicians should avoid exposure to industry marketing. To the contrary, physicians' intentional avoidance of drug reps' marketing makes status quo bias a virtue. Avoiding market prompting on principle reduces the likelihood of prescribing products whose beneficial uses are supported by the best-available data and decreases the likelihood of timely responding to FDA boxed warnings.

Is there specific evidence of product-associated benefit reasonably associated with product promotion? From 1980 to 2000, cardiovascular mortality in the United States declined by 50%. Ford et al. (2007) estimated that half of that decline owed to using industry's products. Is it implausible to suppose that marketing was at least associated with increased prescribing of the products? Clearly, significant increased patient benefit (reduced cardiovascular mortality) has been associated with increased product use.

It is true that FDA approval alone typically does not allow a marketing claim that a newly approved product is cost-effective when compared with other already approved, "in-class" therapeutic options. FDA's standard is "safe and effective," not "safer and more effective." Arguably, a newly approved product is cost-effective compared with doing nothing when there are no other approved options. Sometimes, the benefit may not be proportional to the cost, the quality adjusted life-years (QALYS) too few.

Hep C infection is a widely prevalent, debilitating disease and a major cause of hepatocellular carcinoma. The recently approved Hep C drug, Sovaldi® is 90–95% effective in clearing the infection. The cost/pill is $1000 or approximately $85,000 for a 12-week course of treatment. Gilead, maker of the drug, has been attacked for the price, yet it is cheaper than a liver transplant, cheaper yet than a simultaneous liver-kidney transplant for patients with hepatorenal syndrome (HRS1) which diverts away from the kidney wait list a high-quality organ.

Provenge® is approved for use in castration-refractory non-metastasized prostate cancer on the basis of having proved a significant association between drug use and a mean increased survival of 4 months. Medicare has already agreed to pay for it. Generally, PBMs follow Medicare reimbursement decisions for their non-Medicare contracts. Despite that the FDA's standard is "safe and effective," its approval decision for this novel "immune" therapy has been criticized on what appeared to be cost-effectiveness grounds. The argument is that the drug's mechanism is uncertain, that the benefit is so slight—mean increased survival of 4 months—and that it should not count as "effective." So, of course, government shouldn't pay for it. Should an oncologist prescribe it nevertheless? Should an oncologist prescribe it off-label for patients who are eager to pay out-of-pocket and want their chance at perhaps longer, perhaps much longer, than an extra months? Off-label drug use is very common in oncology and in pediatrics too. Comparatively few drugs have been FDA approved for use in children.

Avoiding rep marketing of newly approved drugs or any branded product on precautionary principle grounds privileges status quo bias as the default without justification. It confuses physicians' therapeutic-familiarity, "comfort," with dedication to the goals of medicine (Wootton 2006). COI entrepreneurs have successfully mass-marketed social distancing from drug reps. They have catered to the economizing interest in "thinking fast." It is not clear that *mindfulness* in medical practice (Epstein 1999) would buy it. Physicians should fight to recover their privilege of "thinking slow" (Kahnemann 2011). Discretionary medical judgment is nothing without it.

Physician Industry Consultants Will Put Their Mouths Wherever Their Money Is

This theory is well-exemplified by a widely cited study by Stelfox et al. (1998). The authors used a survey to correlate the presence or absence of cardiologists' connections with makers of cardiovascular drugs and their published opinions on the dangers of calcium-channel blockers (CCBs) whose safety was controversial at the time. The dangers associated with shorter-acting, immediate-release CCBs was well appreciated. What was controversial: dosing and whether the risks of short-acting CCB should be extrapolated to longer-acting CCBs?

Stelfox and colleagues found that cardiologists who consulted for industry were significantly more likely to discount the dangers allegedly associated with CCBs than those without connections to industry. The authors concluded that "the medical profession needs to develop a more effective policy on conflict of interest." They proposed that journal editors erect a stronger bias filter for manuscript submissions by COI-burdened industry consultants. This suggestion would bear fruit. In 2005, *JAMA* instituted a discriminatory policy burdening industry-sponsored submissions with the cost of securing a sign-off from academic statisticians. Evidence for doubting the validity of this imposition will be discussed below.

Remarkably, the Stelfox study found that the cardiovascular drugs made by the companies for which some of the cardiologists consulted, bore no consistent relationship to the consultants' opinions about the risks of CCBs. Consultants to companies manufacturing cardiovascular medications unrelated to calcium-channel blockers were as likely to discount the risks of CCBs as those consulting for companies that produced these drugs.

How does the theory of risk-emphasizing relationship bias (COI) adopted by Stelfox et al. *explain* this? Is it plausible that cardiologists who consult with makers of any cardiovascular drugs publish favorable comments about all cardiovascular drugs, including CCBs made by companies who do not pay them? Or, might it merit investigation to determine whether the risk-discounting, industry-consulting cardiologists were overall more knowledgeable than the non-consulting cardiologists and had pharmacological justification for claiming that CCBs (properly dosed, longer-acting agents) are not inherently more dangerous than other drugs used to treat hypertension?

Lumping all CCBs together and associating favorable commentary regarding them as a therapeutic class with industry consulting misses important pharmacological distinctions. Industry consultants may have had access to pertinent information from industry scientists that was not yet generally available. They may have already appreciated the relevant distinctions. That confounder was not controlled.

Since "CCB" is ambiguous, industry-consulting experts might have assumed that others would appreciate the ever-present puzzle about dosing and ambiguity between short-acting and longer-acting agents. Knowledgeable people make the mistake regularly. On the other hand, naïve COI-suspicious auditors may blow off any medical subtleties and assume the consultants' risk-discounting was driven by financial COI.

Stelfox and colleagues prejudiced their inquiry by biased risk framing (COI). A corruption narrative is powerful. It substitutes an easier inquiry (into conflicted relationships) for a harder one that would have plumbed pharmacological nuances. The substitution of an easier inquiry for a harder one colored the value of industry-consulting cardiologists' consulting opinions. It may also have delayed use of safe and effective drugs with some, albeit unknown, patient detriment (e.g., forgone benefit).

Subsequent studies have corroborated the safety and efficacy of properly dosed, longer-acting CCBs (Opie and Schall 2002; Epstein et al. 2007). Extrapolating risks from shorter-acting, immediate-release CCBs to longer-acting compounds was inappropriate. Sometimes a sense of smell detects a skunk in the woodpile, but sometimes it detects a fox, which smells similarly. Overreliance on "the smell test" shows willingness to take to the bank the risks of "thinking fast."

Mukherjee (2010: 276) succinctly describes the central weakness in using statistical methods to identify previously unappreciated risk factors. By their very nature, these methods, though powerful, are descriptive and associational, not mechanistic and causal. They rely on a degree of foreknowledge. To run a classic case-control trial to identify an unknown risk factor (e.g., whether one is more likely to get biased advice from paid industry consultants), paradoxically, the investigator must already know the question to ask. Biased framing (COI) does that job. Confirmation bias steps in to do the rest.

Is "Drug Reps Off-Campus" Wise Social-Distancing Policy?

To determine the risks of policies restricting physicians' access to drug reps, Chressanthis et al. (2012) studied the prescribing decisions of 58,647–72,114 physicians. Decisions were statistically analyzed using prescription data from IMS Health. A consulting firm database was used to determine access-to-reps levels for 300,000 physicians. (Reps have a strong financial incentive accurately to log this data because their pay is linked to it.) Physicians were sorted into four categories of access: very low, low, medium, and high. Only high-prescribing doctors of diabetes and lipid-lowering drugs were studied. Low prescribers were ruled out because drugs reps rarely target them.

The drugs were:

- An innovative drug for type 2 diabetes (Januvia-sitagliptin)
- An older diabetes drug with a new FDA-required black box warning for cardiovascular safety (Avandia-rosiglitazone)
- A combination lipid therapy that had reported negative outcomes in a clinical trial (Vytorin-simvastatin+ezetimibe)

The authors found:

- Physicians with very low access to reps had the lowest overall adoption of new, first-in-class therapy and took 1.4 and 4.6 times longer to adopt than physicians with greater access.
- In responding to the FDA's boxed warning for Rosiglitazone, physicians with very low access were 4.0 times slower to reduce their use of this treatment than those with greater access.
- Physicians with very low access to reps were significantly slower to change their prescribing based on negative industry news about combination lipid therapy.
- Overall, cardiologists were the most responsive to information changes relative to primary care physicians.

The authors conclude that limiting access to reps can have the unintended effect of reducing appropriate responses to negative information about drugs just as much as responses to positive information about innovative drugs. When new information became available, primary care physicians who had reduced access to drug reps were more likely to prescribe less effective and potentially more dangerous drugs. A survey of 2996 physicians found that >50% are working in organized provider systems. Forty-two percent of those reported never seeing industry reps (compared with 25% of independent practitioners). Of those physicians working in organized provider systems who never see reps, 80% report that not seeing reps was due to organizational polices (Quantia and Capgemini Consulting 2014).

Supporters of restrictive drug rep access policies might object that reps have seeded physicians' information environment with so much bad data that their making apparently beneficial compensatory updates appears attractive. Perhaps it is better to cut off the whole desultory exercise with a "nudging" social-distancing policy (Thaler et al. 2006) guided by principles of "libertarian paternalism" (Sunstein and Thaler 2003). The premises of these thinkers have received harsh criticism (Wright and Ginsburg 2012). A fuller discussion of freer versus more restricted information markets goes beyond my purposes.

Conclusion

Appreciation for physicians' competing interests is not new. About 2500 years ago, Plato observed that the physician "professes medicine," but he practices an additional art not professed by him—the art of getting payments. Everyone practices the latter art—because no one takes on the troubles of strangers, to straighten them out, but everyone expects payment for that—for the opportunity cost thereby incurred, to use the language of economics.

Plato's analysis recognized that money payments are not the only kind in which people take an interest. Indeed, money payments, while most common, are the lowest kind. Payments additionally include social recognition/status, affirmation from social sources that matter. Professionals commonly have accepted a payment mix—money but also high status, patient approval, social appreciation for pro bono services, and self-approval for doing good, removing harm, and at least avoiding inflicting needless harm. Plato also mentioned the type of payment required to motivate the best of the best—those who had lost interest in money and in further social recognition. Such person could only be motivated to service by avoiding having less qualified, less excellent persons serve in their stead. These exalted souls could not live with themselves were they to allow that to happen. Thus the ultimate sanction and payoff was avoidance of self-disapproval for what one allowed to happen.

The COI cascade seriously misfires by fixating on monetary payoffs while ignoring all the other things that people care about, things that can bias their judgment and lead to wrongdoing. Wrongdoing should be the focus, not the temptations and motivations that sometimes result in it.

Today, more than 50% of physicians are practicing within some type of healthcare system. Their compensation is tied to various performance metrics—process measures, outcome measures, patient volume, and satisfaction measures. They have much lower access to drug reps than in the past. More than 1/3 of physicians are part of an integrated health network, the most restrictive environment for rep access. Seventy percent of physicians who have been out of medical school for 10 years or less are employed by an organized health system. More than 50% of physicians have their prescribing decisions constrained by a formulary and a care pathway or require prior authorization. Prescribing restrictions are greater for in-patients than out-patients.

Conversations with drug reps, where allowed, are rule-constrained. It is not enough for a physician to be convinced about a drug's appropriateness to have it dispensed successfully to a patient. The drug must be on formulary and listed as available on the patient's electronic health record. It is not easy for a physician to discover which drugs are on formulary. He will receive notification in the event he prescribes off-formulary. Nor may a rep detail a product to a physician if it is not already on formulary. Suggestions for formulary additions must be submitted to a Pharmacy and Therapeutics Committee, not to an individual physician. Lobbying an individual physician to submit a formulary addition request is a violation which carries a campus-exclusion penalty for the rep—as long as 90 days.

Corruption fears have resulted in social-distancing policies of drug reps from physicians, and they have given us sharply constrained information markets. There is some evidence that this may be disadvantageous to patients. Information market restrictions do promote powerful political and corporate interests. Interests that cannot reach these levers of power are disadvantaged. Whether there are compensating benefits in reduced expenditures on drugs and devices, which account for roughly 10% of healthcare expenditures, remains to be seen. The prices of some generics have increased sharply. We can only hope that our institutional policies eventually will be revised in a manner that better respects physicians as sophisticated consumers of product marketing and holds them accountable for their choices not their motivational states.

Acknowledgment I am indebted to Thomas P. Stossel, MD, for the PowerPoint slides and to David Barton for the Excel spreadsheet chronographing the COI cascade.

References

AdvaMed Code of Ethics. 2009. http://advamed.org/res.download/112. Accessed 11 Aug 2016.

Barton, D., T. Stossel, and L. Stell 2014. After 20 Years, industry critics bury skeptics, despite empirical vacuum. . *International Journal of Clinical Practice*, June 2014, 68(6): 666–673.

Ben-Shahar, O., and C.E. Schneider. 2010. The Failure of Mandated Disclosure. University of Chicago Law & Economics, Olin Working Paper No. 516; University of Michigan Law & Economics, Empirical Legal Studies Center Paper No. 10-008.

Brennan, T.A., D.J. Rothman, L. Blank, D. Blumenthal, S.C. Chimonas, J.J. Cohen, J. Goldman, J.P. Kassirer, H. Kimball, J. Naughton, and N. Smelser. 2006. Health Industry Practices That Create Conflicts of Interest: A Policy Proposal for Academic Medical Centers. *Journal of the American Medical Association* 295(4): 429–433.

Brody, H. 2007. *Hooked: Ethics, the Medical Profession, and the Pharmaceutical Industry*. Lanham: Rowman-Littlefield.

———. 2010. *Major New Study Published in PLoS Medicine*. Wednesday, October 20, 2010. http://brodyhooked.blogspot.com/2010/10/major-new-study-published-in-plos.html. Accessed 15 Mar 2015.

Cain, D., G. Loewenstein, and D.A. Moore. 2005. The Dirt on Coming Clean: Perverse Effects of Disclosing Conflicts of Interest. *The Journal of Legal Studies* 34: 1–25.

Chressanthis, G.A., et al. 2012. Can Access Limits on Sales Representatives to Physicians Affect Clinical Prescription Decisions? A Study of Recent Events With Diabetes and Lipid Drugs. *Journal of Clinical Hypertension* 14(7): 435–446.

DeAngelis, C.D., and P.B. Fontanarosa. 2008. Impugning the Integrity of Medical Science: The Adverse Effects of Industry Influence. *Journal of the American Medical Association* 299(15): 1833–1835.

Dukeminier, J., S.M. Johanson, J. Lindgren, and R.H. Sitkoff. 2005. *Wills, Trusts, and Estates 158*. 7th ed. Rockville: Aspen Publishers, Inc.

Epstein, R.M. 1999. Mindful Practice. *Journal of the American Medical Association* 282(9): 833–839.

Epstein, B.J., K. Vogel, and B.F. Palmer. 2007. Dihydropyridine Calcium Channel Antagonists in the Management of Hypertension. *Drugs* 67(9): 1309–1327.

Erde, E. 1996. Conflicts of Interests in Medicine: A Philosophical and Ethical Morphology. In *Conflicts of Interest in Clinical Practice and Research*, ed. R.G. Spece, D.S. Shimm, and A.E. Buchanan, 12–41. New York: Oxford University Press.

Estate of Auen. 1994. 35 Cal.Rptr.2d 557.

Ford, E.S., et al. 2007. Explaining the Decrease in U.S. Deaths from Coronary Disease, 1980–2000. *The New England Journal of Medicine* 356: 2388–2398.

Gieser, N. 2009. *Financial Conflict-Of-Interest Disclosure and Voting Patterns at FDA Advisory Committee Meetings*. http://www.fda.gov/downloads/AdvisoryCommittees/AboutAdvisoryCommittees/UCM165328.pdf. Accessed 11 Aug 2016.

Haayer, F. 1982. Rational Prescribing and Sources of Information. *Social Science & Medicine* 16(23): 2017–2023.

Hayward, R.A., S.M. Asch, M.M. Hogan, T.P. Hofer, and E.A. Kerr. 2005. Sins of Omission: Getting Too Little Care May be the Greatest Threat to Patient Safety. *Journal of General Internal Medicine* 20: 686–691.

Hirsch, L.J. 2009. Conflict of Interest, Authorship and Disclosures in Industry-Related Scientific Publication: The Tort Bar and Editorial Oversight of Medical Journals. *Mayo Clinic Proceedings* 84(9): 811–821.

Huddle, T.S. 2008. Drug Reps and the Academic Medical Center: A Case for Management Rather Than Prohibition. *Perspectives in Biology and Medicine* 51(2): 2521–2560.

IMS Institute for Health Informatics. 2014. *Medicine Use and Shifting Costs of Health Care: A Review of the Use of Medicines in the United States in 2013*. http://www.imshealth.com/deployedfiles/imshealth/Global/Content/Corporate/IMS%20Health%20Institute/Reports/Secure/IIHI_US_Use_of_Meds_for_2013.pdf. Accessed 15 Mar 2015.

Institute of Medicine of the National Academies Website. 2009. *Conflict of Interest in Medical Research, Education and Practice*. Released April 28, 2009. http://www.iom.edu/CMS/3740/47464/65721.aspx. Accessed 15 Mar 2015.

Jones, D.J., A.N. Barkum, Y. Lu, et al. 2012. Conflicts of Interest Ethics: Silencing Expertise in the Development of International Clinical Practice Guidelines. *Annals of Internal Medicine* 156: 809–816.

Kahnemann, D. 2011. *Thinking Fast and Slow*. New York: Farrar, Strauss and Giroux.

Kao, A.C., et al. 2001. Physicians Incentives and Disclosure of Payment Methods to Patients. *Journal of General Internal Medicine* 16: 181–188.

Keitz, S.A., K.M. Stechuchak, S.C. Grambow, C.M. Koropchak, and J.A. Tulsky. 2007. Behind Closed Doors: Management of Patient Expectations in Primary Care Practices. *Archives of Internal Medicine* 167: 445–452.

Koch, C., and C. Schmidt. 2010. Disclosing conflicts of interest – Do Experience and Reputation Matter? *Accounting, Organizations and Society* 35: 95–107.

Kuran, T., and C. Sunstein. 1999. Availability Cascades. *Stanford Law Review* 51: 683–768.

Langbein, J. 2005. Questioning the Trust Law Duty of Loyalty: Sole Interest or Best Interest. *Yale Law Journal* 114: 929–990.

Lebel, C., and C. Beaulieu. 2011. Longitudinal Development of Human Brain Wiring Continues from Childhood into Adulthood. *Journal of Neuroscience* 31(30):10937–10947. doi: 10.1523/JNEUROSCI.5302-10.2011.

Lesko, R., S. Scott, and T.P. Stossel. 2012. Bias in High-Tier Medical Journals Concerning Physician-Academic Relationships with Industry. *Nature Biotechnology* 30: 320–322.

Lichtenberg, F. 2006. The Effect of Using New Drugs on Admissions of Elderly Americans to Hospitals and Nursing Homes: State-level Evidence 1997-2003. *PharmacoEconomics* 24(Suppl 3): 5–25.

Lurie, P., M. Cristina, C.M. Almeida, N. Stine, A.R. Stine, and S.M. Wolfe. 2006. Financial Conflict of Interest Disclosure and Voting Patterns at Food and Drug Administration Drug Advisory Committee Meetings. *Journal of the American Medical Association* 295(16): 1921–1928.

Mackey, T.K., and B.A. Liang. 2013. Physician Payment Disclosure under Health Care Reform: Will the Sun Shine? *Journal of the American Board of Family Medicine* 26(3): 327–331.

Madoff, R.D. 1997. Unmasking Undue Influence. *Minnesota Law Review* 81: 571.

McDaniel, S.H. et al. 2007. Physician Self-Disclosure in Primary Care Visits: Enough About You, What About Me? *Archives of Internal Medicine* 167(12):1321–1326.

Mukherjee, S. 2010. *The Emperor of All Maladies: A Biography of Cancer*. New York: Scribner.

Opie, L.H., and R. Schall. 2002. Evidence-Based Evaluation of Calcium Channel Blockers for Hypertension. *Journal of the American College of Cardiology* 39: 315–322.

Pearson, S.D., and T. Hymans. 2002. Talking about Money: How Primary Care Physicians Respond to a Patient's Question about Financial Incentives. *Journal of General Internal Medicine* 17: 75–78.

Phillips, L.S., W.T. Branch, C.B. Cook, J.P. Doyle, et al. 2002. Clinical Inertia. *Annals of Internal Medicine* 135: 825–834.

PhRMA Code. 2008. http://www.phrma.org/principles-guidelines/code-on-interactions-with-health-care-professionals. Accessed 11 Aug 2016.

Policy and Medicine. http://www.policymed.com/physician-payment-sunshine-act/page/11/. Accessed 15 Mar 2015.

Prasad, V., A. Vandross, C. Toomey, et al. 2013. A Decade of Reversal: An Analysis of 146 Contradicted Medical Practices. *Mayo Clinic Proceedings* 88(8): 790–798.
Psaty, B.M., and S.P. Burke. 2006. Institute of Medicine on Drug Safety. *The New England Journal of Medicine* 355: 1753–1755.
Quantia & Capgemini Consulting. 2014. *Working the System: 4 Trends Driving New Opportunities for Engaging Organized Physicians*. http://ebooks.capgemini-consulting.com/Whitepaper%20-%20Quantia%20-%20Capgemini%20Consulting%20-%20Working%20the%20System.pdf. Accessed 11 Aug 2016.
Rohloff, A., and S. Wright. 2010. Moral Panic and Social Theory: Beyond the Heuristic. *Current Sociology* 58(3): 403–419.
Rothman, K. 1993. Conflict of Interest: The New McCarthyism in Science. *Journal of the American Medical Association* 269(21): 2782–2784.
Scalise, R.J. 2008. Undue Influence and the Law of Wills: A Comparative Analysis. *Duke Journal of Comparative & International Law* 19: 41–106.
Scannell, J., A. Blanckley, H. Boldon, and B. Warrington. 2012. Diagnosing the Decline in Pharmaceutical R&D Efficiency. *Nature Reviews Drug Discovery* 11: 191–200.
Shekelle, P.G., E. Ortiz, S. Rhodes, et al. 2001. Validity of the Agency for Healthcare Research and Quality Clinical Practice Guidelines: How Quickly Do Guidelines Become Outdated? *Journal of the American Medical Association* 286(12): 1461–1467.
Sherif, M. 1937. An Experimental Approach to the Study of Attitudes. *Sociometry* 1: 90–98.
Shimm, D.S., and R.G. Spece Jr. 1996. Introduction. In *Conflicts of Interest in Clinical Practice and Research*, ed. R.G. Spece Jr. et al., 1–11. New york: Oxford University Press.
Shojania, K.G., M. Sampson, M.T. Ansari, et al. 2007. How Quickly Do Systematic Reviews Go Out of Date? A Survival Analysis. *Annals of Internal Medicine* 147: 224–233.
Sorrell v. IMS Health Inc.. 2011. No. 10–779 131 S.Ct. 2653.
Spurling, G.K., P.R. Mansfield, B.D. Montgomery, J. Lexchin, J. Doust, et al. 2010. Information from Pharmaceutical Companies and the Quality, Quantity, and Cost of Physicians' Prescribing: A Systematic Review. *PLoS Medicine* 7(10): 1–22.
Stelfox, H.T., G. Chua, K. O'Rourke, and A.S. Detsky. 1998. Conflict of Interest in the Debate over Calcium-channel Antagonists. *The New England Journal of Medicine* 338: 101–106.
Stell, L.K. 2002. Two Cheers for Physicians' Conflicts of Interest. *The Mount Sinai Journal of Medicine* 71(4): 236–242.
———. 2005. Against the Flow: Why Physicians Should Listen to Drug Reps. *ASBH Exchange* 8: 1–5.
———. 2009. Drug Rep Off Campus! Promoting Professional Purity by Suppressing Commercial Speech. *The Journal of Law, Medicine & Ethics* 37(3): 431–443.
Stossel, T.P. 2005. Regulating Academic-Industry Relationships—Solving Problems or Stifling Progress? *The New England Journal of Medicine* 353: 1060–1065.
Stossel, T.P. 2008. Has the Hunt for Conflict of Interest Gone Too Far? Yes. *British Medical Journal*, 336(742):476 PMID:18309999;PMCID:PM2258346.
Stossel, T.P., and L.K. Stell. 2011 Time to 'Walk the Walk' About Industry Ties to Enhance Health. *Nature Medicine* 17(4): 437–438.
———. 2002. *Conformity and Dissent*. University of Chicago Public Law & Legal Theory Working Paper No. 34.
Sunstein, C.R. 2005. *Laws of Fear: Beyond the Precautionary Principle*. Cambridge: Cambridge University Press.
Sunstein, C.R., and R.H. Thaler. 2003. Libertarian Paternalism is Not an Oxymoron. *University of Chicago Law Review* 70(4): 1159–1202.
Thompson, D.F. 1993. Understanding Financial Conflicts of Interest. *The New England Journal of Medicine* 329(8): 573–576.
Timbie, J.W., D.S. Fox, K. Van Busum, and E.C. Schneider. 2010. Five Reasons That Many Comparative Effectiveness Studies Fail to Change Patient Care and Clinical Practice. *Health Affairs* 31(10): 2168–2175.

Todd, J.S. 1991. Professionalism at its Worst. *Journal of the American Medical Association* 266(23): 3338.

United States v Caronia. 2012. 2d Cir.

Vera-Badillo, F.E., R. Shapior, A. Ocana, E. Amir, and I.F. Tannock. 2013. Bias in Reporting of End Points of Efficacy and Toxicity in Randomized, Clinical Trials for Women with Breast Cancer. *Annals of Oncology* 24: 1238–1244.

Wazana, A. 2000. Physicians and the Pharmaceutical Industry: Is a Gift Ever Just a Gift? *Journal of the American Medical Association* 282: 373–380.

Wingrove v Wingrove. 1885. 11 P.D.81: 82–83.

Woolley, K.L., R.A. Lew, S. Stretton, J.A. Ely, N.J. Bramich, J.R. Keys, J.A. Monk, and M.J. Woolley. 2011. Lack of Involvement of Medical Writers and the Pharmaceutical Industry in Publications Retracted for Misconduct: A Systematic, Controlled, Retrospective Study. *Current Medical Research and Opinion* 27(6): 1175–1182.

Wootton, J. 2006. *Bad Medicine: Doctor Doing Harm Since Hippocrates*. Oxford/New York: Oxford University Press.

Wright, J.D., and D.H. Ginsburg. 2012. Behavioral Law and Economics: Its Origins, Fatal Flaws and Implications for Liberty. *Northwestern University Law Review* 106(3): 1033–1088.

Health Measurement, Industry, and Science

Leah McClimans

Patient-reported outcome measures (PROMs) are now common endpoints in clinical trials. PROMs measure latent variables such as mobility, health status, and quality of life typically by asking patients multiple questions using a Likert scale, e.g., "Does your health now limit you in lifting or carrying groceries?" Available answers might include yes, limited a lot; yes, limited a little; and no, not limited at all. Measures such as these gained traction in the 1970s when health measurement started to expand to include patient-reported outcomes alongside more traditional measures of mortality and morbidity. This addition was at least in part due to the development of health technology and improved standards of living, which led to a shift in attention from the cure of acute disease to the management of chronic illness (Cano and Hobart 2011). More recently, the popularity of PROMs can be traced to quality improvement initiatives that emphasize clinical effectiveness and patient centeredness. In the 2009 *FDA Guidance for Industry Patient-Reported Outcomes Measures: Use in Medical Product Development to Support Labeling Claims*, we see a marriage of these two motivations. PROMs provide patient-reported evidence of the effectiveness of drugs targeting the management of chronic illnesses and diseases, for example, in 2010 the FDA approved a patient-reported outcome claim for Actemra, a drug used to treat adults with rheumatoid arthritis (DeMuro et al. 2013).

Despite their popularity with industry and government agencies, PROMs face serious criticisms regarding their measurement properties, e.g., validity—the degree to which a questionnaire actually measures what it was intended to measure (Streiner and Norman 2008). The FDA *Guidance for Industry* goes some way toward recognizing the importance of PROM's measurement properties. For instance, they emphasize that these properties are part of what should be evaluated when determining a measure's suitability for use in medical product labeling. But this emphasis does not go far

L. McClimans (✉)
College of Arts and Sciences, University of South Carolina, Columbia, SC, USA
e-mail: mccliman@mailbox.sc.edu

© Springer Science+Business Media B.V. 2017

D. Ho (ed.), *Philosophical Issues in Pharmaceutics*, Philosophy and Medicine 122,
DOI 10.1007/978-94-024-0979-6_6

enough. As critics, including this author, have argued, the methods that are typically used to establish the psychometric properties of these measures are themselves problematic, e.g., the methods to establish validity are themselves not valid (Hunt 1997; Hobart et al. 2007; McClimans 2010a, b). The result is that even when PROMs appear to meet the FDA criteria, they can still be unethical and unscientific.

Given this claim one might imagine that changes in our psychometric methods—what some refer to as new psychometric methods (Hobart et al. 2007)—would be welcome. But in this paper, I argue that we ought to proceed with caution. Science is not value-neutral and neither are the suggested changes that would make PROMs allegedly more scientific. Indeed, I will argue that by improving the ostensible scientific basis of these measures, we may make it more likely that PROMs will show a treatment effect. While this difficulty is in the best interests of industry, it may not be in the best interests of the public.

Problems with PROMs

Before discussing the kinds of changes and outlook that "new psychometric methods" suggest, it is helpful to understand the context in which they have arisen. Specifically, it is useful to examine some of the criticisms of PROMs that have led to the suggestion of new psychometric methods. The essence of the problem with PROMs can be summarized by a single deficiency that has multiple consequences; namely, PROMs lack a theory that provides a representation of the measurement interaction—the relationship between the construct and its instrument. This has been my argument throughout a number of papers, and, with regard to the general contours of my argument, I am not alone (McClimans 2010a, b, 2015; McClimans and Browne 2012). Consider Donna Lamping's 2008 Presidential address to the International Society of Quality of Life Research, where she identified the need for a theoretical framework as one of three challenges facing the future of PROMs. Or take Jeremy Hobart et al. (2007 *Lancet Neurology*) where they lament the lack of explicit construct theories in their article criticizing the current state of PROMs. In a final example, Sonja Hunt, in her 1997 editorial for *Quality of Life Research*, argues that the surfeit of poorly designed measures suggests that we do not know what quality of life *is* (Hunt 1997). In what follows, I provide a brief overview of three consequences that result from this lack of theory: problems with validity, interpretability, and responsiveness.

Validity

Without a robust theory, establishing the measurement properties of PROMs is often a viciously circular endeavor. Consider construct validity. Construct validity asks whether the questionnaire at hand performs in a way that the underlying theory suggests it should. An increasingly common criticism of construct validation in the context of classical test theory—the dominant psychometric paradigm—is its inability to determine if a measure represents its object of inquiry, i.e., its inability to provide evidence of validation (Hobart et al. 2007).

Construct validity is typically tested by assessing a measure's internal and external validity. Internal construct validity is tested by examining the extent to which the questions or items within a measurement scale are statistically related to one another based on the responses given by a sample population. But this process does not tell us anything about the construct itself, e.g., quality of life or mobility. It tells us only that certain questions tend to occupy the same conceptual space. External construct validity is examined via convergent and divergent validity testing. Here, multiple measurement scales deemed similar to and different from one another are applied to a sample population, and the scores derived from respondent answers are correlated. These correlations determine whether the scale being validated correlates higher with scales that measure similar constructs than with those measuring dissimilar constructs. Once again, this process does not tell us what construct a measure actually assesses. It tells us only that some scales are correlated (or not) with other scales. But without some kind of theory, it is not clear what any of these measures actually measure.

Interpretability

Over the last 15 years, the discussion of how to interpret change in patient-reported outcomes has received considerable attention. Interpretability refers to the clinical significance of increases or decreases on a particular scale or measure over time. For instance, if I score a 30 on the Beck Depression Inventory (BDI-II), we know that I have scored in the middle of the scale—the BDI-II has 63 total points. But imagine two months later I score 42. What does this 12-point increase mean from a clinical point of view? Should my drug regime change? If so, how should it change? PROMs that have been developed using classical testing theory (CTT) only provide ordinal level information, i.e., we know that someone who scores 42 is more depressed than someone who scores 30, but we do not know the degree of that difference. PROMs are thus difficult to interpret.

This has led to the development of methods to enhance their interpretability. One popular method is the identification of a minimal important difference (MID). An MID is the smallest change in respondent scores that represent clinical, as opposed to merely statistical, significance and which would, ceteris paribus, warrant a change in a patient's care (Jaeschke et al. 1989). One popular method for determining a measure's MID is to map changes in respondent outcomes onto some kind of control. These are referred to as "anchor-based" approaches. The idea is to determine the minimal amount of change that is noticeable to patients and to use this unit of change as the MID. The method asks the control group of patients to rate the extent of their symptom change over the course of an illness or intervention on a transition rating index (TRI). TRIs are standardized questionnaires that ask patients questions, such as "Do you have more or less pain since your first radiotherapy treatment?" Typically patients are given seven possible answers ranging from "no change" to "a great deal better" (Fayers and Machin 2007). Those who indicate minimal change, i.e., those who rate themselves as just "a little better" than before the intervention, become the patient control group. The mean change score of this group is used as the MID for the PROM.

This approach of acquiring an MID via a patient control group assumes that respondents who rate their symptom change as "a little better" on a transition question should ceteris paribus also have comparable change scores from the PROM. Put differently, similarities in respondent answers to transition questions ought to underwrite similarities in respondents' magnitude of change over the course of an intervention or illness. But qualitative data from interviews with patients suggests that this assumption is ill founded (Taminiau-Bloem et al. 2011; Wyrwich and Tardino 2006). Whether one understands the magnitude of change over the course of an illness as large or small is a matter of interpretation. As I have argued elsewhere, respondents' answers to TRI ought to be understood against the background of what makes for a good quality of life, e.g., the magnitude of change to which the answer "a little better" refers depends heavily on the significance that, say, worry has within the respondent's vision of the good (McClimans 2011). Thus, it is possible to have an outcome that indicates a large magnitude of change and to interpret this change as minimal.

Consider Cynthia Chauhan, a patient advocate during the deliberations on the FDA guidelines for the use of PROMs in labeling claims. In response to the deliberations, Chauhan cautioned those present, "…not to lose the whole person in your quest to give patient-reported outcomes free-standing autonomy…" (Chauhan 2007). To make her point, she discussed the side effects of a drug called bimatoprost, which she uses to forestall blindness from glaucoma. One of the side effects of bimatoprost is to turn blue eyes brown. Chauhan has "sapphire blue" eyes, in which, she says, she has taken some pride. As she speaks of her decision to take the drug despite its consequences, she notes that doing so will affect her identity in that she will soon no longer be the sort of person she has always enjoyed being, i.e., she will no longer have blue eyes. Moreover, she points out that although the meaning that taking this drug has for her is not quantified on any outcome measure, it nonetheless affects her quality of life (Chauhan 2007).

We can imagine that, even if the bimatoprost is only minimally successful and Chauhan's resulting change score from the PROM is low, she will nonetheless have experienced a significant change—she will not be the same person she was before. But this significance is tied to the place that her blue eyes had in her understanding of herself and what she took to be a good life; ceteris paribus we would not expect a brown-eyed person to summarize their experience in the same way. Thus, it would not be surprising if Chauhan's answer to the transition question was "quite a bit," while the magnitude of her change score was minimal.

I suggest that what examples such as this illustrate is that our understanding of clinical significance ought to be closely linked to our understanding of the construct given the cohort of respondents for whom the measure is targeted. To put this point slightly differently, understanding change in PROMs requires that researchers have a grip on what quality of life or perceived health status means in the context of a particular PROM and the population it serves. In other words, we need a theory of the construct that the PROM aims to measure, i.e., a collection of sentences, propositions, statements, or beliefs and their logical consequences, and these can include statistical and general laws.

Responsiveness

As with validity and interpretability, responsiveness too needs a construct theory. Responsiveness refers to an instrument's sensitivity to change, although just what kind of change a responsive instrument should be able to detect is somewhat controversial, i.e., clinically important changes, changes due to treatment effects, or changes in the true value of the underlying construct (Terwee et al. 2003). But regardless of the kind of change an instrument is meant to identify, the development of a measure requires information about the appropriate distance between units of change. Take a simple example. In the USA, infants are measured in grams because measuring in kilograms is not sensitive enough—the extra grams that an infant weighs can make a difference to their prognosis. But older children and adults are typically weighed to the nearest kilogram because the extra grams are negligible to most of their health outcomes. This decision is in part theory driven including (1) our theoretical understanding of mass, (2) the role body mass plays in our understanding of health outcomes, and (3) the application of this theoretical understanding to different populations, e.g., infants and adults.

What level of sensitivity to a change in quality of life, health status, or mobility, for instance, should PROMs employ? The sensitivity of a scale in the context of PROMs is determined by the number and kinds of questions posed to respondents. For example, single item scales—scales that only ask one question—are limited in sensitivity since they must divide rich variables (e.g., spasticity) into only a few levels (Hobart et al. 2007). But just how finely *should* we divide a variable? In part, the answer to this question can be understood statistically. Questions that are considered "too close" to one another will have overlapping standard errors, but this statistic can be manipulated by increasing the sample size, i.e., the greater the sample size the smaller the standard error around the item estimates.[1] By increasing the sample size, one can increase the precision of the measure.

But as with questions about the appropriate sensitivity of measures of body mass, the responsiveness of a PROM requires a theory of the construct being measured, how that construct relates to other areas of interest, and how our theoretical under-

[1] Standard errors are a way of telling from a statistical perspective if x is significantly different from y. If standard errors overlap, this tells us that, in the case of PROMs, two items are similar enough to be indistinguishable.

standing relates to different populations. If we are trying to establish the effectiveness of a new drug using a PROM as one of the endpoints, then we need some theory that provides a representation of the measurement interaction in the context of the patient cohort as well as an understanding about how the construct in question relates to the condition or illness that is targeted by the new drug. These theoretical considerations cannot be achieved with statistics alone. Indeed, as I will argue below, determining the correct responsiveness of a scale must include considerations of value, in particular harms and benefits.

Solutions to PROMs

Cano and Hobart have been two of the most vocal and consistent critics of the use of traditional psychometric methods to develop PROMs. They have also been two of the most ardent supporters of the use of "new psychometric methods." In this section, I focus on Cano and Hobart's (2011) suggestion for correcting PROMs' current limitations.

While Hobart et al. agree that most of the PROMs in use lack theoretical development, they trace this error to CTT. The problem with CTT is that it does not provide the theoretical resources needed to model the measuring instrument, in this case a PROM. CTT theorizes that a person's observed score on the scale is the sum of the unobserved score to be estimated, i.e., the person's true score, plus measurement error (Hobart et al. 2007). Consider a physical functioning scale with a scoring range of 11–44, where higher scores indicate more limited functioning. Imagine that someone's observed score was 23. CTT tells us that this observed score is the result of their true score plus measurement error. Respondents' true scores are what we would like to know, but to get them, we need some idea of what, e.g., quality of life scores look like for this particular cohort of respondents (say, respondents with lung cancer). At the same time, we need some idea of what counts as instances of measurement error. For instance, does response shift count as an instance of measurement error or part of a person's true score? Response shift is defined as the change in the meaning of one's self-evaluation of a target construct (Schwartz and Sprangers 1999). A classic example of response shift is when a respondent becomes accustomed to her disease/illness/disability and recalibrates her internal standard of measurement. Is this recalibration best understood as measurement error—as much of the quality of life literature treats it—or is it best understood as a legitimate aspect of the quality of life construct, i.e., to be incorporated in our theory of quality of life? CTT leaves us unable to answer this question.[2]

The problem with CTT is that it does not give us a theoretical ideal for the true score as, say, the measurement of time has a theoretical ideal, i.e., the second is defined as the duration of exactly 9,192,631,770 periods of the radiation corre-

[2] For a longer discussion of the difficulties that CTT has with distinguishing between true scores and measurement error, see McClimans 2017.

sponding to a hyperfine transition of cesium-133 in the ground state (BIPM (Bureau International des Poids et Measures) 2006). Nor are CTT's target constructs sufficiently embedded within theories that would allow for approximations of measurement error as, for example, time is embedded within physical theory, e.g., the definition of the second assumes that cesium is in a flat space-time, but the cesium fountains (primary standards) that metrologists build are subject to gravitational redshift. Relativity theory helps us to estimate the error associated with these phenomena (Tal 2011). In other words, CTT does not provide a theoretical representation of the measurement interaction, i.e., the relationship between the construct of interest and its instruments (McClimans 2015).

In place of CTT, Hobart et al. argue for the use of new psychometric methods, particularly Rasch methodology. How is Rasch different from CTT? One important difference is that Rasch has an explicit mathematical model that provides a representation of the measurement interaction. Rasch measurement theory says that a person's response to an item is determined by the difference between a person's location on the ruler (i.e., how much ability they have) and an item's location on the ruler (i.e., how much ability an item requires). Thus it provides a representation of the measurement interaction. In particular, Rasch states that the higher a person's ability with respect to the difficulty of an item, the higher the probability that a person will answer "yes" to an item. The Rasch scale runs from plus to minus infinity with the zero point at the place where the difficulty of the items in the survey is equal to the ability of the sample population. Each item is located on the ruler relative to the point at which there is equal probability of respondents answering "yes" or "no" to that particular item. In Rasch, the probability of answering "yes" or "no" is modeled as a logistic function. The mathematical equation that governs this function is the model that represents Rasch measurement theory.

A second difference between Rasch and CTT is that Rasch prioritizes its mathematical model over the data. CTT defines what it is measuring, say, quality of life, by the items that are generated, usually from a qualitative sample. It is a descriptive approach. The Rasch model, however, operationally defines the construct one is trying to measure as a relationship between a person's ability and the probability they will answer "yes" to an item. When applied to a sample population, this model provides the characteristics and regression weights for selecting items and determining their difficulty, i.e., their place on the ruler (Stenner and Smith 1982). For example, if I ask respondents if they are balanced when seated and 95% say yes and if I ask if they can climb a flight of stairs and only 60% say yes, then balancing when seated is considered less difficult (i.e., requires less ability) than climbing stairs. Climbing stairs will appear farther down one end of the ruler than balanced when seated. Likewise, those who answer yes to more difficult items are considered to have more ability.

Within Rasch, the validity of a measure is determined by how well the data (i.e., respondent answers to survey questions) fit the predictions of the mathematical model applied to a sample (i.e., is balanced sitting less difficult than climbing stairs?) These predictions are made explicit in the construct specification equation in terms of the amount of variance that we should expect to find around the mean with respect to the balancing and climbing questions. If the model correctly describes the probability distributions of people responding to questions about walking and climbing, then we can say that the observed rating scale data satisfy the measurement model, i.e., this is a valid measure. If the model does not correctly describe the observed data, then so much the worse for the data. Within the confines of Rasch, data that do not fit the model cannot be measured.

For Hobart et al., it is not only the validity of PROMs that Rasch improves but also their interpretability. As I discussed above, PROMs developed using CTT are notoriously difficult to interpret. The clinical significance of a 10-point increase on a particular scale is unclear in part because CTT can only deliver ordinal level data. They are also difficult to interpret because they have dubious validity, and if we do not know what something measures, it is difficult to interpret the significance of changes in scale scores. Rasch instruments, on the other hand, provide each item a precise location on the scale. If a mobility scale is validated, then an improvement from −1.5 to .5 indicates an improvement from, say, having the ability to climb the stairs to having the ability to walk on uneven ground. Every increase or decrease in ability is tagged on a Rasch scale to items of various difficulties and thus provides estimates of clinical significance for each move up or down the ruler.

Industry

PROMs that are potentially invalid, difficult to interpret, and of questionable sensitivity can be, as Hobart et al. argue, an impediment to accurate estimates of effect size and detection of clinical change (Hobart et al. 2007). Indeed, they suggest that the failures of clinical trials to yield larger numbers of effective treatments may be due to the lack of scientific rigor of their measuring instruments. It is not surprising that pharmaceutical companies keen to demonstrate the effectiveness of their products while using measures that will satisfy the FDA guidelines are eager to explore methods that will improve their success. And it is not only industry that wants to see the acceleration of medical product development. The FDA also shares this goal.

Public-private partnerships, such as Critical Path Institute (C-Path) created under the auspices of the FDA's critical path initiative program, aim to create drug development tools (DDTs): new data, measurement and method standards to accelerate the pace and reduce the cost of medical product development, etc. (Critical Path Institute 2015). They do so by coordinating collaborations among scientists from the FDA, industry, and academia. Essentially, C-Path puts industry scientists and academics together to develop tools that will enhance the ability of industry to develop medical products. The FDA then provides iterative feedback on the tools they create hopefully ending in the approval of the DDT for use in specific product development. PROMs are one of the four types of clinical outcome assessments eligible to qualify as a DDT (Food and Drug Administration 2007).

The FDA's DDT qualification program along with C-Path is one way to build on the FDA guidelines for the use of PROMs in medical product labeling in order to streamline the process for an instrument's acceptance by the FDA. It is also an opportunity for industry and academics to work together to further their individual ends and, in doing so, flesh out the FDA guidelines, i.e., the FDA is not specific in its guidelines regarding what psychometric methods should be used to establish validity, interpretability, etc. DDTs provide industry and academics with the opportunity to develop new standards for measurement and methods, thus opening up room for new psychometric methods, such as Rasch. Indeed Hobart et al. explicitly call for such developments in their work.

Partly through the work of Sergio Sismondo, the philosophical and bioethics community has learned to have a healthy skepticism of industry/academic partnerships. Much of Sismondo's work focuses on violations of publication ethics through ghost-managed research (see Sismondo's chapter "Hegemony of Knowledge and Pharmaceutical Industry Strategy" in this volume; Sismondo and Doucet 2010; Sismondo and Nicholson 2009; Sismondo 2007). He identifies the entangled nature of ghost management as practically expedient, but ethically troublesome. It is practically expedient because at least at first gloss (almost) everyone involved wins: pharmaceutical companies get more market value out of their publications if well-respected academics put their names on the manuscripts; academics get publications in notable journals; and the journals get well-cited manuscripts, which if published will produce revenue in the form of offprints purchased by industry (Sismondo and Doucet 2010). But ghost management is ethically troublesome—we might even say corrupt—because it reveals how extensively clinical research is driven by market concerns, which in turn begs questions about (1) the justification of subjecting human subjects to research and (2) the integrity of that research. It also intimates a kind of sad desperation among academics for high impact publications and involvement in large clinical trials. As Sismondo points out, what they are doing is unethical, but ambitious academics may have few other options (Sismondo and Doucet 2010).

Although my objective in this last section is not to reveal the kind of widespread corruption that Sismondo does in his work, I do want to suggest that the collaboration of industry and academics to develop DDTs should be critically evaluated. In what follows, I suggest how a PROM developed using Rasch—for the sake of argument, a DDT—could be co-opted to provide evidence of clinical change. Thus, not only should we critically evaluate the collaborative partnerships that C-Path facilitates, but also the use of Rasch as a value-neutral improvement to the scientific rigor of PROMs.

As I discussed earlier, Rasch, unlike CCT, makes use of a more robust measurement theory. As such, it tells us what to make of respondent answers to survey questions, e.g., when respondents answer yes to more difficult items, then they have more ability than those who answer yes to easier items. It also provides us with a ruler with specific item locations. Recall that the Rasch scale runs from plus to minus infinity, with the zero point at the place where the difficulty of the items in the survey is equal to the ability of the sample population. Each item is located on the ruler relative to the point at which there is equal probability of respondents answering "yes" or "no" to that particular item. In sum, Rasch provides a formal theory that tells us where to locate items and where to locate people. But Rasch does not provide an attribute theory that guides us in choosing the content of the scale, i.e., the items or questions.

To be sure, there are constraints in the items that are chosen, the most obvious being that the data resulting from them must coincide with the Rasch model. And as I discussed earlier, if the standard error estimates of adjacent items overlap, then those items are taken to be too similar to one another. Although this latter constraint is not an absolute constraint, since increasing the sample size will decrease the standard error estimates and possibly preserve the questions under consideration. In any case, I want to put aside these two constraints and instead focus on the lack of an attribute theory within the Rasch model.

Rasch lacks a theory regarding the *content* of its target construct. Moreover, unlike the measurement of time, these target constructs are not enmeshed within a robust science such as physics. For example, Rasch does not tell us what is important about a particular construct (e.g., mobility) and neither does psychology. Thus, it is up to researchers who develop such scales to try out different questions if and until the survey data yields a fit with the Rasch model. But without theoretical guidance regarding the content of the construct of interest, how can we determine the adequate sensitivity of a scale? It seems that in this regard, Rasch is no better than CTT and possibly worse.

How might the use of Rasch be worse than CTT when it comes to the sensitivity of a scale? The problem is that Rasch makes it too easy to create a measure that is calibrated to detect clinical change. Consider the following example. It is possible to take survey data from questionnaires such as the European Quality of Life Five Dimensions (EQ-5D) and model it using Rasch. Imagine that when we do so, we find, not surprisingly, that the EQ-5D's five questions are relatively insensitive to change because they divide wide variables into only a few levels, i.e., mobility, self-care, usual activities, pain discomfort, and depression/anxiety. Earlier we discussed a similar problem regarding sensitivity in the context of CTT. In Rasch language, the EQ-5D is too easy, i.e., even respondents without a lot of ability can answer all the questions positively. For instance, eye problems, sleep problems, sexual functioning, memory problems, problems communicating poststroke, and fatigue are a few of the deficits to which the EQ-5D is generally insensitive.

Now, suppose that you were looking at the EQ-5D data because you were interested in whether or not it was the appropriate measure to use in a clinical trial to establish the effectiveness of a drug. You have the mean pretreatment scores of your target population and you know that pretreatment they already have has more ability than the EQ-5D is able to measure. If you want to show a clinical improvement, then you need a measure that is more sensitive. In the language of Rasch, you need a measure that can target a higher-functioning population, i.e., respondents with more ability. Because you already know the mean pretreatment scores, you have an idea where on the ruler you need to develop the scale in order to measure the change you anticipate. Moreover, the more responsive the rulers (i.e., the closer together each step on the ruler), the more likely you will find a clinically significant change.

I want to be very clear: I am not suggesting that anyone is disingenuously using Rasch to demonstrate clinical change. What I am suggesting is that Rasch represents an opportunity to increase the likelihood of finding clinical benefit, while the choice to use Rasch is presented as a matter of scientific rigor. I am not alone in recognizing that Rasch represents this opportunity. Indeed Hobart et al. admit that one criticism of more sensitive measures is that they will increase type 1 errors (false positives) (Hobart et al. 2007). But while they more or less dismiss this concern since blunt instruments are equally problematic, I think it is worth taking seriously.

One reason to do that is because science is a value-laden enterprise. Indeed, as Heather Douglas writes in *Science, Policy, and the Value-Free Ideal*, social and ethical values are necessary to any science that has a public role, i.e., any science that has a role in policy, medicine, technology, etc., as health measurement certainly does. Douglas's argument is twofold. First, she reminds us that our evidence always underdetermines what we should believe (Douglas 2009). We can see her point, if we attend to Rasch measurement scales. Here we see that our knowledge of a construct, including respondent data from items thought to be related to the construct, underdetermines how many questions we ought to ask and at what difficulty level we should target our efforts, including how sensitive the scale should be and if certain areas of the scale should be more sensitive than others. Put another way, there is always an element of uncertainty in the use of scientific evidence. This uncertainty is overcome only when scientists use their judgment to determine which standard, characterization, claim, or theory is indicated (Douglas 2009).

For the second part of her argument, she claims that when science has a public role, when, for instance, a study has the potential to affect public policy or medical treatment options, then the use of expert judgment draws on social and ethical values. When science has the potential to affect others—and it clearly does in the context of health measurement—then the values employed in using one's judgment should be connected to an individual's perception of what is at stake should one make a mistake. Scientists ought to evaluate the social and ethical consequences of error (Douglas 2009: 87). In other words, when considering how sensitive a scale should be, researchers should contemplate the social and ethical consequences of creating an overly sensitive scale that increases the likelihood of type 1 errors. Some of the consequences might be loss of public trust, overmedication, rising healthcare costs, and industry (dis?)satisfaction. Equally, researchers should consider the consequences of creating less sensitive scales that increase the likelihood of type 2 errors (false negatives). Some of these consequences might be increased cost and time to medical product development and increased patient suffering due to the delay in medical product development.

For Douglas, the solution to scientific disagreements that stem from differences in social and ethical value orientations is to make the values on which decisions or judgments are based more transparent (Douglas 2009). In the context of health measurement, we might begin by simply acknowledging their existence. When Hobart et al. criticize CTT as unscientific and suggest Rasch as a replacement in the name of scientific rigor, we might soften the critique by recognizing that the choice to use CTT over Rasch is not only a lack of sophistication and knowledge as they sometimes seem to suggest, but also a value choice of prioritizing expediency and simplicity. Moreover, even if Rasch does provide the basis for more scientific measurement scales—as I believe it does—supporters need to recognize the value-laden decisions that still characterize these scales. Without recognition of values we employ under conditions of uncertainty, we cannot evaluate them. If we do not evaluate them, then I worry that similar to the case of ghost management, we might find ourselves building measures to tailor the marketing needs of pharmacy.

References

BIPM (Bureau International des Poids et Measures). 2006. The International System of Units (SI). Sèvres: BIPM. http://www.bipm.org/en/si/si_brochure/. Accessed 16 Apr 2015.

Cano, S.J., and J.C. Hobart. 2011. The problem with health measurement. *Patient Preference and Adherence* 5: 279–290.

Chauhan, C. 2007. Denouement: A Patient-Reported Observation. *Value in Health* 10: S146–S147.

Critical Path Institute. 2015. http://c-path.org/about/. Accessed 7 Apr 2015.

DeMuro, C., M. Clark, L. Doward, E. Evans, M. Mordin, and A. Gnanasakthy. 2013. Assessment of PRO Label Claims Granted by the FDA as Compared to the EMA (2006–2010). *Value in Health* 16(8): 1150–1155.

Douglas, H. 2009. *Science, Policy, and the Value-Free Ideal*. Pittsburgh: University of Pittsburgh Press.

Fayers, P., and D. Machin. 2007. *Quality of Life: The Assessment, Analysis and Interpretation of Patient-Reported Outcomes*, 2nd ed. Chichester: Wiley.

Food and Drug Administration. 2007. Drug Development Tools Qualification Programs > Clinical Outcome Assessment Qualification Program. http://www.fda.gov/Drugs/DevelopmentApprovalProcess/DrugDevelopmentToolsQualificationProgram/ucm284077.htm. Accessed 7 Apr 2015.

Hobart, J.C., S.J. Cano, J.P. Zajicek, and A.J. Thompson. 2007. Rating Scales as Outcome Measures for Clinical Trials in Neurology: Problems, Solutions, and Recommendations. *Lancet Neurology* 6(12): 1094–1105.

Hunt, S.M. 1997. The Problem of Quality of Life. *Quality of Life Research* 6(3): 205–212.

Jaeschke, R., J. Singer, and G.H. Guyatt. 1989. Measurement of Health Status. Ascertaining the Minimal Clinically Important Difference. *Contemporary Clinical Trials* 10(4): 407–415.

McClimans, L. 2010a. Towards Self-Determination in Quality of Life Research: A Dialogic Approach. *Medicine, Health Care, and Philosophy* 13(1): 67–76.

———. 2010b. A Theoretical Framework for Patient-Reported Outcome Measures. *Theoretical Medicine and Bioethics* 31(3): 225–240.

———. 2011. Interpretability, Validity, and the Minimum Important Difference. *Theoretical Medicine and Bioethics* 32(6): 389–401.

———. 2017. Measurement in Medicine and Beyond: Quality of Life, Blood Pressure and Time. In *Reasoning in Measurement*, ed. N. Mößner and A. Nordmann. London: Routledge.

McClimans, L., and J.P. Browne. 2012. Quality of Life Is a Process Not an Outcome. *Theoretical Medicine and Bioethics* 33(4): 279–292.

Schwartz, C.E., and M.A. Sprangers. 1999. Methodological Approaches for Assessing Response Shift in Longitudinal Health-Related Quality-of-Life Research. *Social Science & Medicine* 48(11): 1531–1548.

Sismondo, S. 2007. Ghost Management: How Much of the Medical Literature Is Shaped Behind the Scenes by the Pharmaceutical Industry? *PLoS Medicine* 4(9): e286.

Sismondo, S., and M. Doucet. 2010. Publication Ethics and the Ghost Management of Medical Publication. *Bioethics* 24(6): 273–283.

Sismondo, S., and S.H. Nicholson. 2009. Publication Planning 101. *Journal of Pharmacy & Pharmaceutical Sciences* 12(3): 273–279.

Stenner, A.J., and M. Smith. 1982. Testing Construct Theories. *Perceptual and Motor Skills* 55(2): 415–426.

Streiner, D.L., and G.R. Norman. 2008. *Health Measurement Scales: A Practical Guide to Their Development and Use*. Oxford: Oxford University Press.

Tal, E. 2011. How Accurate Is the Standard Second? *Philosophy in Science* 78(5): 1082–1096.

Taminiau-Bloem, E.F., F.J. Van Zuuren, M. Visser, C. Tishelman, C.E. Schwartz, M.A. Koeneman, C. Koning, and M. Sprangers. 2011. Opening the Black Box of Cancer Patients' Quality-of-Life Change Assessments: A Think-Aloud Study Examining the Cognitive Processes Underlying Responses to Transition Items. *Psychology & Health* 26(11): 1414–1428.

Terwee, C.B., F.W. Dekker, W.M. Wiersinga, M.F. Prummel, and P. Bossuyt. 2003. On Assessing Responsiveness of Health-Related Quality of Life Instruments: Guidelines for Instrument Evaluation. *Quality of Life Research* 12(4): 349–362.

Wyrwich, K.W., and V.M. Tardino. 2006. Understanding Global Transition Assessments. *Quality of Life Research* 15(6): 995–1004.

Part II
Dispensing

Patient-Centered Care or Drug-Centered Care: The Influence of Pharmaceutical Marketing on Medical Science and Public Health

Howard Brody

Introduction

"Patient-centered care" has become a popular buzzword, leading to concerns that it is being used as pure public relations without substance. When it does have substantial content, it can best be understood in contrast to other sorts of "care" (Duggan et al. 2006). When contrasted with "doctor-centered care," patient-centered care refers to allowing a robust role for patients to be involved in decisions regarding their own healthcare, rather than simply being expected to follow the professionals' directives and recommendations.

Alternatively, patient-centered care could be contrasted with "disease-centered care." It treats patients as fellow human beings in distress and in need of aid and not merely as bodies that encase diseased organs in need of technical fixing.

Patient-centered care then refers to a style of care that takes seriously the human beings who are sick and incorporates a full appreciation for their feelings, thoughts, values, and social relationships.

The major question I wish to discuss in this chapter is whether there remains a further contrast to patient centeredness in medicine and healthcare, which we could term *drug-centered care*. Drug-centered care is what results from the influence of pharmaceutical industry marketing practices over the thinking and activities of both physicians (and other health professionals) and patients. From the standpoint of a drug company, seeking to make a profit in a capitalist economy by selling a useful product, drug-centered care makes a lot of sense. But we must ask whether it makes sense from the standpoint of medical science or public health. To aid in this inquiry, I shall look in more depth at two case studies—medications for depression and for type 2 diabetes.

H. Brody (✉)
The University of Texas Medical Branch, Galveston, TX, USA
e-mail: hbrody13@gmail.com

© Springer Science+Business Media B.V. 2017
D. Ho (ed.), *Philosophical Issues in Pharmaceutics*, Philosophy and Medicine 122,
DOI 10.1007/978-94-024-0979-6_7

Patient-centered care, understood as something substantive, can be defended on at least three ethical grounds—it has been shown to produce superior health outcomes in patients; it better respects the personal dignity and basic rights of patients; and it instills virtuous attitudes in health professionals (Duggan et al. 2006). If these ethical goods are to be achieved, we have to ask whether drug-centered care promotes these outcomes or interferes with them.

Pharmaceutical Industry Marketing

It has been estimated that in 2004, the US pharmaceutical industry spent $57.5 billion on marketing (Gagnon and Lexchin 2008). This huge expenditure produced a tremendous return for the drug industry at a highly profitable time and was used for (Brody 2007; Kassirer 2005; Angell 2004; Avorn 2004):

- A large sales force of "detail people" or sales representatives ("drug reps"), who speak with physicians individually in their practices, often bringing lunch for the entire office staff
- A data system that keeps the reps informed as to the prescribing habits of each physician in their territory, so that their marketing approach can be precisely calibrated
- Samples of drugs for free distribution to patients from physicians' offices
- Dinners for physicians held at fancy restaurants, featuring a talk on a drug-related topic by a drug rep or (preferably) a physician hired by the company
- Print advertisements in medical journals
- Reprints of journal articles favorable to a company's drug, distributed by the drug reps
- Magazine and television advertisements direct to consumers (legally permitted in only two nations, the USA and New Zealand)
- Financial support for continuing medical education (CME) programs, often featuring drug-advertising displays and receptions for attendees, as well as support for the educational presentations
- Financial rewards to high-prescribing physicians, such as payment for drug talks or expenses to attend CME conferences
- "Seeding trials," which masquerade as legitimate scientific studies of drugs already approved for marketing, but function actually as programs that reward practitioners for prescribing the drug by paying them to enroll patients in the "study" and to record minimal data
- Generous contributions to disease-specific patient advocacy groups, such as the American Diabetes Association and the National Alliance on Mental Illness, to encourage industry-friendly policies and educational materials
- Internet-based marketing strategies, that, for example, might use a pop-up ad for a drug when a physician goes to a medical reference website to look up the disease for which that drug is often prescribed (Dolan 2013)

Besides these practices, which can be labeled clearly as marketing, the reach of the marketing division within the drug firm commonly extends further. For example, it appears to be common business practice to involve marketing staff in all planning for scientific clinical trials, to assure that the trial design and results maximally facilitate later marketing efforts. Clinical trials of promising drugs have been stopped prematurely by the company based on marketing estimates that sale of the new drug would not be as profitable as had been hoped (Psaty and Rennie 2003). While companies do not readily reveal such figures, the most reliable estimates indicate that the average company spends two to three times as much on marketing as on research and development (Reinhardt 2001).

The Influence of Marketing

Marketing is most effective when the potential customer hears the same message from multiple, reinforcing sources. The ideal launch of a new drug might go something like this: The physician, attending a meeting of her specialist society, hears an exciting talk about a new drug nearing approval by the Food and Drug Administration (FDA) but not yet on the market. Within months, ads in widely read medical journals announce that the drug is finally available. Then, the company's rep visits the office and leaves a stack of glossy reprints extolling the drug's virtues as well as a supply of samples. Now a patient walks into the office asking for that drug by name, based on a television ad he had seen. It is hard to imagine this physician *not* getting the powerful message that this new drug is important and useful and that failing to prescribe it to patients would be poor medical practice.

K. Applbaum, a medical anthropologist, described the aim of drug companies' marketing as "controlling the channels" (Applbaum 2009, 2010). Chemicals that might serve as drugs follow a complex route, from discovery, scientific testing, and approval for marketing to manufacture, prescription by a physician, and actual consumption by a patient. Along this route, many sorts of people serve as gatekeepers, who can facilitate or impede the drug's flow along the route. These gatekeepers include scientists, government regulators, insurers, pharmacy benefit managers, and many others besides physicians and patients. These various gatekeepers have disparate interests, many at odds with one another.

According to Applbaum, the drug firm has two strategic aims. The first is to control the channel. The gates have to be configured so that the drug can flow smoothly all along the route. Applbaum notes that when a firm, or the industry as a whole, effectively controls the channel, it hardly matters what specific drug is put into it or whether the drug is helpful or harmful; the drug will be prescribed and taken and the companies will profit.

The second strategic aim is not to appear to be controlling the channel at all. Many of the gatekeepers are more likely to cooperate with the company's aims precisely to the degree that they believe that they are acting autonomously and are free from industry manipulation. Smoothly controlling the channel requires that each gatekeeper believes that he is acting independently in his own interests. The very complexity of the channel serves this strategic aim well; it simply does not seem possible to the average onlooker that any single actor could possibly exert so much control, at so many places, simultaneously. (Of course, when one has $57 billion to spend, many things become possible.)

Applbaum studied in depth one of the case examples I will refer to later, the newer classes of antidepressant drugs. An example he gives of "controlling the channels" is the creation of simple-to-administer questionnaires designed to allow primary care physicians to determine whether a patient is sufficiently depressed to warrant drug therapy. From the standpoint of the physician, the questionnaire seems a valuable aid. It saves time and increases the efficiency of an office visit. Psychiatrists commonly criticize primary physicians for missing the diagnosis of depression, so this instrument reassures the physician that she is being appropriately thorough. Also, since the diagnosis of a treatable depression can now be made (or so it appears) in the office, there is less need to refer patients out of the practice either to psychiatrists or to other mental health professionals.

From the standpoint of the drug company, things look a bit different. If psychiatrists prescribe a new antidepressant, then the potential market is a certain size. If primary care physicians as well as psychiatrists can be persuaded to prescribe the drugs, the potential market increases greatly. If primary care physicians can also be persuaded that a patient who in former days would have been viewed as simply reacting to bad life circumstances, or going through a bad spell, can be reclassified as having a depression that deserves drug treatment, the market is further expanded. If the questionnaire is designed to set the threshold for diagnosing depression lower than is strictly warranted, it is unlikely that the primary physician will have the expertise to challenge this.

For such a questionnaire to have the desired effect, other aspects of the channel have to be favorable. The physician, for example, must be persuaded that whereas older antidepressants were somewhat dangerous to administer, the newer classes of drugs are considerably safer and so can be prescribed with relative impunity. In this way, the various aspects of the overall marketing campaign all fall into place.

The twin strategies of controlling the channels, and not appearing to be controlling the channels, have important implications for other parties along the pathway, especially physicians and patients. The role of the pharmaceutical industry becomes inherently less trustworthy. Even if, in any specific instance, a drug firm is communicating candidly, other actors must remain suspicious that some level of subterfuge is intended to conceal an effort to exert control over other parties' behavior. Another important implication, which I will develop more below, is that one highly efficient way to control a channel is to *change the way both physicians and patients think* about a medical condition. If these other actors can be persuaded to think about a disease in a drug-prescribing-friendly fashion, the industry can be assured of sales, even while other actors fondly believe that they are doing nothing more than acting in accord with their own self-interests (or professional responsibilities) and are not a party to marketing manipulation at all.

The case studies of antidepressant drugs and drug therapy for type 2 diabetes illustrate how what at first seems to be the desirable "patient-centered care," or even the less desirable "disease-centered care," can easily be rechanneled into drug-centered "care."

Case Study: Antidepressants

Antidepressants represent an attractive market for pharmaceutical marketers for two reasons. First, most often, antidepressants are taken long term, often for life. By contrast, even though there is a pressing need for new antibiotics, to replace older drugs to which bacteria have developed resistance, very few new antibiotics have been introduced in recent years. Though the scientific research needed to gain approval of an antibiotic costs the company at least as much, if not more, than research for a new antidepressant, patients typically take an antibiotic for only two to three weeks. The potential profit from a longer-term drug makes it much more attractive for companies to focus their research efforts in that direction, regardless of clinical need.

The second reason that antidepressants constitute a more lucrative potential market is that there is no definitive blood test or other organic marker for the diagnosis of depression. That makes the diagnosis a more elastic category and provides more of an avenue for drug marketing to expand and extend the diagnosis. (We will see below, in discussing type 2 diabetes, that having a blood test does not necessarily pose an insurmountable barrier, however.)

Even with $57 billion to spend, it is unlikely that the pharmaceutical industry would manage to convince us that something is black when we are quite sure that it is white. So most successful marketing campaigns start out by finding ways to reinforce something the target audience already wants to believe. In the case of antidepressants, this worked well for both psychiatrists and patients. During the last decades of the twentieth century, psychiatrists wanted to emerge from the perceived shadows of the domination of Freudian psychoanalysis, which incidentally was dismissive of most drug therapy. In the minds of both physicians and the general public, psychoanalysis stood for arcane mumbo jumbo akin to witch doctoring. Psychiatrists wanted above all else to prove that they were every bit as respectable as other medical specialists—and in the language of twentieth-century medicine, this meant what came to be known as *biological psychiatry*. Mental illness had to be just like any other disease of any other organ system—it had to be based on some purported organic lesion that was detectable by the right sorts of chemical tests and imaging studies. Mental illnesses had to be diseases of the brain. If a drug that changed the chemistry in the brain in a targeted and known manner made a mental illness better, that fact helped to convince everyone that mental illness was "real" according to the coin of the medical realm and therefore that psychiatrists were "real" doctors.

Patients for their part also wanted to hear that mental illness was caused by a chemical imbalance in the brain and could be fixed by the right pill. First, many people in society still regarded mental illnesses, especially those that used to be called neuroses, as simple failure of will. Presumably, if only the individual tried hard enough, he'd snap out of it. A disease model that absolved the individual of personal responsibility for mental illness, analogous to the way we typically regard the victim of pneumonia as not having done anything to bring on the disease, was fervently sought. Finally, Americans are generally impatient and hardly want to hear, as psychoanalysis often seemed to say, that years of patient probing into one's life history and feelings would be needed for a meaningful response. A pill that promised to reverse the condition in just a few weeks was far preferable.

It would perhaps be reassuring to say that the specialty of psychiatry was simply caught napping and was blindsided by a clever industry ploy; but the truth is more complicated. Robert Whitaker, whose history of recent psychiatry offers a scathing critique, names a particular date when psychiatry's leaders made an explicit decision to get into bed with the pharmaceutical industry. The American Psychiatric Association (APA) formed a task force in 1974 to explore common interests with the pharmaceutical industry, including joint public relations efforts. In 1980, the APA adopted a policy of accepting drug company funds for specific educational presentations at its convention (Whitaker 2010: 268–282). This created the near-circus atmosphere that a senior psychiatrist deplored in 2002, reporting on an international psychiatric congress in Berlin that featured, among a plethora of drug displays, a picturesque babbling brook and a 40-foot rotating tower. Presentations about drug treatment for mental disorders were thrust into the attendees' faces; sessions describing alternatives to drugs were hidden away in back rooms (Torrey 2002). Following this sort of negative publicity, the APA announced in 2009 that it would no longer accept drug funding for its annual conferences (Tanne 2009).

As Applbaum argued, once the industry has succeeded in controlling the channels, it hardly matters what drug is put into the pipeline. The selective serotonin reuptake inhibitor (SSRI) class of antidepressants is a good illustration. The ideal of biological psychiatry is that a particular psychiatric condition is caused by a single, unique, and discrete form of chemical malfunction in the brain. The ideal psychoactive drug, in turn, targets that single chemical pathway and leaves all other brain functions untouched—the proverbial magic bullet. The problem from the start was that the SSRIs refused to conform to this pattern. As David Healy, psychiatrist, drug industry critic, and experienced investigator of these drugs, has reported, SSRIs actually appeared to affect a wide array of symptoms (Healy 1997). The first such drug, buspirone, was marketed in the late 1980s as an antianxiety drug, and that effort flopped miserably. The company claimed that it was an excellent tranquilizer but was non-habit forming. But physicians would have none of it. They had just been through a phase of chastisement over the too liberal use of supposedly safer tranquilizers like Valium and Librium, only to find out how addictive these drugs were in practice. Everyone now *knew* that tranquilizers were habit forming.

Therefore, when the Eli Lilly Company sets out to market its new SSRI drug, fluoxetine (Prozac), they knew exactly what to do and what to avoid. They promoted the drug as an antidepressant, only one that was safer than the older class of tricyclic antidepressants, which were known among other things for occasional side effects involving heart rhythm and so could be fatal in overdose. Everyone knew that antidepressants were nonaddictive, so that part of the marketing campaign worked well. Then, over time, all that was needed was to convince physicians that what had previously been termed anxiety was, in reality, a disguised form of depression.

Two inconvenient facts created potential problems for this marketing campaign—fluoxetine did not work very well against depression, and it had its own nasty set of side effects. (In fact, it's reported that when Prozac was first submitted for marketing approval in Germany, the drug agency there turned it down, unimpressed with any of the evidence of efficacy.) (Healy 2003: 204) One of the stunning stories about pharmaceutical marketing during this era is for how long the industry was able to conceal these inconvenient truths. Research studies that indicated high success rates with SSRIs were promptly published, often in multiple journals, while equally well-done studies showing disappointing results were quietly buried (Turner et al. 2008). Study design was carefully manipulated to avoid recording serious adverse reactions. The most worrisome though fortunately rare reaction attributed to SSRIs was akathisia, an agitated state, often occurring within the first few weeks of therapy, in which patients might become homicidal or suicidal. Some of the questionnaires used to measure adverse reactions in trials of SSRIs seemed specifically designed to avoid revealing any signs of akathisia. For years, companies insisted that any patient who committed suicide shortly after starting an SSRI did so as a result of the underlying depression and not because of the drug—an explanation that ceased to hold water when non-depressed patients, started on SSRIs for other conditions, also occasionally became suicidal.

Another secret that the drug industry managed to keep for many years is that patients who attempted to discontinue taking SSRIs often experienced a nasty withdrawal reaction. This reaction happened to be good for drug sales, as psychiatrists swayed by company marketing invariably attributed these symptoms to worsening depression and used them as evidence that the patient needed lifelong drug therapy. Even better, the psychiatrist might decide that stopping the SSRI had unmasked a coexisting psychiatric illness such as bipolar disorder, and henceforth this patient needed to be placed on two or three psychiatric drugs, not just one. Whitaker, in his book, reviewed both the extensive clinical evidence supporting such a withdrawal syndrome and the biochemical mechanisms that rendered such a syndrome a logical outcome of drug treatment (Whitaker 2010). But as long as the dominant narrative circulating in the medical community was the one generated by the industry marketers, voices such as Whitaker's were easily drowned out.

Perhaps the single most successful aspect of the marketing of the SSRIs was selling the general public, as well as the medical community, on the serotonin theory of depression. Many lay people can explain to you today precisely how antidepressants work—that the depressed person has too little serotonin in the synapses between nerve cells, due to too rapid reuptake of the chemical by the cells; and SSRI drugs slow the reuptake process and so restore serotonin to its proper levels. It sometimes seems as if "How's your serotonin level?" is as likely to occur in casual conversation as "How's everything?" Like much of the rest of the dominant, industry-promoted narrative, the serotonin theory of depression turns out scientifically to be mostly a mirage. At best it's a serious oversimplification; at worst it's a plain falsehood (Healy 2003; Leo and Lacasse 2008). But the theory serves so many useful purposes for both physicians and the general public that everyone is loath to let go of it, regardless of what the scientific evidence shows.

Case Study: Diabetes

Just as everyone knows that too little serotonin causes depression, everyone knows what is wrong with you if you have diabetes and what should be done about it. Diabetes is popularly known as "sugar" in some communities, and that's accurate because the basic problem in diabetes is a too high blood sugar level. This is true in both sorts of diabetes, type 1 (juvenile onset) and type 2 (adult onset). The difference is that in type 1, the problem is too little insulin, and no treatment is possible without injecting insulin. In type 2, however, some of the problem is resistance to insulin in the body's cells, and this can sometimes be treated with oral medications, without the need for insulin injections. But in any type of diabetes, the mainstay of successful treatment is to restore the blood sugar to a normal or at least near-normal level.

If anyone doubted this narrative about the cause and treatment of diabetes, their doubts would be removed by watching any of the advertisements on daytime TV aimed at the US Medicare population with type 2 diabetes. These ads make clear that it is an absolute necessity that anyone with the diagnosis of type 2 diabetes possesses and frequently uses a home glucose monitor. Your physician would, of course, tell you this (and, by implication, any physician who does not is not worth listening to). And don't worry about the finances, because Medicare will happily pay for most of the cost of both the monitor and the test strips. (Indeed, the device companies find that selling the test strips is such a lucrative business that they will sometimes give away the monitor for free.)

Now, watching these ads, one would never know that when randomized trials are conducted, in which half the subjects with type 2 diabetes are given monitors and the other half not, no difference is found in either control of blood sugar or other health outcomes (Farmer et al. 2012: e486). Nor would one know that a number of major studies over several decades have shown with a high level of confidence that normalizing blood sugar in type 2 diabetes is not a particularly good strategy for achieving the most important health goals (Boussageon et al. 2011: d4169).

Why is diabetes bad for you? It is true that once a certain very high level is reached, the hyperglycemic state itself causes one to feel bad and function poorly, so at that level, it does in fact make a real health difference to lower blood sugar. Many people with mild to moderate type 2 diabetes, however, seldom if ever reach that level of hyperglycemia.

The much more serious problem with type 2 disease is the complications. People with diabetes have a much higher rate of serious and potentially fatal complications, including heart attacks, stroke, kidney failure, blindness, and blood vessel disease requiring limb amputation. So it's reasonable to ask what the scientific evidence shows about the relationship between strict control of blood sugar and preventing these serious complications.

The largest and most extensive study to address this question was the UK Prospective Diabetes Study (UKPDS), which enrolled a large number of patients and followed them for ten years, a feat unlikely ever to be repeated. The main UKPDS investigators confidently expected to show a close correlation between blood sugar control and reduced rate of complications, and so that's how they reported their results, even though they had to play fast and loose with a number of statistical analyses to get that outcome (UK Prospective Diabetes Study Group 2008). Others, looking at the UKPDS body of data more critically, have pointed out that the data simply do not support that conclusion (Montori and Fernandez-Balsells 2009). Particularly with regard to the complications that involve the larger blood vessels (heart attack, stroke, amputation), there was no good evidence that reducing the blood sugar lowered the rate of complications. One medicine, metformin, turned out to be very good at preventing bad complications. Other medicines, however, that lowered sugar as much or more had no such preventive effect, so it seemed that metformin exercised its benefits in a way separable from its glucose-lowering properties.

If one looks at UKPDS in this critical manner, it seems a good example of why association needs to be distinguished carefully from causation. There is a good deal of evidence from epidemiologic studies that there's a close association between elevated glycohemoglobin (the definitive blood test that shows that one has had a high blood sugar over an extended time period) and higher rates of complications from diabetes. It would then seem eminently logical that treatment that lowers glycohemoglobin level (by better controlling blood sugar) would reduce the rate of complications. But evidence doesn't seem to support this conclusion. Since elevated glycohemoglobin is a marker for more severe diabetes, it seems that those with more severe diabetes are more likely to suffer complications and vice versa. But drugs that lower blood sugar generally seem ineffective in turning severe cases into mild cases. By contrast, attempting to control diabetes through diet and exercise seems more promising in terms of reducing complications, but this is not the place to discuss that.

In light of the lessons we have learned from a closer scrutiny of the UKPDS' data, we'd naturally expect that when other studies used medications to lower blood sugar levels in type 2 diabetes, the same outcome would occur. And to date, this is precisely what these further studies have shown. In general, studies that use medication to more strictly control blood sugar do not significantly reduce the rate of complications. However, such a regimen almost always causes harm, because the strict-control patients suffer more frequent episodes of low blood sugar (hypoglycemia). And hypoglycemia can be very serious, leading to seizures and coma; and the risks of and damage from hypoglycemia increase as one gets older.

What's striking about these further studies is that diabetes experts, who might be expected to say, "Oh well, that's what UKPDS suggested, so no big surprise here," act astounded each time such a study is reported (Kishore et al. 2012). It seems that the comfortable narrative of diabetes = sugar is so deeply entrenched in both medical and popular thinking that even physician-scientists cannot quite let go of it, no matter how much contrary evidence accumulates.

Medical historian Jeremy Greene has studied where this compelling narrative comes from in his book, *Prescribing by Numbers: Drugs and the Definition of Disease* (Greene 2008). During the middle of the twentieth century, beginning with the famous Framingham Study of heart disease risk, medical science happened upon a new concept—the idea of the *risk factor*. In older times, either you were healthy and the doctor left you alone or you were sick and the doctor treated you. Now suddenly there was a new way that you could be healthy and yet need the doctor's care. You might have a risk factor which, if unattended to, made it more likely that you'd suffer some really serious illness later on but that could be mitigated with the proper medical attention, including drugs.

Everyone at the time thought that risk-factor medicine was a marvelous advance. A common criticism of the older style of medicine was that it failed to attend to prevention. Risk-factor medicine, by contrast, seemed to be preventive medicine *par excellence*. What's not to like?

As Greene reviews the history, what failed to get noticed at this time (or that more accurately, seems meaningful only in hindsight) was the very close cooperation between the pharmaceutical industry and medicine in making the turn toward risk-factor interventions. What the industry most wanted, and medical science appeared eager to supply, was "prescribing by the numbers"—developing and encouraging the widespread use of simple blood tests to identify a threshold level of a risk factor and then persuading doctors that a healthy patient who had such an abnormal lab result ought immediately to be placed on a medicine, ideally, an expensive medicine that needed to be taken for the rest of one's life.

"Prescribing by the numbers" is very near a dream world for a pharmaceutical marketer. According to the recently adopted mantra of patient-centered care, the ideal research study randomizes subjects to the drug or no-drug condition and then follows them long enough to see how many develop an outcome that actually matters in people's lives, such as death or a condition like a heart attack that causes significant disability and hospitalization (Washington and Lipstein 2011). Such a study (like UKPDS) takes a long time and is very expensive. Moreover, it is hard to predict what will be the actual outcome of such a study; and drug marketers hate that uncertainty.

Consider by contrast the sorts of studies drug companies love, aimed at what evidence-based experts call "surrogate markers." Once you have decided that high blood pressure or high blood sugar indicates conditions that count as risk factors for later heart disease, you then do a study to show that a drug lowers blood pressure or blood sugar. Typically a drug will show an effect on the surrogate marker within a few weeks, making for a much shorter and hence cheaper study. Based on preliminary data, it's much easier to predict a positive outcome involving surrogate markers than it is if one were studying actual clinically relevant outcomes. And, there's a bonus: It's not uncommon that a drug will lower a surrogate marker promptly but show a serious adverse reaction only if taken for some months. If the company is lucky, it can get in and out with its study in time to "prove" success with the surrogate marker, but not run the study long enough to pick up any unwelcome signs of bad side effects.

The recent career of rosiglitazone (Avandia) nicely illustrates the success of "prescribing by the numbers." Rosiglitazone is one of a class of drugs introduced with much fanfare for type 2 diabetes in the late 1990s, because they promised a new biochemical approach to drug therapy. They were approved for use because they worked well on the surrogate marker, glycohemoglobin, or blood sugar, even though no studies had been sufficiently lengthy to see whether any of these drugs actually reduced the long-term complication rates. Indeed, no evidence was ever accumulated to show that rosiglitazone successfully prevented major diabetic complications. Ideally, such a drug would not have been used at all for type 2 diabetes, or else it would have been a third-line drug for use in selected recalcitrant cases. Yet with aggressive marketing, Avandia was quite widely used, despite its high cost.

With time, however, worrisome evidence accumulated that rosiglitazone actually increased the risk of serious heart disease. At first the manufacturer simply hid the evidence and attacked independent investigators who attempted to publish such findings (Nissen and Wolski 2007; Moynihan 2010). In the end, the company had to admit the problem and submit to a major warning label. The drug was released in 1999 and by 2006 had reached a sales peak in the USA of $2.5 billion. Once the negative heart disease information was published in 2007, US sales plummeted, but even then Avandia racked up $1.2 billion in worldwide sales in 2009 (Noble 2013). Since the drug would have faced stiff competition from cheaper generic versions after it went off patent in 2012, the company got to enjoy almost a full lifetime of generous profits from a largely useless and actually dangerous drug. Whatever the effect on unfortunate patients, as a business plan, Avandia succeeded splendidly.

The Public Health and Inverse Benefit

The Avandia case is a prime example of what a colleague and I call the inverse benefit law. Avandia sales jumped markedly following publication of a study in 2006 in *The Lancet*, showing the drug to slow the progression from the state designated as "prediabetes" to full-fledged type 2 diabetes (Gerstein et al. 2006). This was presented to the medical community as a major breakthrough. Apparently one was supposed to forget that both prediabetes and diabetes are diagnoses that rely solely on numerical values of blood tests. It is questionable that a condition "prediabetes" even exists apart from the arbitrarily labeling of lab values. It has never been demonstrated that preventing the progression of prediabetes into diabetes (as opposed to waiting for diabetes to develop and then treating that condition) is associated with any superior outcomes in terms of serious complications.

In short, success was declared for the drug in the absence of any data whatsoever that patients received any benefit. At the same time, as was later demonstrated, all the patients given Avandia suffered a higher risk of later developing heart disease.

To explain the phenomenon of which this Avandia study is a single example, sociologist Donald Light and I proposed an *inverse benefit law* (Brody and Light 2011). According to this "law," intended as a heuristic rather than as a quantitative predictor, the public health consequences of the use of a drug worsen proportionately to how aggressively a drug is marketed.

The inverse benefit law is illustrated in the figure:

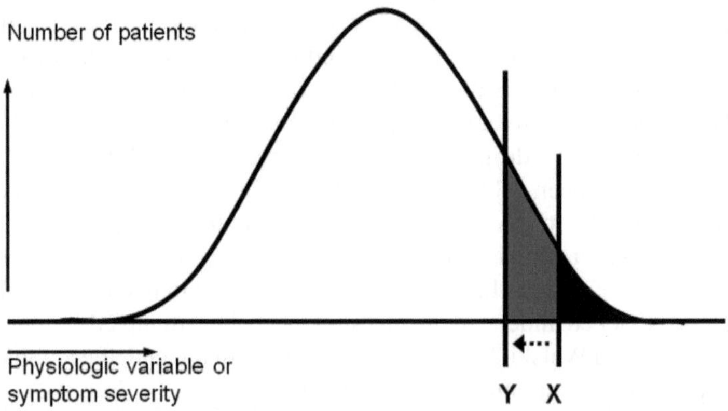

The example of blood sugar level works particularly well to illustrate the law. A physiological variable such as blood glucose is distributed normally through the population, generating the familiar bell-shaped curve. Let's say that the line in the figure marked X is the cutoff level for the diagnosis of diabetes. If a drug is supposed to treat diabetes, the area shaded in black represents the population of patients who are candidates to receive the drug. The likelihood that the drug will cause adverse reactions is distributed (we assume evenly) throughout the entire population. That is, if a drug like Avandia causes more risk of developing heart disease, it does so across the entire population, regardless of one's blood sugar level.

If a drug is a good drug for treating diabetes—we have seen that Avandia isn't—two good things happen when it is prescribed for the group of people in the black-shaded area. First, there's a decent chance they will be helped, since they actually have the disease for which the drug is recommended. Second, relatively few people are at risk for developing adverse reactions, since such a small part of the total population are exposed to the drug. The ratio between harm and benefit, therefore, is likely to be favorable.

The black-shaded area may be a good sign for the public health advocate, but it is bad news for the industry marketer. So long as such a small percentage of the population is a candidate for the drug, profits will be severely limited. The marketer's goal is to pull off a "left shift," that is, to move the threshold for prescribing the drug from X to Y in the figure. Claiming that Avandia is good for treating "prediabetes" as well as diabetes is an example of such a shift, if we assume that the line Y corresponds with the lower blood sugar level at which the condition "prediabetes" is diagnosed.

The "left shift" from X to Y is a huge success for marketing due to the shape of the bell curve. The actual difference in blood sugar value between X and Y may be slight, yet the population considered candidates for drug treatment (now both the gray- and black-shaded areas) has much more than doubled. By contrast, the left shift has been a public health disaster. The patients in the gray area are much less likely to derive actual benefit from drug therapy compared to the black area. Since they begin with a lesser chance of developing serious complications, the drug has a corresponding lesser chance of preventing those complications. Yet a much greater number of people have now been exposed to the risks of adverse reactions. The harm-benefit ratio has shifted considerably to the negative side.

In summary, drugs generally prove to be a public health benefit when science is able to identify a specific population that benefits from the drug, and prescriptions of the drug are then targeted at that specific population. Drug marketing almost inevitably seeks to extend the use of the drug beyond that population, which in turn almost always constitutes a threat to the public health. Marketing, one might say, turns good drugs into bad drugs. Marketing fails to serve patient-centered care and even falls short in promoting the less desirable disease-centered care. Instead, marketing creates an unfortunate parody of good medicine—drug-centered care.

Conclusion

When drug-centered "care" takes the blatant form of what amounts to bribery or corruption, as when physicians are given financial incentives to overprescribe, it may be extremely difficult to find effective ways to rein in the corrupt practices and to restore some semblance of professional integrity to the arrangement (Gøtzsche 2013). But from a conceptual viewpoint, there is, at least, little mystery about what is happening, what is wrong with it, and whose interests are served.

By contrast, when drug-centered "care" takes the more insidious form of altering the way we think about illness, it may be much harder to detect, let alone correct, the problem. At first blush, the altered way of thinking is almost always presented to us as a true scientific advance or alternatively as mere common sense. ("Everybody knows" that diabetes is basically a disease of too high blood sugar, and "everybody knows" that depression is caused by a deficiency of serotonin.)

The danger of drug-centered "care" extends even beyond the realm of disease and extends to how we think about life generally. The term "disease mongering" was coined to label the tendency of industry marketing not to be content merely by redrawing the lines at which a disease is diagnosed, as shown in the figure. An even more effective way of extending a drug's potential market is to relabel as disease something that previously was not viewed as a disease at all, but merely as a common problem of living or as one variant of normal behavior (Moynihan and Cassels 2006).

As men age, for example, they commonly find themselves having less energy and perhaps less interest in sex, but now they are informed that these may be serious symptoms of a testosterone deficiency (handily redesignated "low T") that requires drug management (Braun 2013). The women married to these men may find them less desirable as sexual partners, but recently there was a serious effort to label a condition of their own as "female sexual dysfunction," also requiring pharmaceutical intervention (Tiefer 2006). We used to imagine that grief after the loss of a spouse or parent might go on for a long time without triggering the prescription for a medication; a recent psychiatric guideline tells us to mark our calendar for day 13 after the death, and if grief has not resolved by then, one has to start inquiring as to whether a pathological reaction exists that might benefit from a drug (Friedman 2012).

There are sound ethical and policy reasons to replace doctor-centered care and disease-centered care with patient-centered care. But drug-centered "care" is a retrograde step which threatens the integrity of both medical science and medical practice, along with posing a serious danger to the public's health. How severe that threat to the public health has become is illustrated by calculations that indicate that *taking prescription drugs as prescribed* (not errors or overdoses) is now either the third or fourth leading cause of death in the USA (Light 2012; Gøtzsche 2013). How both the medical profession and the public have come to regard it as normal, indeed desirable, that so many people are taking so many prescription drugs requires thoughtful reexamination.

References

Angell, M. 2004. *The Truth about the Drug Companies: How They Deceive Us and What to Do about It.* New York: Random House.

Applbaum, K. 2009. Getting to Yes: Corporate Power and the Creation of a Psychopharmaceutical Blockbuster. *Culture, Medicine and Psychiatry* 33: 185–215.

———. 2010. Shadow Science: Zyprexa, Eli Lilly, and the Globalization of Pharmaceutical Damage Control. *BioSocieties* 5: 236–255.

Avorn, J. 2004. *Powerful Medicines: The Benefits, Risks, and Costs of Prescription Drugs.* New York: Knopf.

Boussageon, R., T. Bejan-Angoulvant, M. Saadatian-Elahi, et al. 2011. Effect Of Intensive Glucose Lowering Treatment On All Cause Mortality, Cardiovascular Death, And Microvascular Events In Type 2 Diabetes: Meta-Analysis Of Randomised Controlled Trials. *British Medical Journal* 343: d4169.

Braun, S.R. 2013. Promoting "Low T"—A Medical Writer's Perspective. *Journal of American Medical Association Internal Medicine* 173: 1458–1460.

Brody, H. 2007. *Hooked: Ethics, the Medical Profession, and the Pharmaceutical Industry.* Lanham: Rowman and Littlefield.

Brody, H., and D.W. Light. 2011. The Inverse Benefit Law: How Drug Marketing Undermines Patient Safety and Public Health. *American Journal of Public Health* 101: 399–404.

Dolan P.L. 2013. Physician EHRs Emerge as Hot Advertising Venue for Drugs. *American Medical News*. http://www.amednews.com/article/20130121/business/130129987/4/. Accessed 12 Sep 2014.

Duggan, P.S., G. Geller, L.A. Cooper, and M.C. Beach. 2006. The Moral Nature of Patient-Centeredness: Is It "Just the Right Thing to Do"? *Patient Education and Counseling* 62: 271–276.

Farmer, A.J., R. Perera, A. Ward, et al. 2012. Meta-analysis of Individual Patient Data In Randomised Trials Of Self Monitoring Of Blood Glucose In Patients With Non-Insulin Treated Type 2 Diabetes. *British Medical Journal* 344: e486.

Friedman, R.A. 2012. Grief, Depression, and the DSM-V. *The New England Journal of Medicine* 366: 1855–1857.

Gagnon, M.A., and J. Lexchin. 2008. The Cost of Pushing Pills: A New Estimate of Pharmaceutical Promotion Expenditures in the United States. *PLoS Medicine* 5(1): e1.

Gerstein, H.C., S. Yusuf, J. Bosch, et al. 2006. Effect of Rosiglitazone on the Frequency of Diabetes in Patients with Impaired Glucose Tolerance or Impaired Fasting Glucose: A Randomised Controlled Trial. *Lancet* 368: 1096–1105.

Gøtzsche, P.C. 2013. *Deadly Medicines and Organized Crime: How Big Pharma Has Corrupted Healthcare*. London: Radcliffe Medical Press.

Greene, J.A. 2008. *Prescribing by Numbers: Drugs and the Definition of Disease*. Baltimore: Johns Hopkins University Press.

Healy, D. 1997. *The Antidepressant Era*. Cambridge, MA: Harvard University Press.

———. 2003. *Let Them Eat Prozac*. Toronto: James Lorimer.

Kassirer, J.P. 2005. *On the Take: How Medicine's Complicity with Big Business Can Endanger Your Health*. New York: Oxford University Press.

Kishore, P., S.H. Kim, and J.P. Crandall. 2012. Glycemic Control and Cardiovascular Disease: What's a Doctor to Do? *Current Diabetes Reports* 12: 255–264.

Leo, J., and J. Lacasse. 2008. The Media and The Chemical Imbalance Theory of Depression. *Society* 45: 35–45.

Light, D.H., ed. 2012. *The Risks of Prescription Drugs*. New York: Columbia University Press.

Montori, V.M., and M. Fernandez-Balsells. 2009. Glycemic Control in Type 2 Diabetes: Time for an Evidence-based About-face? *Annals of Internal Medicine* 150: 803–808.

Moynihan, R. 2010. Rosiglitazone, Marketing, and Medical Science. *British Medical Journal* 340: c1848.

Moynihan, R., and A. Cassels. 2006. *Selling Sickness: How the World's Biggest Pharmaceutical Companies are Turning Us All into Patients*. New York: Nation.

Nissen, S.E., and K. Wolski. 2007. Effect of Rosiglitazone on the Risk Of Myocardial Infarction and Death from Cardiovascular Causes. *The New England Journal of Medicine* 356: 2457–2471.

Noble, P. 2013. *FDA Eases Restriction to Avandia Access. Thomson Reuters/Life Science Connect*. https://lsconnect.thomsonreuters.com/fda-eases-restrictions-avandia-access/. Accessed 13 Sept 2014.

Psaty, B.M., and D. Rennie. 2003. Stopping Drug Research to Save Money: A Broken Pact with Researchers and Patients. *Journal of the American Medical Association* 289: 2128–2131.

Reinhardt, U.E. 2001. Perspectives on the Pharmaceutical Industry. *Health Affairs (Millwood)* 20: 136–149.

Tanne, J.H. 2009. American Psychiatric Association Says No to Industry Funding for Symposiums. *British Medical Journal* 338: b1426.

Tiefer, L. 2006. Female Sexual Dysfunction: A Case Study of Disease Mongering and Activist Resistance. *PLoS Medicine* 3(4): e178.

Torrey, E.F. 2002. The Going Rate on Shrinks: Big Pharma and the Buying of Psychiatry. *The American Prospect* 13: 15–16.

Turner, E.H., A.M. Matthews, E. Linardatos, et al. 2008. Selective Publication of Antidepressant Trials and Its Influence on Apparent Efficacy. *The New England Journal of Medicine* 358: 252–260.

UK Prospective Diabetes Study Group. 2008. Tight Blood Pressure Control and Risk of Macrovascular and Microvascular Events in Type 2 Diabetes: UKPDS 38. *British Medical Journal* 317: 703–713.

Washington, A.E., and S.H. Lipstein. 2011. The Patient-Centered Outcomes Research Institute—Promoting Better Information, Decisions, and Health. *The New England Journal of Medicine* 365: e31.

Whitaker, R. 2010. *Anatomy of an Epidemic: Magic Bullets, Psychiatric Drugs, and the Astonishing Rise of Mental Illness in America*. New York: Crown.

Is There a Legitimate Concept of Drug-Centered Care?

Kenneth A. Richman

Brody introduces drug-centered care in contrast to patient-centered care, disease-centered care, and doctor-centered care (Brody 2017). Patient-centered care sounds like a good idea, and disease-centered care sounds potentially justifiable in some circumstances. Doctor-centered care names a real but clearly misguided approach. Drug-centered care, on Brody's concept, is a flavor of disease mongering.[1]

Unless it involves fish, "mongering" is a term of abuse. Hate, war, fear, rumors, and disease are entities that get mongered. To the uninitiated, drug-centered care as disease mongering can seem like an absurd conspiracy theory—too outrageous to be a real phenomenon. But the plausibility of disease mongering as an industry strategy was well illustrated by Ray Moynihan's widely believed April Fools' announcement in the British Medical Journal (Moynihan 2006). The piece announced that a team led by a neurologist named Leth Argos [sic] had discovered a debilitating illness called "motivational deficiency disorder (MoDeD)" which, in extreme cases, "can be fatal, because the condition reduces the motivation to breathe" (Moynihan 2006: 745). Not coincidentally, Dr. Argos is associated with a company that has just the pill to ban the indolence caused by motivational deficiency—Indolebant. The announcement is sufficiently realistic to make many readers think they are coming up with the disease mongering critique themselves in response to a real example rather than having it prompted on purpose by a thinly but cleverly veiled parody.

[1] "Disease mongering can include turning ordinary ailments into medical problems, seeing mild symptoms as serious, treating personal problems as medical, seeing risks as diseases, and framing prevalence estimates to maximize potential markets" (Moynihan et al. 2002: 886). Disease mongering can also involve displacement of social problems or violations of social norms. "We must turn from the inappropriate use of the disease model of emotional distress and understand that individuals' psychological pain arises within social systems as well as within their own brains" (Skomorosky 2015). See also Szasz 1960.

K.A. Richman (✉)
School of Arts and Sciences, MCPHS University, Boston, MA, USA
e-mail: kenneth.richman@mcphs.edu

© Springer Science+Business Media B.V. 2017

D. Ho (ed.), *Philosophical Issues in Pharmaceutics*, Philosophy and Medicine 122,
DOI 10.1007/978-94-024-0979-6_8

Brody's critique of disease mongering is legitimate. He fills in the details of how we have moved beyond the sociohistorical situatedness of medical concepts to where "[t]he social construction of illness is being replaced by the corporate construction of disease" (Moynihan et al. 2002: 886). This is no surprise, of course. Our current system holds pharmaceutical manufacturers responsible to shareholders. By accepting capital from shareholders, corporations take on a duty to make decisions that are good from a business standpoint. In late capitalism, this often means creating new "needs."[2] These "needs" can be for bottled water, fine Bavarian motorcars, or medications. It is easier to be appalled by these practices when it comes to pharmaceuticals as compared to other consumer products, and pharmaceuticals are rightly the focus of Brody's attention. From a broader view, however, it would be disingenuous to be appalled by one of these things and not the others. Or at least we need reasons for thinking that one of these things is not like the others.

One reason to think that the pharmaceutical industry is different from the grocery and automobile industries is the status of medicine as a profession. Professions define primary duties for practitioners (Thompson 1993). The primary duty in medical practice is promoting health in patients. When this primary duty faces undue influence by a secondary goal, such as monetary gain, we have a classic example of conflict of interest.[3] Some conflicts of interest are expected and unavoidable. Such conflicts need to be managed. Other conflicts of interest can and should be avoided or eliminated.

Notice, however, that drug discovery, pharmaceutical manufacturing, and marketing are not professions in the same sense as medicine, nursing, or pharmacy. Drug reps influence patient care, but they are not (at least not in virtue of their role as drug reps) healthcare professionals. Their practices may be objectionable in many ways, but they are not guilty of failing to deal with their own professional conflicts of interest. Industry employees simply do not have conflicts of interest in the same way that members of professions do.

To mitigate healthcare providers' conflict between profit and patient care, it might be necessary to take drug discovery and development out of the business realm and into the realm of the health professions. Such a "never mix, never worry" approach would require radical changes from the current system. We can imagine moving drug discovery, testing, and distribution, as well as dissemination of information about new drugs, into public sector institutions such as the National Institutes of Health. Special care would be needed to set up reasonable incentives and controls to promote relevant values. These values would include, *inter alia*, good science, innovation, efficiency, and justice. We could spend a great deal on such a system

[2] U'Ren describes the commodification of mental health services and the concomitant creation of psychiatric needs in U'Ren 1997.

[3] "A conflict of interest is a set of conditions in which professional judgment concerning a primary interest (such as a patient's welfare or the validity of research) tends to be unduly influenced by a secondary interest (such as financial gain)" (Thompson 1993: 573).

before we even approached the amount of money our society currently puts into the pharmaceutical industry. The public might be less comfortable knowing where money is being spent compared to the present system in which some factors affecting drug prices are hidden. However, the current mix of profit seeking and medicine introduces social and economic toxicities that could be minimized in an alternative scheme.

Even given the legitimacy of Brody's critique, we might still wonder whether there are acceptable ways to let the availability of treatments guide medical decisions. We can call a treatment recommendation, request, or decision "drug centered" when it is prompted by the availability of a drug rather than belief in the presence or risk of a symptom or disease. The following claims might provide reasons for thinking that drug-centered treatment decisions can in some cases be legitimate:

- Drugs are no different from other commodities.
- The wholesale rejection of drug-centered care is based on inadequate conceptions of what it means to be healthy.
- Drug-centered care may not be good treatment, but it could contribute to permissible enhancement.
- Drug-centered care may not be good treatment, but it can contribute to permissible improvements in quality of life.

The drug-centered approaches supported by these lines of reasoning will not be the same as the drug-centered care that Brody rejects, but they might have some *prima facie* reasonableness that Brody's target does not. Without committing to the truth or falsehood of any of these claims, the remainder of this chapter will focus on exploring these options.

Drugs Are No Different from Other Commodities

It is common to design university courses around available books, to create a wardrobe around available clothes, or to design an ornamental garden based on plants available at the nursery. These choices are shaped by the relevant industries. These industries are not free of moral responsibilities, but (hoarding, budgeting, and labor issues aside) it seems reasonable to choose a sweater simply because it is available and attractive.

Books, clothes, and gardens might seem irrelevant because they affect health only indirectly. Dietary decisions, however, affect health directly. When I cook dinner for my family, I make dietary decisions based on available ingredients. These ingredients are themselves shaped by what is in the store or what is in our share box from the community-supported farm. How is treating according to available drugs different from cooking according to available ingredients?

Perhaps the relevant comparison is not what to cook for the main course but whether to have dessert. Cheesecake is not a need but a luxury. Because choosing to eat cheesecake too often can be bad for our health, it would be contrary to duties of medical beneficence to cause people to believe that cheesecake is a necessary or expected part of dinner. However, those who make and sell cheesecake in diners along New Jersey's Routes 1 and 9 do not (at least not *qua* diner owners or cheesecake mongers) have such a duty. They have a duty to refrain from deception, but this duty is consistent with pages of high-calorie entrees that come with soup, salad, soft drink, and dessert. If these menus threaten the health of the residents of the Garden State beyond tolerable limits of *caveat emptor*, it is up to the state to set limits.

Such limits would constitute state paternalism (in this case, Garden State paternalism) and reflect widely accepted duties of government. As a matter of fact, New Jersey took action against the dangers of runny eggs in 1992 (Osgood 1992).[4] The system for regulating drugs in the USA is similarly paternalist (Flanigan 2012). Some drugs (such as acetaminophen) are available over the counter, some (such as Prozac) by prescription, and some (such as crack cocaine) are banned.

There may also be a difference between self-pay treatments and treatments where the cost is distributed among members of a community or a group covered under a common insurance contract. We can accept that health insurance ought to cover all and only treatments that are "medically necessary" without thinking that other treatments (teeth whitening, removal of benign moles, etc.) should be universally withheld. But even then, communities with sufficient resources may and perhaps should support treats. A case in point is that cake and ice cream are both allowed purchases for those receiving SNAP benefits (Supplemental Nutrition Assistance Program, also known as food stamps) in the USA.[5]

In short, drugs are not unique in their status as commodities that affect health and are peddled by those who have no special obligation to medical beneficence. Recommending or requesting a drug that is not medically necessary may be no more or no less problematic than choosing a food that is not nutritionally justified.[6]

---

[4] New Jersey's ban on eggs over easy was later repealed, but Brookline, Massachusetts, successfully banned commercial use of trans fats (Abel and Guilfoil 2007). In a similar move, New York City has attempted to enact a ban on extra-large soft drinks (Grynbaum 2012).

[5] I have absolutely no objection to the following: "Items such as birthday and other special occasion cakes are eligible for purchase with SNAP [food stamp] benefits as long as the value of non-edible decorations does not exceed 50 percent of the purchase price of the cake" (Supplemental Nutrition Assistance Program 2014).

[6] On the general principle that one philosopher's modus ponens is another philosopher's modus tollens, the comparison can be used to defend either greater permissiveness or greater paternalism for both food and drugs.

The Rejection of Drug-Centered Care Is Based on Inadequate Conceptions of What It Means to Be Healthy

What counts as treatment will depend at least in part on how we understand health and illness. Dominic Murphy divides theories of health into naturalist theories and constructivist theories (Murphy 2015).[7] A naturalist concept of health such as Boorse's theory of health as species-typical functioning (Boorse 1977) treats claims about whether a state of some organism is healthy as made true by biological facts—or at least facts that are part of the natural world rather than socially constructed. Naturalist approaches provide clear grounding for suspicions about disease mongering. Having a drug that is useful or provides a socially or individually desirable effect is not evidence that the drug is promoting health. If the "channels"[8] tell us that a drug is promoting health, there is an objective way to test this claim.

Constructivism holds that our disease attributions are based on "some shared, usually culturally specific, conception of human nature" (Murphy 2015). Constructivist objections to disease mongering will be based on claims that the purported disease is inconsistent with the authentic view of a particular culture or that recognizing the disease being mongered would be morally objectionable. Foucault, for example, can object to the medical community's explanations of hysteria on the basis that they create unacceptable power relationships (Foucault 1964: 158). A strict constructivist does not have the same type of objective foundation for rejecting drug-centered disease claims that the naturalist has.

Embedded Instrumentalism

There are other ways of thinking about health that do not rest entirely on the biological and yet are also not entirely constructivist. One such approach I have termed "embedded instrumentalism" (Richman 2004). Embedded instrumentalist theories take health to be a state of an individual that allows him or her to reach or strive for his or her actual goals. This general approach allows great variability among sets of goals and, hence, among the corresponding conditions that would count as healthy for the individuals who hold those goals, although the range of variability differs among specific versions of embedded instrumentalism. Nordenfelt (e.g., 1995) has been the most prolific proponent of a version of embedded instrumentalism. Here I want to consider drug-centered care in the context of embedded instrumentalism with the distinction between health of individuals *qua* organisms and health of

[7] Murphy also identifies an additional distinction—between conservative and revisionist theories—that cuts across naturalist and constructivist theories (Murphy 2015). This second distinction is less relevant to our discussion here.

[8] See Brody (2017) p. 191.

individuals *qua* persons that I have defended elsewhere (Richman and Budson 2000; Richman 2004).

On the embedded instrumentalist account I developed with Andrew Budson, one set of goals determines the range of states that count as healthy for an individual considered as an organism, and another set of goals determines the states that count as healthy for an individual considered as a person—an agent in the world. The goals relevant to health *qua* organism are determined by biology. As far as health *qua* organism goes, the naturalist objections to drug-centered care hold. If you tell me that a drug provides health benefits, facts about biology can be brought forward to evaluate your claim.

For health of the individual *qua* person, the relevant facts are not facts about biology but facts about goals that will vary from person to person. Many of these goals will be the result of social influence. Some of that influence will come from marketing, and some of that marketing will come from drug companies. To adjust for inconsistent goals, self-defeating goals, and goals problematically based on false beliefs, I invoke the concept of the idealized objectified subjective interest of the individual (Richman 2004: 45). Adjusting for the influence of marketing may require an additional filter to eliminate inauthentic goals. However, even with all of these adjustments, someone could still authentically adopt a goal based on the availability of a drug that can facilitate reaching or striving for that goal.

Viagra might provide an example. Reduced sexual performance for men over a certain age appears to be species-typical. Before drug interventions, as now, there were men for whom engaging in sexual intercourse was an unrealistic goal. The availability of a pharmaceutical intervention has made this a realistic goal for some of these men. Assuming that the goal is not eliminated by the filters mentioned above (consistency with other goals, authenticity, etc.), adopting this goal is a drug-centered decision, and taking that drug contributes to health of the individual *qua* person. Availability of this drug is another fact about the environment that serves as the platform for our choices, including our choices of which goals to adopt.

Self-Actualization

Taken as a theory of requirements for health, Abraham Maslow's famous hierarchy of needs bears some similarities to the embedded instrumentalist theory just discussed. At the bottom of the hierarchy are biologically determined needs such as food, water, and shelter. Farther up are emotional and social needs. At the top of the pyramid are needs of self-actualization. "Self-actualization… may be loosely described as the full use and exploitation of talents, capacities, potentialities, and the like" (Maslow 1954: 126). Under Maslow's understanding of self-actualization as required for full health,[9] whatever facilitates achieving this "full use and exploitation" will count as treatment:

> What is psychotherapy, or for that matter any therapy or growth of any kind? Any means of any kind that helps to restore the person to the path of self-actualization and of development along the lines that his or her inner nature dictates. (Maslow 1954: 115)

Although an individual's inner nature may dictate the path in a general sense, it seems fair to say that the options for how that general path becomes manifest will often be limited and facilitated by what opportunities are available. That a particular drug is available may provide a relevant opportunity. The characteristics of self-actualized people, including accuracy of perception, ethics, and resistance to cultural influences, among others (Maslow 1954: 128), provide the filters required in the previous theory. Thus, both in the context of health *qua* persons and in the context of Maslow's broad sense of psychological health, we can imagine situations in which a decision can be drug-centered without being morally or medically suspect on that basis.

Drug-Centered Care and Enhancement

We might think that any action properly called medical treatment or healthcare should aim to mitigate or prevent a disease or symptom. Along these lines, Murphy and others have suggested that the concept of health *qua* persons is more a theory of well-being than a theory of health (Murphy 2015). On such a view, drug-centered decisions (starting, as they do, with the availability of a drug rather than a biological need) would be something other than medical treatment or healthcare. This would also cast doubt on Maslow's claim that self-actualization is required for health rather than being something extra. Rather than focusing on well-being, I will consider whether drug-centered care might contribute to enhancement or quality of life, which corresponds more directly with how I carve up these concepts.

So let us suppose, *contra* my own view discussed above, that an intervention is only healthcare if it is biologically based. Enhancement will then be something like what moves us to better-than-health status on health-related measures. On this limited understanding of enhancement, not all enhancements will be pharmaceutical or even medical. They may, perhaps, even include simply learning. (Bostrom and Savulescu 2009: 3)

Enhancement is a bit like dessert: it is not necessary, but it can be nice; it is hard to pass up; and if we make it available to one person, we should probably make it available to all. Like dessert, we might be willing to share the cost of healthcare for other members of our community but unwilling to share the cost of enhancement.

[9] The title of Chapter 11 of Maslow 1954 is "Self-actualizing People: A Study of Psychological Health" (125). Cf: "…self-actualization has been found to relate positively to measures of psychological adjustment and negatively to measures of psychopathology" (Ivtzan et al. 2013: 120).

Unlike with dessert, more enhancement is by definition always better—at least along some measure of better.

When it comes to healthcare in the biology-limited sense, we can distinguish the introduction of acceptable and useful drugs from inappropriate, manipulative moves to market drugs that are not actually beneficial. We can draw a similar distinction in the area of enhancement. Some drugs or nondrug interventions provide benefits that can move us beyond health by enhancing us in ways that we authentically find good or useful; others will be chosen for inauthentic reasons having to do with external influences, such as marketing rather than for reasons of self-actualization.[10]

Authenticity becomes a key concept here. Consider Charles Taylor's idea that:

> Being true to myself means being true to my own originality, and that is something only I can articulate and discover. In articulating it… I am realizing a potentiality that is properly my own. (Taylor 1992: 29)

In a strict sense, no desire that is the result of marketing would count as "properly my own" because it does not spring from "my own originality." On the other hand, if I learn about a drug, consider the facts, and integrate a desire for that drug into my goals without being deceived or coerced, this could be an authentic choice.

I suggest, then, that a biology-only view of healthcare leaves open the possibility of legitimate drug-centered decisions that contribute to enhancement. This is supported by the fact that there are any number of functions that could be enhanced. Attention, memory, empathy, physical endurance, and eyesight are just a few. If there is more than one measure for which enhancement is permissible, the decision to pursue enhancement on one measure rather than another could very well be based on the availability of a drug—a drug-centered decision.

Drug-Centered Care and Quality of Life

The distinction between treatment and enhancement may not be very clear. Indeed, as John Harris remarks:

> [t]he overwhelming moral imperative for both therapy and enhancement is to prevent harm and confer benefit. Bathed in that moral light it is unimportant whether the protection or benefit conferred is classified as enhancement or improvement, protection or therapy. (John Harris, quoted in Bostrom and Savulescu 2009: 7)

Looked at in this way, we might decide to focus on quality of life rather than health or enhancement as the relevant measure or status in determining whether a drug or other treatment is beneficial.

[10] Eric Parens explores concepts of authenticity in the context of enhancement in Parens 2009, distinguishing authenticity as gratitude for the self as given from authenticity as creativity—as an internal drive to change or grow. The question of whether a desire for some enhancement is authentic sits in the realm of authenticity as creativity (Cf Wargo 2011).

Healthcare providers cannot be charged with preventing harms and conferring benefits in all aspects of life; their duties are limited to addressing health-related harms and benefits. Clearly, however, there are health-related interventions that can promote quality of life. Some of these would count as healthcare under any view; others will be health-related only in the sense that they draw on skills characteristic of healthcare providers or affect functions that can also be affected in different ways by healthcare or by disease. Teeth whitening, removal of benign moles, and certain ways of enhancing cardiovascular fitness may be health-related quality of life interventions. As with enhancement, the availability of a drug to improve quality of life in one area may reasonably drive a decision to take that drug.

Conclusion

Each of the four lines of reasoning pursued here has provided a way to describe drug-centered care that we could accept as unobjectionable or at least not obviously corrupt. This does not weaken the core of Brody's critique of disease mongering. It does suggest that there might just be some babies that should not be thrown out with the Big Pharma bathwater. That is, there can be situations in which it is perfectly legitimate to make a decision prompted primarily by the availability of a drug.

References

Abel, D., and J.M. Guilfoil. 2007, June 1. Brookline OKs Trans Fat Ban. The Boston Globe. http://www.boston.com/news/local/articles/2007/06/01/brookline_oks_trans_fat_ban/. Accessed 9 Apr 2015.

Boorse, C. 1977. Health as a Theoretical Concept. *Philosophy of Science* 44: 542–573.

Bostrom, N., and J. Savulescu. 2009. Human Enhancement Ethics: The State of the Debate. In *Human Enhancement*, ed. J. Savulescu and N. Bostrom, 1–22. New York: Oxford University Press.

Brody, H. 2017. Patient-centered Care or Drug-centered Care? The Influence of Pharmaceutical Marketing on Medical Science and Public Health. In *Philosophical Issues in Pharmaceutics: Development, Dispensing, and Usage*, ed. D. Ho. Philadelphia: Springer.

Flanigan, J. 2012. Three Arguments Against Prescription Requirements. *Journal of Medical Ethics*. Published online July 26, 2012. http://jme.bmj.com/content/early/2012/07/25/medethics-2011-100240.full.pdf+html. Accessed 9 Apr 2015.

Foucault, M. 1964. *Madness and Civilization: A History of Insanity in the Age of Reason*, 1st ed. New York: Vintage Books.

Grynbaum, M. 2012, May 30. New York Plans to Ban Sale of Big Sizes of Sugary Drinks. *The New York Times*. http://www.nytimes.com/2012/05/31/nyregion/bloomberg-plans-a-ban-on--large-sugared-drinks.html. Accessed 11 Aug 2016.

Ivtzan, I., H.E. Gardner, I. Bernard, M. Sekhon, and R. Hart. 2013. Wellbeing Through Self-Fulfilment: Examining Developmental Aspects of Self-Actualization. *The Humanistic Psychologist* 41: 119–132.

Maslow, A.H. 1954. Motivation and Personality. New York: Harper and Row.

Moynihan, R. 2006. Scientists Find New Disease: Motivational Deficiency Order. *British Medical Journal* 332(7544): 745.

Moynihan, R., I. Heath, and D. Henry. 2002. Selling Sickness: The Pharmaceutical Industry and Disease Mongering. *British Medical Journal* 324(7342): 886–891.

Murphy, D. 2015. Concepts of Disease and Health. In *Stanford Encyclopedia of Philosophy*. http://plato.stanford.edu/archives/spr2015/entries/health-disease/. Accessed 9 Apr 2015.

Nordenfelt, L. 1995. *On the Nature of Health: An Action-theoretic Approach*, 2nd ed. Dordrecht: Kluwer Academic Publishers.

Osgood, C. 1992, January 17. N.J. Runny Egg Law "Bad Yolk" on Diners. *Sun Sentinel*. http://articles.sun-sentinel.com/1992-01-17/news/9201030706_1_soft-boiled-eggs-benedict-new--jersey. Accessed 9 Apr 2015.

Parens, E. 2009. Toward a More Fruitful Debate About Enhancement. In *Human Enhancement*, ed. J. Savulescu and N. Bostrom, 181–198. New York: Oxford University Press.

Richman, K.A. 2004. *Ethics and the Metaphysics of Medicine: Reflections on Health and Beneficence*. Cambridge, MA: MIT Press.

Richman, K.A., and A.E. Budson. 2000. Health of Organisms and Health of Persons: An Embedded Instrumentalist Approach. *Theoretical Medicine and Bioethics* 21(4): 339–352.

Skomorosky, A. 2015, March 30. *Don't Blame it on Depression*. Slate.com. http://www.slate.com/articles/health_and_science/medical_examiner/2015/03/germanwings_co_pilot_mental_illness_suicide_is_linked_to_depression_but.html. Accessed 9 Apr 2015.

Supplemental Nutrition Assistance Program (SNAP). 2014. *Eligible Food Items*. http://www.fns.usda.gov/snap/eligible-food-items. Accessed 9 Apr 2015.

Szasz, T. 1960. The Myth of Mental Illness. *The American Psychologist* 15: 113–118.

Taylor, C. 1992. *The Ethics of Authenticity*. Cambridge, MA: Harvard University Press.

Thompson, D.F. 1993. Understanding Financial Conflicts of Interest. *The New England Journal of Medicine* 329(8): 573–576.

U'Ren, R. 1997. Psychiatry and Capitalism. *Journal of Mind and Behavior* 18(1): 1–11.

Wargo M. 2011. *Authenticity and Ethics in Human Augmentation*. MA Thesis, Boston University Department of Philosophy, Boston.

Prescription Paternalism: The Morality of Restricting Access to Pharmaceuticals

Robert M. Veatch

Although physicians are clearly experts on the facts of health and illness, they are not experts on the value judgments that necessarily are incorporated into prescription-writing. Two policy options are explored: (1) A libertarian view holds that all medications that pose danger primarily to the patient himself or herself should be available without prescription for use by mentally competent adults. (Restrictions on drug access to children do not constitute paternalism as usually defined nor do restrictions on those that pose a danger to others—amphetamines and alcohol, for example.) (2) A more modest option is to replace the act of prescription-writing with a requirement that adults who purchase pharmaceuticals possess a certificate from a physician testifying that the purchaser has an adequate knowledge of the agent and possibly that the purchaser has a condition for which the agent is typically recognized as useful.

Luke Braddock, a 45-year-old business professional left his New England home on a Friday afternoon with his wife and two children for a weeklong Florida vacation. On arriving at their Clearwater hotel late that evening, Mr. Braddock realized that he had left his medication at home. He took daily propranolol, 20 mg., to help control a ventricular arrhythmia. His physician had told him he should not miss any doses since an abrupt stop could trigger a rebound cardiac arrhythmia that could well be fatal.

He knew that he could not reach his physician in New England and that, even if he could, that physician could not prescribe in Florida. He explored his options: (1) go to the hospital emergency room, wait perhaps hours, and pay an enormous fee to get an additional week's supply of the propranolol; (2) go to a local pharmacy and try to persuade a pharmacist to give him enough capsules to get through the week; or (3) take a chance and go without the drug for a week.

The first option was unattractive. It would involve getting a taxi to the hospital, making use of the ER personnel, spending perhaps many hours, and generating significant costs.

R.M. Veatch (✉)
The Kennedy Institute of Ethics, Georgetown University, Washington, DC, USA
e-mail: veatchr@georgetown.edu

© Springer Science+Business Media B.V. 2017

D. Ho (ed.), *Philosophical Issues in Pharmaceutics*, Philosophy and Medicine 122,
DOI 10.1007/978-94-024-0979-6_9

> The second option might not work. Mr. Braddock knew that it would be illegal for the pharmacist to provide the drug without a prescription, and he wasn't sure it would be ethical to ask. He chose the third option. Fortunately, there were not adverse events. Mr. Braddock asked himself why it was necessary to have a prescription from a physician to gain access to drugs, especially a drug that was not subject to significant abuse and was reasonably safe.

This true story is, no doubt, repeated daily by those who need or want pharmaceuticals that are, by law, restricted to prescription access. Some drugs are surely more dangerous and more subject to abuse than propranolol. Some consumers are surely less well educated about the risks and benefits of pharmaceuticals that they might want to access. This example raises the larger question of whether the widespread policy of restricting access of lay people to pharmaceuticals that have been prescribed by physicians or other authorized healthcare professionals is philosophically sound and morally justified. Many chemicals, often not conceptualized as pharmaceuticals, are as dangerous (or more dangerous) than many drugs, and yet in a liberal society, they are made available "over the counter" for sale directly to consumers without a physician's review and approval. Alcohol is certainly responsible for more abuse and harm to the consumer (not to mention other parties) than almost any pharmaceutical. Cigarettes are a drug delivery system with enormous potential for harm, yet they are available from informed and usually uninformed sales people (or even from vending machines without any interaction with a human seller). Ammunition for firearms is sold without certification from a recognized expert that the consumer will use the item properly. Even rat poison can be bought over the counter in spite of its lethal potential for misuse.

This raises the question why American and most other liberal societies make these lethal chemicals available so readily and yet restrict even many benign pharmaceuticals to cases in which an intermediary has decided that the consumer should have access. This is surely the height of paternalism, a feature that is generally frowned upon in liberal society. Consistency requires that pharmaceutical access be at least as open as these other toxic agents.

After spending some time defining paternalism and teasing out some of its implications for access to pharmaceuticals, we will consider two arguments for questioning the current restrictive policy: what we will call the liberal critique and the empirical critique. We shall suggest that current policies regarding legend drugs are unacceptably paternalistic and violate the ethical principle of autonomy, which is in other contexts considered fundamental. This we will call the liberal critique. Then we shall examine whether the empirical evidence supports the assumption that patients are, in fact, better off with a restrictive policy that strives to protect them from themselves.

Having found the current restrictive policy indefensible, we will explore two alternatives: a libertarian policy that would permit unrestricted purchase of pharmaceuticals (while still permitting restriction of sales to children and the mentally incompetent as well as restriction of agents that pose significant third-party risks) and a policy that would require that anyone who wants to purchase a drug deemed potentially dangerous would have to show a certificate from a physician indicating that the physician has judged that the patient has an adequate understanding of the drug, its effects, its possible dangers, and treatment alternatives. As part of that discussion, we will consider a modification in which the physician must also certify that the patient has a condition that is generally viewed as amenable to the drug the individual desires to purchase.

The Paternalism Problem

The central problem of policies that restrict access to pharmaceuticals to prescription is paternalism. The primary purpose of such policies is to protect consumers from bad decisions they might make about the use of drugs if they have unrestricted access. We first need to be clear on exactly what is meant by paternalism and why it is a problem.

The Definition of Paternalism

Paternalism describes an action or policy whereby one person acts for the purpose of benefitting another against that person's substantially autonomous choice (Dworkin 1972, 1988; Gert and Culver 1979). From this definition, it is clear that only persons who have the capacity to make substantially autonomous choices about the issue at hand can be the subjects of paternalism. Paternalism is always an action that violates autonomy, and it is for that reason that it is morally suspect.

Hence, taken literally, one cannot act paternalistically toward small children or mentally incompetent adults who do not possess enough autonomy to be violated. Similarly, actions or policies undertaken to benefit others or protect them from harm do not count as paternalism. This is true even if the one who is the target of the action also benefits in some ancillary way.

Similarly, actions taken to protect the welfare of those who are so addicted to a chemical that their actions are not substantially autonomous are not rightly considered paternalistic (Faden et al. 1986). There is no substantial violation of autonomy.

The term "substantial" is important. It recognizes that people and their actions are not rigidly classified as either autonomous or nonautonomous. Autonomy varies by degree and by the nature of the decision being made (Buchanan and Brock 1989). Autonomy is sometimes said to be a "threshold concept" (cf. Buchanan & Brock: 26–29). People are thought either to possess enough autonomy to make choices on their own, in which case they can be said to be "substantially autonomous," or they fail to reach the threshold, in which case they can be said to be substantially nonautonomous.

The Implications

A psychiatrist who decides to breach confidentiality of a patient who plausibly claims he is about to do serious harm to another is not acting paternalistically if the motive for the breach is the protection of the third party.[1] This is true even if it is plausible that breaching confidentiality in a way that will stop the patient from harming the other will also benefit the patient by protecting him from the punishment that would likely result if he carried out the threat.

Similarly, in the psychiatry example, if the patient is so psychotic that his actions could be said to be not substantially autonomous, breaching confidentiality even for the purpose of benefiting the patient would not count as paternalism since it would not violate the patient's autonomy. This may not be the obvious use of the language of paternalism. One might claim that since the word *paternalism* derives from the root word that means "father," it is odd that one cannot be paternalistic to a child or other person who is not a substantially autonomous agent. Nevertheless, the moral bite of the concept of paternalism is that it involves the violation of another's autonomy. Acting to benefit another who cannot autonomously choose some other course of action cannot, by definition, compromise that individual's autonomy. Hence, paternalism, at least as the word is used here, has to involve acting for the benefit of another in such a way that the other person's autonomy is violated.

Sometimes actions to benefit someone who is not substantially autonomous are said to be *weak paternalism* (Beauchamp 1977: 67). This term is better reserved for a narrower kind of case in which real paternalism with a violation of autonomy occurs. If a militant anti-paternalist encounters someone acting strangely in a way that endangers the actor's safety, the anti-paternalist might be inclined to intervene to protect the stranger from himself. Someone standing on a bridge threatening to jump might trigger an inclination to restrain the stranger. In such a case, a militant anti-paternalist might believe that, if the stranger knows what he is doing and is substantially autonomous, he has the right to jump, but if he is so deranged as to be substantially nonautonomous, he should be restrained. If one were to intervene long enough to assess the mental state of the stranger, one might discover he, indeed, knows exactly what he is doing and can give rational reasons for his action. The militant anti-paternalist would at that point presumably step aside, but during the time when he is intervening in order to do an assessment, he might well feel justified in his intervention.

In this example, the intervention would really be paternalistic in the sense of violating the autonomy of a substantially autonomous agent, but it is justified because it was necessary to rule out the possibility that the stranger was deranged, thus nonautonomous. Sometimes the phrase *weak paternalism* is used in this case to refer to the instance when real paternalism occurs, but it is necessary to determine the mental state of the subject of the paternalism. On the other hand, if the stranger continues to be restrained after he is determined to be substantially autonomous,

[1] See Tarasoff 1974.

any action for the purpose of benefiting that individual can be called *strong paternalism*.

The Liberal Critique of Restricting Access to Pharmaceuticals

The most obvious problem with policies that restrict access to pharmaceuticals to cases in which a physician or other health professional has signaled approval by writing a prescription is that these are blatant cases of paternalism. The primary purpose of the policy is to protect the potential consumers of drugs from the harm that they can do to themselves. Even if the consumers are sufficiently autonomous, the policies restrict their freedom to purchase and use medicinals that they believe are beneficial. We can call this the *liberal critique* of restricting access to pharmaceuticals to prescription. It is a policy designed to benefit the consumers by protecting them from themselves in spite of the fact that the consumers are presumed to be substantially autonomous agents.

One gets a sense of how controversial this might appear when one compares it with similar policies applied to alcohol, tobacco, guns, rat poison, household cleaners, and so forth. Any of these latter items can pose serious threats to the one using them, yet we generally permit unrestricted access to substantially autonomous beings. We restrict the right to purchase for children and others who are not competent, but we let competent adults make these purchases without supervision. Moreover, we presume the adult is competent to make the purchase unless there is evidence to the contrary.

In fact, the case for restriction of alcohol, guns, and poisons seems much stronger than for pharmaceuticals. There are much greater third-party risks from alcohol and arguably from guns and poisons than from almost all drugs. Since the "harm to others" principle provides a basis for restricting access, it would seem reasonable that these substances be more properly the target of controls. On the other hand, almost all pharmaceuticals, if they are dangerous at all, are dangerous primarily to the consumer.

There are, of course, some exceptions in which pharmaceuticals pose a danger to third parties. Methamphetamines, narcotics, and possibly agents that have a side effect of making one drowsy could all be restricted on the grounds of risk to third parties, but these constitute a very small subset of pharmaceuticals. Most, like propranolol, if they are dangerous at all, are primarily of danger to the consumer.

One might respond by pointing out that some pharmaceuticals can be dangerous to some subset of people who for genetic, psychological, or other reasons are at risk in ways they do not know about. This is also true, however, for alcohol and other agents that are freely available even to the subset who do not know that they are at unique risk. Even over-the-counter medications, generally thought to be particularly safe, can pose serious health risks to some subsets of the population with unusual risk factors. Aspirin is dangerous for people at risk for hemorrhagic stroke or gastrointestinal bleeding that can be rapidly fatal.

The reasonable conclusion is that public policy restricting access to pharmaceuticals is out of line with similar policies that permit ready access to competent adults for other agents that pose a danger to the consumer (in addition to posing third-party risks). It makes sense to restrict a small group of drugs that pose real risk to third parties—alcohol, amphetamines, and addictive drugs like narcotics and nicotine. It also makes sense to control substances that can render one nonautonomous—like addictive agents. But a wholesale restriction on access to prescription drugs seems unjustified from the point of view of one who generally accepts the moral premises of a liberal society that permit competent adults to use their own judgment to make purchases even if some of those purchases can end up harming the purchaser. A much better case can be made to restrict alcohol, tobacco, and rat poison than to restrict the typical legend drug. Perhaps we need a finer-grain distinction that captures the in-between group of drugs. I will suggest some options below. Treating all pharmaceuticals currently restricted to prescription access as more restricted than alcohol, tobacco, and rat poison seems indefensible.

Of course, not everyone is a pure anti-paternalist. Some would accept the moral legitimacy of relying on calculations of benefit to justify violations of autonomy. That is to say, some accept the moral legitimacy of paternalism, even strong paternalism, in cases in which the potential risk to the substantially autonomous person is great enough to outweigh the offense of the autonomy violation. Even the strong paternalist, however, must recognize that the one acting paternalistically must be confident that his or her calculation of the potential harm to the substantially autonomous agent is reliable. One cannot legitimately act paternalistically if there is good reason to doubt that the paternalistic one lacks the ability to know that the intervention will, on balance, really be beneficial. Thus, the strong paternalist must have sufficient justification of his or her belief that intervention will be better for the target of the paternalistic action.

The Empirical Critique

One necessary condition for justifying paternalism is that one must be confident that the paternalistic policy will, in fact, produce good for the target of the policy. If the paternalist is not confident that his or her action will benefit the recipient, then the restriction is not justifiable.

Is the Patient Better Off with Current Restrictions?

The case of the man on vacation in Florida without his propranolol illustrates the problem. There are lots of reasons why patients would benefit from being freed from restriction on access. Getting the drugs would almost certainly be much less expensive and less time consuming. At least if one needs a pharmaceutical when one does not have access to one's physician, a policy of open access would certainly provide *prima facie* benefits to the patient.

There would be more subtle benefits of open access as well. If the patient made purchases without the requirement of a physician's prescription, the patient would surely take more responsibility for the use of drugs. Under the present system, patients presume they are in good hands and that their doctor can figure out what is best for them. This, unfortunately, is not always the case. There is good reason to believe that often physicians are not in a position to be capable of determining exactly which drug use is most beneficial to the patient. If determining what is most beneficial is a subjective matter, based on the patient's beliefs and values, there is no reason to assume that the physician can make these evaluations accurately. In short, doctors really cannot know what is best for their patients. They, no doubt, can more accurately predict the effects of the drug than the patient can, but they are not in a position to know whether those effects are beneficial from the patient's perspective and, assuming they are beneficial, whether some other drug, dose level, route of administration, or manufacturer would be deemed better based on the patient's unique and subjective value framework. Moreover, having the patient as an active participant in the doctor-patient relation may offer great benefits. If the patient perceives that he or she is the final decision-maker with the authority to purchase or refrain from purchasing pharmaceuticals, more careful assessment of the risks, benefits, and alternatives is likely to result.

Of course, some terrible things might happen if patients are free to purchase pharmaceuticals on their own. They may not adequately understand the nature of a drug's actions and may fail to grasp the unique risks that their disease, body type, and genetic makeup could pose. Any rational patient will, of course, need to rely on the physician and other health professionals as the authorities on the diagnosis and prognosis of their illnesses, the presentation of a reasonable range of plausible alternative therapies, and the like. One would hope that wise patients armed with the right to buy what are now prescription drugs would exercise caution and avail themselves of the opportunity to get expert input on the medical facts. Nevertheless, sometimes the patient will surely be better off without the current restrictions, especially if the patient is prudent enough to make judicious use of professional advisors. Although patients are sometimes described as having lost autonomy because of disease and disability, a very large portion of patient/physician interactions involve healthy patients receiving physical exams or therapies for chronic diseases that in no way can be said to compromise autonomy. The vacationing patient who forgot his propranolol is an example. On the other hand, occasionally patients' autonomy is significantly compromised, but in those cases, normally a surrogate—parent, spouse, or significant other—should be in the lay decision-making role. Thus, even if we recognize the fact that physicians possess knowledge about medical facts of certain pharmaceuticals that patients usually lack, they have no basis for making the value judgments that can better be made by substantially autonomous patients or their surrogates. It seems that physicians ought to play a significant role in patients' access to drugs, but that role should be limited to education about the relevant medical facts and should exclude necessary inputs about nonmedical facts as well as about the value judgments needed to make a proper therapeutic choice.

In many cases, it is probably wiser to permit the patient to exercise discretion in whether to consult an advisor. The uninsured poor have a very limited range of choices. If funds are limited, many choose to go without any medical care—physician consultation or prescription—because they cannot afford the full ritual of seeing a doctor, paying for tests, and getting a prescription filled. It is not obvious that the current requirement of getting a prescription from a physician to gain access to legend drugs is in the patient's best interest.

The Doctor Can't Know What Is Best

There is an even more subtle problem. Relying on a professional medical expert to prescribe what he or she thinks is best runs afoul of a mistake in the model for medical decision-making in the modern era. The model of modern medical decision-making is that there is a class of well-trained experts who can diagnose, review treatment alternatives, and choose what is best for the patient. This has, in fact, been the dominant ethic of physician ethics since the Hippocratic era (Veatch 1978), but in the twentieth century, an adjustment was made. The clinician was still deemed authoritative in matters of diagnosis, prognosis, and the expected effects of treatment alternatives but was expected to test his or her judgments about these matters of medical fact with standard conventions of modern science: peer review, randomized clinical trials, and other devices to test the individual clinician's understanding of the facts against the consensus of experts and the best scientific knowledge (Veatch 2012). What remained was the assumption that "doctor knows best," that is, that once the diagnosis and prognosis were established (now based on the best standards of medical science), this could lead to a professional judgment about what was best for the patient. Hence, medicine held out the view that there were objective standards of "medically indicated treatments" and among alternative treatments the science could determine a "treatment of choice." Medical textbooks could specify what the appropriate treatment was for any given patient's condition once that patient's condition was specified sufficiently.

This assumption that physicians (aided by standards of peer consensus) could know what was best for the patient was what was behind the model of modern medicine, which included the requirement that treatments be "prescribed" by physicians and that patients—presumably having less knowledge of these medical facts—were not capable of deciding what was appropriate except in the most trivial cases of easily diagnosed problems for which safe, simple, and reasonably effective treatments were available for purchase over the counter.

The Emergence of the Ethically Exotic Case

By the end of the twentieth century, we began to discover that this model of modern medical decision-making was flawed in a crucial way (Veatch 2009). We first discovered the problem in the last decades of the century when considering medical interventions in a few medical situations that could be called "ethically exotic." The care of terminally ill patients, the pregnant woman, or those contemplating contraception or sterilization were the opening challenges to the model of modern medical decision-making. We discovered that a physician with complete medical facts and complete knowledge of treatment alternatives was nevertheless helpless in determining what was best for a terminally ill patient. For the elderly patient with metastatic cancer, was aggressive, experimental treatment or hospice care better? Should patients receive life support, or should it be omitted in order to let the patient die? We discovered that all the knowledge of medical science in the world could not tell us what was best for a patient. It was a value choice about which the patient's physician could not be presumed to be expert.

In the case of a pregnant woman with a genetically afflicted fetus (or simply one who did not want to have a baby), is terminating the pregnancy or carrying it to term best for the patient? Once again perfect knowledge of obstetrics could not tell the answer. Is contraception best for an unmarried teenager who is sexually active? It is not a question a medical education should be able to answer. In the mid-twentieth century, doctors attempted to decide what was best for their patients based solely on the medical facts. They imposed aggressive life support on permanently vegetative patients (In re Quinlan 1976). They decided whether a 30-year-old married woman with fewer than four children should be sterilized (Scrimshaw and Pasquariella 1970). (A rule known as the 120 rule was used. The woman's age multiplied by the number of children she had had to equal 120 before sterilization was acceptable.) Physicians had to give their approval before a woman could have an abortion.

We gradually began to realize how irrational this was. The physician, even with perfect knowledge of the medical facts in these cases, did not have a basis for deciding which treatment was best for the patient. Deciding the best treatment required imposing an evaluative judgment—a religious or philosophical choice—on the medical facts. Moreover, it became clear that physicians had no unique expertise in making these value judgments even if they were obviously expert on all the medical facts. In fact, there was in some cases reason to believe that physicians as a group were systematically biased in their evaluative judgments. For example, data revealed that physicians had an unusually high fear of death (Feifel et al. 1967) perhaps because students with high fear of death or unusually traumatic childhood death experiences chose to go into medicine. And, they tended to make judgments in terminal illness cases that led to aggressive provision of life support.

Whether physicians as a group had tendencies for certain moral and other evaluative biases, there was no reason to assume that an individual physician (or even one aided by peer review and the consensus of the profession) should be deemed expert on the evaluation of terminal illness care choices, decisions about whether to abort a pregnancy, or choices about the use of contraception. Temporarily, society began to recognize that there was a class of medical decisions about which physicians were not expert. Even if they were the definitive authorities on matters of diagnosis and prognosis and had clear expertise in knowing the treatment alternatives and their likely consequences, they had no expertise in moral and other evaluative judgments about which treatment was best. In short, doctors did not know best what to do in these ethically exotic cases.

The Discovery of the Ubiquity of Value Choices in Prescriptions

It turns out that this discovery that those with expertise on matters of medical facts were not also expert on matters of evaluation of treatment alternatives could not be restricted to a few ethically exotic cases. Theorists in clinical decision-making began to realize that the problem of differentiating expertise on medical facts from expertise on medical value judgments was pervasive (Veatch 1972; Emanuel and Emanuel 1992; Goldman 1999). The problem can arise in literally every medical decision.

The logical structure of every medical decision can be reduced to a simple syllogism:

> If one believes that X is the best medical outcome, one ought to do Y when conditions A, B, C… exist.
>
> Conditions A, B, C… do exist.
>
> Therefore, one ought to do Y.

This syllogism reveals that the judgments of medical and other facts (conditions A, B, C, etc.) are in the minor premise. The major premise always must contain an evaluative statement (such as "one ought to do Y") if one is to get to a clinical action in the conclusion. Physicians are clearly presumed expert on the medical portion of the minor premise. (They may be wrong, but they run the risk of medical malpractice if they are.) They are just clearly not experts on the major premise. If one ought to do Y, that judgment comes not from medical science but from someone's general system of beliefs and values. Hence, if one ought to choose aggressive life support (or hospice care) when a terminal illness is diagnosed and no effective treatment is available, that is a value judgment, not a matter of medical science. Likewise, if one ought to abort (or carry to term) a fetus with Down syndrome, that is a value judgment, not a matter of medical science. But, just as clearly, if one ought to take an antihistamine for springtime allergies, it is a value judgment. Likewise, it is a value judgment if one ought to take penicillin for pneumonia, a statin for high cholesterol, or propranolol for a ventricular arrhythmia. Literally every medical choice is a value judgment.

It has taken a long time for the radical implications of this analysis to come to fruition. The conclusion that a clinical decision must incorporate a value judgment about which physicians cannot be presumed to be expert applies not only to the ethically exotic cases but also to all medical choices. If the patient with the ventricular arrhythmia should take propranolol given the alternatives available, that is a value judgment, not a matter of medical fact. Propranolol has some modest risks of undesired effects as well as some documented impact on arrhythmias. Whether the risk of the drug is justified given the potential benefit is a value judgment that must be made taking into account the costs and potential perceived harms of the alternatives. The only plausible conclusion is that in the era of what can be called postmodern medicine, literally every clinical choice involves an evaluative choice about which physicians cannot be presumed to be experts. The implications for the practice of prescription writing are radical.

If clinicians cannot be experts on the major premise of every clinical choice (because it requires a value judgment about which the clinician cannot be presumed to be expert), clinicians are not justified in "prescribing." Prescribing is the act of reviewing the relevant facts—medical and nonmedical—and deciding that a treatment, in the usual case, a pharmaceutical, is best for a particular patient in a particular circumstance. If the clinician has no basis for determining the value premise needed to decide what is best, then the physician cannot justifiably prescribe. Doctor cannot know what is best. It is, of course, true that the physician may still be the expert in knowing the effect of drugs and, hence, which drugs might be considered for achieving the patient's goals. Physicians are left merely being able to offer hypotheticals (if you want to achieve X with the highest probability, use drug Y). They cannot recommend that the patient use a particular drug. Moreover, there may be other legitimate non-paternalistic limits on patient access to pharmaceuticals. There may even be paternalistic ones, but the physician is not in a position to impose or even authorize those limits.

There is an even more basic point. If prescriptions need to be based on value judgments imposed on the relevant facts—both medical and nonmedical—then the physician is not even an expert on all the relevant facts. In particular, the physician cannot be expected to know the patient's nonmedical situation—his or her economic, social, familial, religious, psychological, legal, and historical situation. The physician can know only one portion of the relevant medical facts and, at best, only vicariously the relevant value judgments.

What a clinician can do is diagnose, determine prognosis, and assemble a list of plausible treatment options with their expected effects (what the clinician may think of as "risks" and "benefits"). Even this task will involve some evaluative judgments—determining what counts as a plausible treatment option and how likely an effect must be before it is worth mentioning. What should happen at this point is complicated. Ideally, if time and alternative costs were not a concern, the clinician would presumably explain all this to the patient and let the patient impose his or her own value judgment, filling in the major premise of the clinical decision-making syllogism as well as the relevant nonmedical facts. As a practical matter, since there are costs in time and energy involved, something short of this may occur. The patient may ask the clinician for a recommendation. Doing so, however, is, in effect, asking the clinician to fill in his or her personal value judgments as well as guesses about some presumed nonmedical facts. It makes a difference whether the doctor we are asking about terminal illness options is Jack Kevorkian or a practicing Orthodox Jew who believes in preserving life to the very end. Likewise, the physician's personal values will make a difference in literally every recommendation made. For the patient with chronic allergies, the antihistamine may improve the inflammation but cause drowsiness. The clinician must make the value trade-off if he or she is to prescribe. The rational patient will realize that taking the shortcut of asking for the clinician's recommendation is, in effect, asking for the physician's value judgment. If the patient believes that clinician's values are similar to his or her own, this may make sense as a practical and efficient shortcut, but typically there is no reason to believe that patients share values with their clinicians.

One might be tempted to suggest at this point that the physician ask the patient about what matters to him or her and then explain how a therapeutic choice maximizes those values. At this point the physician is, once again, not making a recommendation but merely stating a hypothetical (if you want X, follow course Y). There are two problems with this reinterpretation, however. First, at best, the physician will be able to gain only a crude approximation of the patient's complex value orientation, nothing like the rich texture needed to evaluate the options. Second, it reduces the physician to a mere technician offering the patient a strategy for achieving the patient's goals no matter how offensive they may be to the physician's own value framework.

The implications for the practice of prescribing are radical. The practice incorporates a logical flaw. It depends on the healthcare professional not only for professional expertise about the medical facts but also for the necessary value judgments about which the professional cannot be presumed expert. In a liberal society, we normally assume that the value judgments are the patient's to make, at least if there are no significant third-party costs. If there are third-party costs—in deciding whether to use amphetamines that may induce irrationally violent behavior or to use narcotics that pose an addiction risk that could jeopardize third-party welfare—then it is the society, not the individual clinician, that should make the value judgments. The practice of prescription writing is built on the model of modern medicine in which we incorrectly assumed that physicians were expert on the value judgments. Once we realize that error, it makes no sense to permit physicians to prescribe. A new social practice is called for.

Alternatives to Prescribing

The thought of eliminating prescription writing and making all pharmaceuticals available over the counter probably seems radical to many. Surely, however, for anyone who accepts the moral power of the ethical principle of autonomy, such an option must be considered. Let us consider two policy options, one a more purely libertarian view that elevates autonomy to a central place, at least taking precedence over strong paternalism that would attempt to protect drug consumers from their own poor choices, and another that retains some features of moral paternalism while coming to terms with the recognition that prescription writing requires a value judgment about which clinicians cannot be presumed expert.

The Libertarian Option

The first possible policy we can call the "libertarian option." It is firmly grounded in the ethical principle of autonomy. It is the policy already in place for chemicals like alcohol, tobacco, and warfarin for pest control. Competent adults would be free to purchase drugs on the open market. They would, of course, also have the moral responsibility to become educated consumers learning about the evidence that the drug is effective (for which they might rely on websites such as those of NIH or WebMD), even if there would be no legal requirement that they become so educated. They would also be able to consult experts and would, for serious drug choices, presumably want to talk to their physician. In doing so, however, they would rely on these sources for knowledge of the medical facts, or, if they asked for advice, they would do so after making a judgment about whether the source's value judgments should be relied upon.

Libertarians since the days of John Stuart Mill (1956 [1859]) have recognized that the individual's autonomy-based rights to consume something like a drug must be limited by the "harm to others" principle. Even a libertarian should accept the need for controls on methamphetamine and probably narcotics—drugs that pose a risk to third parties. This could be the basis for restricting purchase of a drug like pseudoephedrine. In our liberal democracy, we actually tolerate liberty at a higher threshold than libertarians would. We allow individuals to do many things that harm other unwilling people, for example, smoking in public or owning a gun. These activities pose a risk to those around the practitioners, but we tolerate them because of the premium we place on autonomy. If we are consistent, we would permit patients to make drug choices even when they pose similarly modest risks to third parties. Since relatively few drugs pose significant third-party risks, this is not a serious limitation to a libertarian policy of permitting competent adults to purchase drugs without a prescription.

Also, the libertarian option would have to come to terms with the fact that some potential consumers are not competent adults. Such a policy would presumably limit purchases by children and others who are not mentally competent. Not only would children not be permitted to make purchases, but adults as well would not necessarily be permitted to make purchases for them and administer the drugs based on their own value judgments. Just as society imposes limits on the right of adults to buy and give children alcohol and tobacco, so presumably some drug-use choices made by adults for a child would also be constrained.

These constraints on the use of pharmaceuticals by both incompetent and competent persons in a way that poses significant risks to third parties are consistent with libertarian principles, in particular, the notion that it is not paternalism to take action to protect those who are not substantially autonomous or to take action when the purpose is to protect not the consumer of the drugs but rather third parties.

This is a plausible conclusion for anyone who accepts a moral theory that places the principle of autonomy over the principle of beneficence when it is used in such a way to bring benefits to one who, while substantially autonomous, does not want the presumed benefit. Still, many would find this goes too far. Some believe that, at least in extreme cases, autonomy is not a lexically ranked principle that always should take precedence over beneficence, even beneficence directed toward the one being benefited (Beauchamp and Childress 2013: 220–223 and *passim*). We may be worried that the libertarian option at least poses the risk that some drug-use decision-makers will not be acting in an adequately autonomous way because they are not adequately informed or free to choose. We may also be worried that some adequately autonomous decision-makers may make such bad choices that some limited form of paternalism may be justified, that is, we may hold an ethical theory that sometimes permits the ethical principle of beneficence to take precedence over autonomy. If so, we might consider a second, more conservative policy as an alternative to the purely libertarian policy regarding the practice of prescribing.

The Certificate Option

The libertarian option poses a problem for those who are concerned about the possibility that consumers will not be adequately informed in their drug-use choices. Even competent adults may be inclined to purchase pharmaceuticals without the benefit of adequate education about the pharmacology of the agents—the evidence of their effectiveness, the data on potential undesired effects, and the evidence regarding alternatives. Lay people may be unduly influenced by lay press accounts of wonder drugs or by advertising intentionally misleading consumers about the drug's safety and effectiveness.

The Weak Paternalism Version

This is, in effect, a concern addressed by the notion of *weak paternalism*. Some who are, in principle, anti-paternalists worry about dangers to people who are not acting in a mentally competent manner based on adequate knowledge. They would support a limited use of paternalism to intervene long enough to determine if the consumer is, in fact, adequately competent and informed. Once that is determined, the weak paternalist will step aside and let the actor choose a course of action, even if observers continue to believe that the consequences will be bad.

That is the policy that defenders of a "certificate option" would favor. They would restrict the purchase of legend drugs, those currently available only on prescription (or a shorter list of those that should only be available on prescription) to purchasers who have a certificate signed by a licensed physician, asserting that he or she has educated the patient adequately about the drug, its potential effects, and its alternatives.

This would look very much like the piece of paper that is currently called a prescription, but it would be very different philosophically. It would not be a written declaration that the physician believes the drug is best for the patient. It would merely be a testament that the patient is adequately informed. It would remain the patient's decision to determine whether the drug is best. The certificate requirement would be a paternalistic attempt to make sure that the consumer is, in fact, acting in a substantially autonomous manner. Once that was determined, it would be the patient's decision whether to use the drug and how it is used.

A Strong Paternalism Version

This may still be too risky for some who want to balance concern for the well-being of the consumer against the principle of autonomy. They may fear that getting the certificate could lead a consumer to buy and use a drug about which a physician had provided information that the drug poses serious risks and offers very little chance of benefit. If the certificate merely testifies to the fact that the patient has been educated, this educated patient may still make what seem like bad choices that are not in the patient's interest. How could someone who is willing to tolerate some paternalism in the strong form make use of the certificate given the realization that prescribing requires incorporating value judgments about which the physician cannot be presumed to be expert?

We could require that the certificate not only testify that the patient is adequately informed but also that he or she has a condition for which the drug has been judged adequately beneficial. Presumably, it should not be the individual clinician's judgment that the drug is adequately beneficial since that would require the clinician to impose his or her personal, idiosyncratic values. We could rely on a more societal judgment, however. This should not be the judgment that is merely the consensus of the medical profession. Even the profession as a whole may have systematic biases. For example, in the mid-twentieth century, the professional consensus was that sterilization should be available only to patients who were older or already had enough children.

There is no reason why the professional consensus on such matters should be authoritative any more than the individual clinician's judgment is. If we are to be strongly paternalistic in protecting adequately informed consumers of pharmaceuticals from harms to themselves, the judgment about the benefits and harms should be as sound as possible. For example, if the best medical science finds no evidence that a drug is effective against a particular condition and it poses a risk of producing what most would consider a serious side effect, then a defender of strong paternalism might consider a policy that the drug not be sold to someone who wants it for that condition. We might rely on the FDA to compile a list of drugs paired with conditions for which the drug has been found effective. Since the determination of whether there is adequate evidence of a drug's effectiveness always incorporates value judgments, the process should not be in the hands of professional groups. Instead, it should be a democratic decision based on most people's judgment that a drug should be permitted to use to treat a condition. This, of course, reintroduces strong paternalism and the imposition of the value judgments of the group compiling the list of approved uses, but it at least avoids the serious mistake of permitting the medical professional to be the paternalist.

The result would be that the certificate would not only testify to the fact that the patient is educated but also that he or she has a condition for which the drug has been deemed effective enough that its use would be reasonable. Presumably, the worse the collective value judgment about the effects generally deemed bad, the more evidence of effectiveness would be required. A drug with little evidence of benefit might still be made available provided there is also little evidence of harm. (Consider homeopathic remedies as an example.)

We would consider this a strong paternalism. The patient, even when certified adequately educated and mentally competent, would not be able to purchase the drug unless he or she had a condition on the list of approved uses. This would, of course, run the risk of jeopardizing the rational, if idiosyncratic, judgments of those who have unusual values that would lead them to rule out standard treatments and prefer a drug that most would find inadequately supported. A pure libertarian would object to the certificate option in either of its forms, especially in the latter form that imposes a societal judgment on the rationality of the consumer's decision that the drug offers enough promise of benefit to be worth pursuing.

Conclusion

The practice of limiting pharmaceuticals to prescription access is so deeply entrenched that most would not question it. On reflection, however, it is an odd custom given the propensity, especially in American culture, to accept the right of competent adults to engage in practices based on their own judgment of risk and benefit as long as they do not significantly jeopardize the welfare of others. Since most drugs pose very little if any risk to third parties, it seems strange that a liberal society would routinely presume that drugs cannot be purchased on the open market except in the case of those found most innocuous. This practice is even more puzzling given the strong consensus that access to some very potent and dangerous chemicals should be open and without the review and approval of medical professionals. Even for one who is not a libertarian, the appeal to consistency would seem to require that, if admittedly dangerous chemicals like alcohol, nicotine, and rat poison are readily available to adults, most pharmaceuticals should be as well.

The most likely explanation is that for too long Western culture bought into the traditional medical model, which assumed that only physicians could know when a drug was good for a patient. That model changed in the modern period so that it was the consensus of the professional group that would determine whether a drug should count as beneficial. Until late in the twentieth century, the question of when a drug was "indicated" or a "treatment of choice" was conceptualized as a matter of medical science. Physicians who were expert on the science of medicine were also assumed to be experts on the value judgments that were necessary to decide that a treatment was beneficial. That error was slowly discovered, first with a few "ethically exotic" medical interventions surrounding terminal care and fertility decisions, but now the necessity of nonmedical value judgments to determine if a treatment is beneficial is increasingly seen as logically necessary.

Given the critical role of these nonmedical value judgments about which physicians cannot be presumed to be expert, the practice of *prescribing* no longer makes sense. A prescription is merely the recording of some physician's value judgment that a particular medication in a particular dosage form at a particular strength (and perhaps by a particular manufacturer) is better for his or her patient than any alternative. Since the patient may not share that value judgment, logically the reasoning that supports the writing of a prescription is flawed.

An alternative policy is called for. We present two options. The defender of the lexical priority of the principle of autonomy will be a militant antipaternalist and will support the open access to all pharmaceuticals by competent adults except for those posing serious risk of harm to others. That defender of autonomy will plausibly accept limits on the right of competent persons to transfer pharmaceuticals to incompetents including children, but not on the right of those persons to consume the drugs themselves.

For those more willing to balance the duty to produce benefit against autonomy, a second policy should be considered, a certificate policy. Such a policy would come in two forms. One form merely certifies that the patient has been adequately educated to purchase a particular pharmaceutical. This view could be defended on grounds of acceptance of weak paternalism: The right of society to intervene to assure that the actor is, in fact, substantially autonomous. A second form of a certificate policy would also require health professional certification that the patient has a condition that those best able to judge consider to be adequately responsive to the drug in question. The therapeutic suitability of a drug would be determined not by the consensus of the medical profession, which may have biases, but by some broader, interdisciplinary group representing a wide range of societal perspectives. This second form of a certificate would cross the line into strong paternalism in that it would override an individual judgment that a drug should be purchased (and presumably used), even though the broader society considers that there is no adequate evidence that the risks are justified by the potential benefits.

Even this second, more conservative form of a certificate would avoid the no-longer-defensible assumption that health professionals are presumed experts on the value judgments that go into the practice of prescribing. The practice of physician prescribing is not defensible in a postmodern society. Some alternative is needed.

References

Arras, J., and B. Steinbock (eds.). 1999. *Ethical Issues in Modern Medicine*, 5th ed. Mountain View: Mayfield Publishing Company.

Beauchamp, T. 1977. Paternalism and Bio-Behavioral Control. *The Monist* 60(1): 62–80.

Beauchamp, T.L., and J.F. Childress. 2013. *Principles of Biomedical Ethics*, 7th ed. New York: Oxford University Press.
Buchanan, A.E., and D.W. Brock. 1989. *Deciding for Others: The Ethics of Surrogate Decision Making*. Cambridge: Cambridge University Press.
Dworkin, G. 1972. Paternalism. In *Morality and the Law*, ed. R. Wasserstrom, 107–126. Belmont: Wadsworth.
———. 1988. *The Theory and Practice of Autonomy*. Cambridge: Cambridge University Press.
Emanuel, E.J., and L.L. Emanuel. 1992. Four Models of the Physician-Patient Relationship. *Journal of American Medical Association* 267(16): 2221–2226.
Faden R., and T.L. Beauchamp in collaboration with King NNP. 1986. *A History and Theory of Informed Consent*. New York: Oxford University Press.
Feifel H. et al. 1967. *Physicians Consider Death*. Proceedings of the American Psychological Association, 201–202.
Gert, B., and C.M. Culver. 1979. The Justification of Paternalism. *Ethics* 89: 199–210.
Goldman, A. 1999. The Refutation of Medical Paternalism. In *Ethical Issues in Modern Medicine*, 5th ed., ed. Arras et al. Mountain View: Mayfield Publishing Company.
In re Quinlan. 1976. 70 N.J. 10, 355 A. 2d 647, *cert. denied* sub nom., Garger v. New Jersey, 429 U.S. 922 (1976), overruled in part, In re Conroy, 98 NJ 321, 486 A.2d 1209 (1985).
Mill, J.S. 1956. *On Liberty*. New York: The Liberal Arts Press.
Scrimshaw, S.C., and B. Pasquariella. 1970. Obstacles to Sterilization in One Community. *Family Planning Perspectives* 2: 40–42.
Smith, D.H., and L.M. Bernstein, ed. 1987. *No Rush to Judgment—Essays on Medical Ethics*. Bloomington: The Poynter Center/Indiana University.
Tarasoff v. Regents of the University of California Supreme Court of California. 1974. 1974 13 Cal. 3d 177.
Veatch, R.M. 1972. Models for Ethical Medicine in a Revolutionary Age. *The Hastings Center Report* 2(3): 5–7.
———. 1978. The Hippocratic Ethic: Consequentialism, Individualisem and Paternalism. In *No Rush to Judgment—Essays on Medical Ethics*, ed. D.H. Smith and L.M. Bernstein. Bloomington: The Poynter Center/Indiana University.
———. 2009. *Patient, Heal Thyself*. New York: Oxford University Press.
———. 2012. *Hippocratic, Religious, and Secular Medical Ethics: The Points of Conflict*. Washington, DC: Georgetown University Press.
Wasserstrom, R., ed. 1978. *Morality and the Law*. Belmont: Wadsworth.

Conscientious Refusals in Pharmacy Practice

Zuzana Deans

It is common for pharmacists as well as other healthcare professionals to be given the right to conscientiously object to any procedure that falls under their professional remit. This is usually facilitated by a conscience clause within a code of ethics. By their nature, conscientious refusals, or conscientious objections, are usually made in relation to the most polarising issues. In recent years, pharmacists have been criticised for exercising their right to make conscientious refusals, with the media spotlight on the refusal to supply contraception, including emergency hormonal contraception (EHC) (BBC News 2010; Stein 2005; Stokes 2008). When this topic is discussed, there are often two debates running in parallel. One concerns the moral or religious acceptability of the procedure to which the objection is being made (e.g. supply of EHC), and the other concerns whether it is morally justifiable to have a conscience clause in place at all. It is this second debate that is the subject of this chapter.

This chapter offers a critical review of the current thinking on the philosophical and practical ethical questions about the validity of conscientious refusals. It is organised into four main sections. First is an examination of two key concepts on which a conscience clause is commonly thought to be based: conscience and integrity. Following this is a discussion addressing some practical ethical questions: whether the conventional form of the conscience clause is adequate for meeting patients' rights, protecting the pharmacist's integrity and guarding against wrongdoing; whether pharmacists should publicly announce their objections in advance; and how we should decide what the acceptable bases for making an objection should be. Next, consideration is given to whether the provision of conscientious refusal should apply equally to students, who arguably have the right to conscientiously refuse to participate in some aspects of their pharmacy training. The chapter finishes

Z. Deans (✉)
Centre for Ethics in Medicine, School of Social and Community Medicine, Faculty of Health Sciences, University of Bristol, Bristol, UK
e-mail: zuzana.deans@bristol.ac.uk

© Springer Science+Business Media B.V. 2017
D. Ho (ed.), *Philosophical Issues in Pharmaceutics*, Philosophy and Medicine 122,
DOI 10.1007/978-94-024-0979-6_10

by stepping back from the detail of the debate and suggesting some ways in which progress can be made in understanding this complex practical philosophical problem. The arguments presented share their origin with those in the debate about conscientious objections among physicians, and the same basic principles apply to any profession. Where the discussion differs slightly is in how these principles play out in practice, given the particular circumstances of the pharmacy and the obligations pharmacists have that compete with their right to conscientiously object.

Conscience and Integrity

Conscience

When an individual makes a conscientious refusal to perform a certain action, she does so because she believes that to perform that action would be to contradict her core moral beliefs (Wicclair 2011: 4–5) or her religious faith and would go against what her conscience is "telling" her to do. When an individual listens to her conscience, she is responding to a moral judgement and/or feeling that urges her to act or discourages her from acting. Most people are familiar with this feeling, though it is hard to pinpoint exactly what the conscience is and how exactly it relates to whether an action is (or would be) right or wrong. Hill describes the conscience as a "capacity, commonly attributed to most human beings, to sense or immediately discern that what he or she has done, is doing, or is about to do (or not do) is wrong, bad and worthy of disapproval" (Hill 1998: 14). It is most common for the conscience to be referred to in the context of wrong action, as Jenkins explains: "We hear a great deal about a 'bad' or 'guilty' or 'nagging' conscience, and about 'pangs' with which conscience afflicts us when it is 'hurt' or 'outraged'. We hear very little about a 'good' conscience. This indicates that conscience is peculiarly alert to human malfunctioning" (Jenkins 1955: 261–2). When the good conscience is referred to, it is as a wholesome entity or is described as having an absence of negative properties. As Childress puts it, "often the good conscience is described by nouns such as 'peace', 'wholeness' and 'integrity' or as adjectives such as 'quiet', 'clear', 'clean' and 'easy'" (Childress 1979: 318).

Despite the sometimes powerful "gut feeling" that an (in)action would be wrong (or right), the conscience is not generally considered by philosophers to be a reliable and definitive indication of what is in fact right and wrong (see Childress 1979; Hunter 1963; Sulmasy 2008). Indeed, one's conscience can be desensitised over time by consistent wrongdoing, which implies it is a poor evaluator of whether an action is right or wrong.

The moral judgements of the conscience are not solely based on feelings or intuitions; a conscientious refusal is often the result of careful, rationalised thought. This intellectual aspect of moral judgement may make it easier to understand, challenge, persuade or dissuade someone from the position she holds, but it does not necessarily make her conscience a more reliable indicator of what is in fact right or wrong. Individuals with equally strong conscientious positions can hold opposing views, again implying the conscience is not a reliable epistemic tool.

Even though the conscience gives an unreliable indication of whether an action would be morally permissible, a person's appeal to her conscience remains morally valuable because the conscience represents an individual's commitment to doing what is right. This is captured in Sulmasy's definition of conscience, which is perhaps the clearest and most fitting among the various definitions in the literature, and will be adopted for the remainder of this chapter:

> (1) a commitment to morality itself; to acting and choosing morally according to the best of one's ability, and (2) the activity of judging that an act one has done or about which one is deliberating would violate that commitment. (Sulmasy 2008: 135)

Importantly, it is this commitment that matters when a conscientious refusal is honoured. In this way, the conscience clause has little to do with the particular procedure to which the individual is objecting; rather, it is in place partly to respect the individual's integrity. Take the example of pharmacists supplying EHC. In several countries, including the USA, the pharmacy profession as a whole supports the view that supplying EHC in certain circumstances is morally permissible but allows conscientious refusals. The fact that an individual pharmacist's conscience tells her that it is not morally permissible does not in itself persuade the profession that the practice is wrong. Rather, the conscientious refusal is honoured in part because the profession respects the pharmacist's commitment to morality. The profession accepts that for some individuals, the act of supplying EHC would be a breach of this commitment and would violate their integrity.

Integrity

There are several accounts of integrity in the philosophical literature.[1] One of the significant differences between them is the implication for the relationship between integrity and morality. There are two reasons why this relationship is important when defending conscientious objections. First, if integrity is linked to doing the right thing, then protecting it would be instrumentally valuable for the profession. Second, if acting with integrity is morally worthy in and of itself, then, all things being equal, it would be wrong to prevent an individual from exercising her integrity.

Cox et al. (2003) make a convincing case for integrity as a virtue that keeps in check the balance of other virtues, describing integrity as a cluster concept, made up of other virtues. This view is not without its faults, specifically that it does not provide an explicit definition of integrity. Even so, it goes some way towards capturing

---

[1] For further explanation, see Cox et al. 2013, Palanski and Yammarino 2007, and Scherkoske 2013.

intuitions about integrity and how it is applied to individuals where other accounts seem to miss the nuances of what it means to be a person with integrity. By this account, a person with integrity is reflective and responsive. She takes seriously her commitment to morality, which includes continual assessment and adjustment of her values, particularly by considering the perspectives of others. Under this view, a person with integrity can make mistakes and do wrong, but in the main she will do good because the qualities underlying integrity (e.g. self-awareness, sincerity, reflection) usually contribute to good character. I would add to this, more explicitly, that integrity includes the virtue of a commitment to morality. It is worth briefly visiting three other accounts of integrity ("self-integrated integrity", "integrity as identity" and "objective integrity") to explain why integrity is best understood as a virtue.

Under the self-integration account, integrity is a kind of wholeness of the self. It is achieved by a person when her various desires, principles and commitments are in harmony without inner conflict. "The integrated-self picture…emphasizes the fact that people of integrity know who they are, and what they stand for. They have settled reasons for taking the stand they do, and those reasons are their own reasons" (Mendus 2009: 16). In this view, it makes no difference whether a person is harmoniously evil or harmoniously good; it is through wholeness, not goodness, that a person achieves integrity. A major criticism of this view is that to exempt integrity from moral evaluation would be to miss an important aspect of integrity, which is about a commitment to the good. Another criticism, made by Cox et al., is that the proponent of this account is mistaken in the claim that conflicts of commitments, values and desires threaten integrity. This is because integrity necessarily involves a capacity to change and involves a "continual re-making of the self" (Cox et al. 2003: 41). "[T]he view that the person of integrity is a whole integrated self, that the successful integration of self is both necessary and sufficient for integrity, is oversimplified and mistaken. …[B]eing such a 'whole integrated self' is antithetical to integrity since it suggests an end to, or elimination of, the kinds of conflict that integrity is rooted in and thrives upon" (Cox et al. 2003: 18–19). In other words, we should not resist conflict of our own values or regard it as a weakness or sign of a lack of integrity. Instead, it should be accepted that a person with integrity is likely to experience conflict and will possess the necessary virtues (which together constitute integrity) to reflect on and make adjustments as appropriate.

This responsive and adaptable element is also missing from Williams' notion of "integrity as identity", which is to hold steadfastly to one's principles and values as part of one's identity. For Williams (1981), integrity is a matter of honouring one's commitments which, if abandoned, would corrupt the very person one is. A person might compromise on minor matters but will show unwavering commitment to the values with which she most closely identifies. Unlike integrity as a virtue, integrity as identity does not necessarily motivate one to act morally, only to act in which a way that is in keeping with one's character. Thus integrity by identity is descriptive (of consistency with one's character) and nonnormative. This means that a person can have integrity even if she consistently commits heinous acts. One might respond to this by pointing out that this account sets out a duty to be true to one's character (even if that character is otherwise morally flawed) and that it therefore does have a normative element to it. Pitted against the person who is inconsistent, unreflective and with little integrity of any kind, the evil person who is true to her character might attract a degree of admiration for her consistency that is separate from judgements of the acts she commits. However, this single trait of consistency with one's character is insufficient as an account of integrity. A person who met a commitment to maintaining her poor character would be more likely to be described as constant, stubborn, predictable or unchanging than as having integrity. Integrity is a richer, more complex term. It is used to comment on a person's moral character, not a single trait. According to Cox et al., integrity would be more appropriately attributed to a person who overcame her bad character and was *not* true to her previous self (2003: 31). Again, the integrity as identity account seems to miss important aspects of integrity as being aligned with what is morally right. It is only by understanding integrity as a virtue can we explain that a person may be guided by her integrity to do the right thing.

Doing the right thing is essential under the "objective" account of integrity, which requires "that we abide by our moral commitments and that these commitments stem from the moral obligations we actually have" (Ashford 2000: 425). Cox et al. (2013) claim it would therefore be impossible for someone to act with integrity and be morally mistaken. This is far more demanding than the "integrity as virtue" account, in which a person strives to do good and usually succeeds but is sometimes mistaken in her judgement.

In terms of the correlation between integrity and doing the right thing, the "virtue" account sits somewhere between the "identity" and "objective" versions. As Cox et al. describe it, "A person of moral integrity cannot be a moral monster . . . because attributions of integrity, being attributions of an important virtue, presuppose a certain moral success; the qualities that make for a character of integrity only constitute integrity when they succeed in making a person, with some degree of latitude, a good person" (Cox et al. 2003: 69). This means the evil tyrant cannot have integrity, even though he stands steadfastly to his principles. It also means that a person can have integrity despite sometimes faltering in her judgement. Instead, what matters is that a person takes seriously her moral commitments and usually displays praiseworthy virtue. A person with the virtue of integrity holds an honest adherence to her values, ready to adjust these values in response to new knowledge, experiences or realisations. Honesty is a virtue in its own right but also forms part of the cluster of virtues that makes up integrity. This honesty demands that the individual is open to adjusting her values as appropriate by considering the perspectives of others and keeping open-minded about the possibility that she has been mistaken or naive. "The appearance of certainty throughout a person's moral and volitional life may be, and often is, less a sign of integrity than an indication of its lack. Thus, we learn early to beware of the self-righteous and sanctimonious" (Cox et al. 2003: 3).

On the whole, a person with integrity will judge correctly. Importantly, though, integrity is not sufficient for moral action as it may lead to wrong action, and it is not necessary for a morally desirable outcome. This can be seen in the following examples. Sally is a pharmacist who believes that using animals in research is morally abhorrent. Even so, she continues to supply medicines that have been tested on animals because she believes that to refuse to supply them would put her current position of employment in jeopardy. For the sake of argument, assume Sally is mistaken in her beliefs about using animals in research.[2] In this example, Sally is acting against her integrity to do something that is otherwise morally acceptable. In a similar example, George also believes that using animals in research is morally abhorrent, and he refuses to make the supply. In this case, George acts with integrity to refuse to supply medication tested on animals, but, for similar reasons to Sally, he is mistaken in his reasons for the refusal. Sally's actions have resulted in a morally desirable outcome despite her lack of integrity, and George's actions have resulted in a morally undesirable outcome despite his integrity. As someone with integrity, George's misjudgement is likely to be exceptional; usually, his thoughtfulness, sincerity and commitment to morality lead him to make careful, sensitive and well-justified decisions.

So far, it has been shown that neither conscience nor integrity is a reliable indication that the moral judgement a person makes is correct, though integrity as a virtue is generally conducive to doing the right thing. Integrity is thought to be valuable in its own right, such that to force a person to act against her integrity would be to commit a great wrong against her. When conscientious refusals are granted, it is because it is thought to be wrong to force a person to act against her integrity to perform an otherwise good act even in cases in which acting with integrity may lead to wrongdoing or an undesirable outcome (Benn 2007: 348). This implies there is something valuable about integrity over and above it being instrumental for morally worthy acts. Importantly, it might also mean that it would sometimes be wrong to force someone to act against her integrity to prevent her from acting in a morally undesirable way. In other words, it might sometimes be wrong to deny a conscientious objection, even when that conscientious objection amounts to a morally undesirable action. As I have explained elsewhere, "if pressurizing someone to act against her integrity is wrong, then potentially…there are two wrong actions from which to choose the lesser of two evils: pressurizing someone to act against her conscience, or letting her do the wrong thing" (Deans 2013: 51). Take an example of Mike, a pharmacist who is working as a locum in a community pharmacy practice that usually runs a needle exchange service for users of recreational intravenous drugs. Mike has a moral objection to the needle exchange service because he believes that participating in the scheme would be assisting individuals to harm themselves through drug use and that this would be worse than providing clean needles to reduce the risk of infection through sharing. In one version of events, Mike is allowed to refuse to supply clean needles and therefore denies the patient access to safe instruments, increasing the risk of the patient acquiring or transmitting an infection, but his integrity is respected. In another version of events, Mike is pressured to act against his conscience to supply clean needles, thus reducing risk of infection for the patient and others.

[2] Suppose that, despite her efforts to engage with the subject, Sally has not heard all sides of the arguments in the debate and they are arguments that would persuade her to change her mind.

This example is, of course, oversimplified, but it is intended to illustrate that, whether it would be right or wrong to supply the clean needles, there is an additional moral consideration about whether it would be right to pressure someone to act against her conscience. In this example, if Mike is allowed to refuse to supply the clean needles when the profession supports needle exchange programmes, then this is because the profession considers it worse to pressure someone to act against his integrity than for a morally undesirable outcome to occur (i.e. for the needles to not be supplied).

One reason against pressuring someone to act against her conscience is that it could make her feel uncomfortable. This discomfort has been described as "distressing" (Benn 2007: 348) and "excruciating moral anguish"(Cohen 1968: 269), but, as Benn points out, there are other distressing aspects of professional life (e.g. delivering bad news to a patient) that are unavoidable (Benn 2007: 348). The comparison is imperfect, since it conflates distress arising from fundamentally unpleasant situations with distress caused by doing what one believes to be wrong; arguably, healthcare professionals should be prepared to overcome distress caused by unfortunate circumstances but should be spared the particular torment of being pressured into wrongdoing. Whatever status is given to the kind of distress brought about by wrongdoing, granting conscientious refusals in order to save the objector from moral anguish will frustrate the conscientious objector. She would like her objection to be upheld out of respect for her integrity and moral position, not merely in order to prevent her from suffering distress (Deans 2013: 50). Professionals who use the conscience clause to avoid a procedure because it is unpleasant rather than because to perform the procedure would be to contravene their moral principles are met with criticism for misuse of the privilege (Millward 2010: 425).

What, then, might the defence be for respecting integrity, especially in cases in which respecting it comes at the cost of allowing a morally undesirable outcome to occur? Contrary to Giubilini's claim that integrity is an "anaemic" concept that has been assigned too much importance in moral decision-making in healthcare (Giubilini 2014: 162), integrity has weighty intrinsic and instrumental value. First, the agent's moral engagement brings value to a decision. This is to claim that it is not just the consequences of an action that have moral weight but also the intentions and motivations behind the act. It is to recognise that a person who acts with integrity is acting through a self-aware, reflective commitment to doing the right thing. This involves attempting to make well-considered decisions that take into account the possible outcomes of her actions, while also ensuring she is moral in other ways (e.g. by acting with honesty, kindness, fairness and compassion and by considering carefully her responsibilities). This was exemplified by Sally and George: Sally's actions resulted in the most desirable consequences (supply of medicine), but it was George who displayed good moral character. Although ultimately mistaken, George showed a commitment to doing what is right.

Second, to act with integrity is to take seriously one's moral responsibilities. Restricting individuals in making independent judgements according to their integrity may discourage them from engaging with the world around them. Very basically, the healthcare professions, including pharmacy, are driven by moral values, and they require moral agents to navigate the situations in which they and their patients find themselves.

This second aspect, and to a lesser extent the first aspect (on which more later), can be achieved through the common arrangement for applying conscientious objections, namely, the conventional compromise (Brock 2008: 194). The conventional compromise is commonly put in place to allow an individual professional to exercise her right to conscientiously object while minimising the effect on the patient and not denying the patient her right to access services to which she is entitled. Thus, it prevents individuals' judgements leading to morally undesirable outcomes. In this way, integrity is arguably respected to an appropriate degree: enough to respect and encourage genuine moral engagement of professionals while ensuring that a morally desirable outcome is achieved.

Practical Resolutions

Conventional Compromise

When an individual makes a conscientious refusal, she declines to carry out a procedure or supply a medication that would otherwise be provided by the profession. Since patients have the right to access the health service being denied, the professional's rights to act autonomously and with integrity are in direct conflict with the patient's rights (and correspondingly the profession's duty to provide those services).

Savulescu argues that a healthcare professional who conscientiously objects is unjustifiably failing in her duties: "When the duty is a true duty, conscientious objection is wrong and immoral. When there is a grave duty, it should be illegal. A doctors' conscience has little place in the delivery of modern medical care" (Savulescu 2006: 294). This is based on the claim that it is the practitioner's duty to provide healthcare services and that patients are treated unfairly when access to services depends on practitioners' values and when some patients may be unable to seek alternative care. Savulescu claims it is unreasonable for individuals to expect to be able to opt out of activities that are central to their role. Practitioners can expect to be asked to perform particular procedures and should avoid the profession if they are not prepared to provide these services (e.g. obstetricians should be willing to terminate pregnancies, and pharmacists should be willing to supply contraception). By implication, an individual should not sign up for the profession if she feels she cannot carry out these duties because of her values and should resign from her profession if she encounters such conflict.[3] Commenting on the duties of physicians, Savulescu writes, "people have to take on certain commitments in order

[3] This may seem a particularly hard line when applied to those who find their values conflict with practices that have been introduced after they joined the profession due perhaps to innovations in medical treatment.

to become a doctor. They are a part of being a doctor. Someone not prepared on religious grounds to do internal examinations of women should not become a gynaecologist. To be a doctor is to be willing and able to offer appropriate medical interventions that are legal, beneficial, desired by the patient, and a part of a just healthcare system" (Savulescu 2006: 295).

Savulescu's position is known as the "incompatibility thesis" (Wicclair 2008: 172) because the claim is that holding values that prevent one from performing the duties of a healthcare practitioner is not compatible with being a healthcare practitioner. It has, however, been shown that the incompatibility thesis is not problematic if the practitioner refers the patient to a colleague who is willing to provide the service. This is because it is the profession itself, not the individual practitioner, that must provide certain services (Brock 2008: 193). It does not matter which professional provides the service as long as the duty is fulfilled by the profession. Indeed, Savulescu is in agreement that conscientious refusals are sometimes acceptable in order to honour the liberty of the practitioner if there are plenty of other practitioners available to provide the service.

The conventional compromise allows an individual professional to act in accordance with her conscience (to some extent) without denying the patient access to health services. It is based on the principle that the duty to provide healthcare services to the patient is stronger than the right to conscientiously object. Typically, the conventional compromise involves three parts:

1. The professional provides the patient with information about the relevant service or treatment.
2. The professional informs the patient of an alternative means of accessing the service or treatment she needs.
3. This means of access does not present an unreasonable level of burden for the patient (Brock 2008: 194).

This compromise fits well with common sense; the individual pharmacist need not provide the service that she believes so strongly would be wrong to provide, and the patient can receive the treatment without very much inconvenience. There are, however, a couple of problems with this response.

The first difficulty is in interpreting "unreasonable burden". It seems sensible and fair that the referral must not present an unreasonable level of inconvenience for the patient. Unreasonable burdens may include a long journey to another pharmacy, a long waiting time before a service can be accessed (perhaps with worsening symptoms, reduced effectiveness of treatment, a longer period of anxiety or simply the inconvenience of waiting longer than anticipated), financial costs incurred or emotional distress caused by the referral.[4] However, what seems reasonable to one person will not be reasonable to another. The ambiguous nature of claims for rea-

---

[4] In Great Britain, the General Pharmaceutical Council requires pharmacists to ensure that referrals allow patients to access treatment within an appropriate timeframe that will not compromise contraceptive cover or effectiveness of the treatment. In making this assessment, pharmacists are advised to consider factors such as the practice opening hours and the patient's ability to get there (Royal Pharmaceutical Society of Great Britain 2014: 136).

sonableness might also make it hard to reach firm conclusions in a dispute over whether a referral had presented an unreasonable burden. Further, the burdensome consequences of a referral could be long-lasting, unpredictable and difficult to measure or prove.

The second challenge to the conventional compromise is that there is serious doubt as to whether it safeguards integrity at all. This is because the compromise still requires the pharmacist to facilitate the patient to access the treatment or service to which she has an objection. Take the example of a pharmacist, Sarah, who has a conscientious objection to supplying EHC because she believes it is wrong to interfere with the natural process of conception. A patient comes to Sarah with a request for a non-prescription based, over-the-counter supply of EHC. In accordance with the conventional compromise, Sarah refers a patient to a pharmacist on the same street who she knows would be willing to make the supply. The patient follows the referral and is supplied with EHC.

In this case, Sarah has not succeeded in her aim to avoid participating in an act that she believes amounts to wrongfully interfering with the natural process of conception. Although she herself did not make the supply, she was nevertheless involved in the process of the patient obtaining EHC. In other words, by making the referral, the pharmacist is still involved in the provision of the service to which she was objecting. Furthermore, Sarah was involved in a morally relevant way (i.e. she understood the possible outcomes of her referral and was a willing participant) and therefore has a degree of moral responsibility for the supply of the pill to this patient. Even so, Sarah may be less responsible than she would have been had she made the supply herself. This is because her responsibility may have been diluted by the inclusion of another moral actor, namely, the pharmacist to whom the patient was referred. In making the referral, Sarah changed the event into one of cooperation; without Sarah's input, this supply of EHC would not have been made, but likewise without the input from the second pharmacist this supply of EHC may not have been made. This is analogous to the responsibility an individual committee member has for a decision made by a committee, for which a unanimous vote is required (Mellema 1985: 178). For Sarah, as with the member of the committee, her action (the referral) was not sufficient to bring about the supply of EHC. However, if Sarah had made the supply directly, her action (the supply) would have been sufficient and she would have borne full responsibility for the supply.

However diluted a pharmacist's responsibility is, she still bears some responsibility, and a pharmacist seeking "clean hands" would only achieve this by not making the referral.[5] It has been suggested that there is a moral distinction between direct and indirect referrals, such that a pharmacist making a direct referral would be morally complicit in the supply, but if she were to make an indirect referral, she could not be said to be morally complicit (Chervenak and McCullough 2008: e2 as

[5] This is not necessarily a major concern for pharmacists. For instance, research has revealed moral passivity among some pharmacists who do not always engage in moral decision-making even when they regarded something as ethically problematic. Instead, pharmacists admitted shifting the moral responsibility to the prescribing doctor (Cooper et al. 2008b: 443).

discussed in Wicclair 2011: 37–38). A direct referral involves the first pharmacist making specific arrangements for a second pharmacist to make the supply. If Sarah were to telephone the pharmacist on the same street to check his availability, check his willingness to make the supply and let him know to expect the patient and were to give the patient specific instructions about how to reach the second pharmacist, then the referral would be direct. If instead Sarah were to tell the patient to use the Internet to find the address of another pharmacy that might make the supply, then the referral would be indirect. Sarah's indirect supply has elongated the causal chain that leads to the supply of EHC, and some suggest this reduces complicity (Cantor and Baum 2004: 2011). The indirect referral also increases the chance that the patient will not obtain EHC at all (e.g. if she does not get around to finding an alternative pharmacist). This in turn may make the final outcome more or less foreseeable (e.g. a pharmacist in a busy city setting might reasonably suppose the patient could access the medication elsewhere, while a pharmacist in a remote rural setting would know that it is less likely that the patient would find an alternative). This arguably increases or reduces moral culpability, in parallel with legal principles. Note, however, that the less foreseeable the outcome becomes, the less helpful the referral is, and this could be to the point of the referral being obstructive or giving rise to an unreasonable burden for the patient.

Public Disclosure of Conscientious Objections

A practical solution to the inevitable complicity of a referral may be for pharmacists to give advance notice of which services they are unwilling to provide, reducing the number of patients who would otherwise approach the pharmacist for treatments of services to which the pharmacist has an objection and therefore reducing complicity by way of referral. Aside from a strategy of advance notification, it would not be possible to avoid complicity without defying the conventional compromise. It has been suggested by Harter, in relation to physicians, that public disclosure of conscientious objections would reduce instances of delay or discontinuation of patient care (Harter 2015: 225). Elsewhere, I have argued that advance notification is akin to a blanket refusal to provide a service and that this contravenes one of the conditions of the conventional compromise, which is that the patient should not be subjected to an unreasonable burden (Deans 2013: 56–7). While Harper is correct in pointing out that advance public notification could be advantageous to patients, this is only in comparison to situations in which the patient approaches a professional who has an objection and is then referred to another colleague in accordance with the conventional compromise. There may be circumstances in which a referral would be inappropriate because it would present an unreasonable burden for a patient. In such cases, it is less likely that advance notification of a conscientious objection would be in the patient's favour.

It might be objected, however, that a blanket refusal does not contravene the conventional compromise because the obligations associated with the conventional compromise only apply in circumstances in which a patient approaches a pharmacist with a request for provision of a service to which the pharmacist has an objection. If the patient does not approach the pharmacist in the first place, then arguably the obligation to make the supply in cases in which a referral would cause an unreasonable burden to the patient does not apply. This is because, by making a blanket refusal, the pharmacist is simply implementing an absence of service. By giving advance notice, the pharmacist is omitting, rather than refusing, to provide the service. This absence of service, so the objection runs, relieves the pharmacists from any obligations that would otherwise be associated with provision (or refusal) of the service.

This objection is problematic for two reasons. First, it is wrong to regard a blanket refusal of a central pharmacy service as a mere absence of a service. Unlike some other absences, this lack of provision is not morally neutral. When a pharmacist makes a central service unavailable, she is failing to provide the standard professional services. She cannot obliterate the associated obligations by simply removing one of the services, and one of those obligations is to redirect the patient where appropriate. The absence of a central pharmacy service is not like absences of auxiliary, non-pharmacy services, for example, film processing. In the absence of these kinds of services, pharmacists are under no obligation to aid would-be customers because there had never been an obligation to supply the service in the first place.

The second problem with the objection is that it wrongly assumes that the method of communication (e.g. a refusal declared in advance on a poster, instead of a refusal made face-to-face) alters the obligations of the pharmacist. This rests on the false premise that pharmacists only have obligations towards patients with whom they directly interact. This is not true; it is well accepted that pharmacists have obligations towards the public (e.g. this motivates public health initiatives by pharmacists). This is not to say that pharmacists have obligations to actively seek all members of the public who may benefit from their services, but the obligation to serve the public does include an obligation not to discourage patients from seeking a service. This is reinforced by Wicclair's argument that each individual pharmacy practice has an obligation to promote the public health, safety and welfare of the population within its catchment (Wicclair 2011: 136). This obligation is derived from a commitment made when the licence was given, from principles of reciprocal justice (i.e. fair exchange of rights and privileges with obligations) and from other moral obligations (e.g. beneficence). To properly fulfil this obligation, Wicclair argues that a "pharmacy-by-pharmacy" standard should be adopted. This is opposed to a "general public availability" standard, which would see a population served by the pharmacies in the area, without any one particular pharmacy being obliged to fulfil the obligation. Wicclair argues that the pharmacy-by-pharmacy standard is justified on the grounds of fairness (especially as some decisions about which medicines to stock would be business-led. A pharmacy-by-pharmacy standard would avoid any one pharmacy carrying the burden of providing services that were less profitable or less advantageous in business terms). Wicclair also claims the general public availability standard would be unworkable. In instances in which more than one pharmacy could serve the needs of the population, there is no obvious way of deciding which pharmacy this should be. In addition, it would be infeasible to measure whether patients' needs could be adequately met through implementation of a general availability policy because the criteria (e.g. whether requiring a patient to travel for her medication would be excessively burdensome) are controversial and elusive[6] (Wicclair 2011: 140–141). Each pharmacy ought, therefore, to be available to serve the relevant population.

There are other reasons pharmacists may be encouraged to publicly disclose their position. The most obvious is that public disclosure is congruent with an open, honest and transparent profession.[7] More fundamentally, Brownlee argues that public exposure goes some way towards proving that the individual is truly consistent and non-hypocritical. Brownlee classifies conscientious (or, to use her terminology, "personal") objections as a non-communicative disobedience, or personal disobedience, instances of the individual avoiding certain action without true conviction (Brownlee 2012: 27). According to Brownlee, where disciplinary action is relevant, those who do not publicly declare their position openly should be treated more harshly than they tend to be, and those who are prepared to pay the costs of standing up to their beliefs publicly should be treated less harshly then they tend to be (Brownlee 2012: 8–9). Open dialogue, in his view, is a key element in any genuine pursuit of change to policy or law, and this is absent when the objecting individual keeps her position private. This may be evasive, for example, when a pharmacist who privately objects to the supply of methadone to drug users busies herself when she sees that the patient who regularly visits the pharmacy for his methadone prescription is next in the queue, leaving her colleague to serve the patient. Alternatively it could be non-evasive, for example, when a pharmacist with an objection to the use of contraception exercises a conscientious objection by having an initial discussion with the patient and then referring her to a colleague for supply. Her refusal to supply is not affected by whether or not her position is publicly known; she simply desires non-interference in her decision to not make the supply. Brownlee claims that civil disobedience (communicating one's objections to regulations, policies or law) is more deserving of respect than a personal objection or a conscientious objection, because it displays a greater degree of moral conviction (Brownlee 2012: 15–50). But it should be remembered that the two are not mutually exclusive; a pharmacist might quietly refuse to make a supply to a patient and engage in rigorous debate with her colleagues and the regulatory body in an effort to change policy. In cases like these, full disobedience (e.g. not following the conventional compromise with the intention of being called up to a disciplinary panel) will be at the expense of the patient. One reason a pharmacist may not do this is because she holds patient

[6] Interestingly, this is not regarded as an insurmountable problem when similar criteria are applied in the third part of the conventional compromise.

[7] However, it is worth noting that sharing one's personal beliefs is not always regarded as appropriate. For example, the British Medical Association's guidance for doctors who conscientiously object is that they should not share their moral views unless they are explicitly invited to do so (BMA 2015).

welfare and autonomy higher than the principle she would like to adhere to by conscientiously objecting. If, however, the pharmacist regards the principle as more important than the patient's autonomy (e.g. a pharmacist who believes the use of EHC amounts to killing an innocent person), then she may be driven to take a stand. Wicclair cautions against how conscientious objections are used: to prevent the pharmacist in participating in wrongdoing or to prevent another person in participating in wrongdoing, "The point is to allow individuals to refrain from performing actions against their conscience and preserve their moral integrity. However permitting *x* to refrain from acting against *x*'s conscience is not to be confused with enabling *x* to prevent *y* from performing legal actions that are contrary to *x*'s (but not *y*'s) ethical or religious beliefs" (Wicclair 2011: 112). This is important because the conscience clause is designed to allow a professional to preserve her integrity; it is not intended as a direct instrument for wider change, though widespread, organised conscientious refusals could equate to a protest movement within the profession.

Adding Further Restrictions to Permitted Use of Conscientious Objections

So far, it has been shown that while there is good reason to allow pharmacists the right to conscientiously object, restrictions should be in place (the conventional compromise or a variation thereof) to ensure that pharmacists' integrity and autonomy are preserved as much as possible without jeopardising patient welfare. Contemporary policy and literature on this subject suggest some further limitations should be set so that only conscientious objections based on certain grounds should be considered valid. Increasing the limitations would be a further restriction of the pharmacist's autonomy, which would come up against arguments in support of conscientious objections that are based on the premise that a variety of moral and religious beliefs ought to be tolerated. Further conditions for the acceptability of a conscientious objection are that the reasons for the refusal should be reasonable (Card 2007: 13), should not be based on prejudice (Wicclair 2014: 279), should be in keeping with the core values of the profession[8] (Wicclair 2000: 217 and Deans 2013: 53) and should be genuine (Meyers and Woods 2007: 20). This list, and each individual item on the list, presents a tall order. Looking at each in turn, one can see that none of these criteria is easily or completely achievable but nor is it futile to aspire to these requirements.

It is "not unreasonable to ask for reasons", states Card (2011: 62), the "beliefs on which conscientious objection is based must be reasonable and should be subject to evaluation in terms of their justifiability" (Card 2007: 13). Under such a proposal, the reasons behind a conscientious refusal would have to be based on scientific knowledge and, where applicable, true or not implausible non-clinical claims.

[8] A similar criterion of goals of the profession has also been suggested (Wicclair 2006: 244).

Wicclair points out that the professional must have her clinical facts right (Wicclair 2006: 243) and just as important are non-clinical facts.

A position is most clearly unreasonable when the facts are incorrect or when a judgement is based on prejudice. For example, empirical research shows that some pharmacists are mistaken about some aspects of EHC and that views vary on the clinical action of EHC, which may affect the supply of EHC over the counter (Cooper et al. 2008a: 50). There is also evidence that pharmacists make their decisions about whether to supply based on their perceptions of women's propensity to abuse EHC based on their age, wealth or locality (Cooper et al. 2008a: 50). Judgements about the non-clinical context include, for example, assuming the availability of EHC would lead to a rise in sexual promiscuity and irresponsible sexual behaviour (Barrett and Harper 2000: 205).

In many cases, there will be no evidence available to prove or disprove a non-clinical belief. Metaphysical beliefs (including religious) are almost immune to verification. As Marsh points out, "since many moral judgements stem from metaphysical or religious assumptions, this class of refusal [metaphysical] turns out to be rather common" (Marsh 2014: 314). There are very strong epistemic barriers to verifying metaphysical claims. While some metaphysical positions may be more plausible or more reasonable than others, there is great difficulty in validating or invalidating many metaphysical theories. Take, for example, Paul, a pharmacist who believes life begins when the soul is created, which is the point at which an egg and sperm unite to become the two-celled zygote. Paul understands that EHC is not an abortifacient, but he is correct in his belief that the use of EHC could destroy a zygote. Paul believes all humans who present themselves to him at his pharmacy are his patients. Subsequently, Paul believes the zygote constitutes a patient. In line with his professional code of ethics, Paul believes the interests of his patients are of prime importance. As such, Paul has a conscientious objection to supplying EHC.

Paul is unable to prove his claims are correct, but neither can his position be disproven. The problem with making it a requirement of conscientious refusals that the objector gives sound reason is, as Marsh argues, that it is either too easy or too difficult to satisfy (Marsh 2014: 313). Nevertheless, sound reasoning is so fundamental to how we justify decisions that the profession must surely strive to overcome epistemic difficulties wherever possible. Reasonableness tests are notoriously problematic, but in practice, the majority of accounts and narratives are accepted as being either reasonable or unreasonable, and therefore justification for decision-making is deemed achievable.

It has also been claimed that conscientious objections should be in line with the core values of the profession (Wicclair 2000: 217; Deans 2013: 53). The most basic problem with this approach is immediately obvious: how can we identify the core values? Some attempts have been made to do this (Benson et al. 2009), but there remains no definitive understanding of the core values. There will, however, be cases in which it is clear that an individual is acting against the values of the profession. As with the reason-giving requirement, it will be a matter of judgement within a real-life context whether a pharmacist could be said to be acting within the core values of the profession.

The requirement that the conscientious objection be based on core values of the profession is driven by a resistance to the conscience clause being an "anything goes" policy, such that when a pharmacist's beliefs come into conflict with the values of the profession, the pharmacist's beliefs are given priority. This presents something of a contradiction within the profession's policies: the profession sets standards, presumably based on norms and values, which are collectively agreed by its members. At the same time, the profession permits an individual to act in a way that is contrary to those values (Deans 2013: 53). This could perhaps be avoided by recognising that one of the core values of the profession is respect for the integrity of the individual pharmacist. This would not eliminate conflict, but arguably conflicts of this kind inevitably exist within professions and do not present deep philosophical challenges.[9]

A requirement of genuineness has also been proposed, so that the objector has "a sincere scruple-based objection to the procedure" (Meyers and Woods 2007: 20). This is intended to eliminate refusals that are based on a non-moral aversion. This, again, may be hard to verify, though as Marsh points out, this may become evident when the individual is pressed to give justification for their conscientious refusal (Marsh 2014: 318–9).

Finally, Kantymir and McLeod suggest a middle ground of "reasonableness or genuineness plus", which demands that the conscientious objector prove either:

1. That it is reasonable, in particular, by showing that what grounds the objection is as likely or more likely to be true than what grounds the standard of care for patients
2. That it is genuine, *plus* that it satisfies certain criteria

For option (2), the criteria are as follows: patients will still get the care they need in a respectful and timely fashion, any empirical beliefs on which the objection rests are not baseless, and the moral or religious beliefs on which it rests are not discriminatory (Kantymir and McLeod 2014: 21).

The strongest challenges this position faces are related to the applicability of such criteria: the interpretation, verification and practical management of imposing these standards.

Conscientious Objections in an Educational Context

So far, the discussion has been about qualified pharmacists and their obligations as professionals. In this final section, the focus turns to students of pharmacy. Empirical evidence shows that students are supportive of the right for professionals to conscientiously object, but this is limited to students reflecting on the appropriateness of conscientious objections among qualified pharmacists (Hope et al. 2014). There is also some evidence of medical students' views, again only on qualified

[9] I thank Dien Ho for raising this point.

professionals (Strictland 2012; Nordstrand et al. 2014), and little is known about how inclined pharmacy students are to conscientiously object to carrying out procedures as part of their training. Nevertheless, this is an area worth exploring further since there are additional arguments to consider in the educational context. Wicclair puts forward a compelling set of reasons for allowing conscientious objections among students, including that diversity in student population should be encouraged, and that students' views should be taken seriously and should not be dismissed as outliers and that student objections can and do lead to changes in practice (Wicclair 2011: 171–172).

The key difference between the qualified pharmacist and the pharmacist in training is in the purpose of the procedure. A qualified pharmacist aims to meet her obligations towards the patient and the public, while the student pharmacist aims to ensure she receives the training to obtain the skills, knowledge and experience necessary for becoming a competent pharmacist. Where the qualified pharmacist has a conscientious objection, she may be able to ensure her aims are met by a colleague. For a student pharmacist, it would be far less plausible for her to find a substitute to carry out the educational exercise and for the objecting student to still gain the necessary skills and experience. Wherever possible, it is the procedure that would have to be substituted, not the person. For example, a number of medical schools in the USA now use interactive computer programmes instead of dog labs (in which a dog would be anesthetised, cut open and examined while still alive. It would then be killed once the demonstration was over). Similarly, students could examine a member of the same sex instead of a member of the opposite sex where sexual characteristics are not relevant to the educational exercise. This is, however, limited. For example, it is hard to see how a student who objects to participating in reading, lectures and examinations on diseases caused by sexual activity or alcohol (Foggo and Taher 2007; Strickland 2012: 24) could gain a comprehensive education. In addition, Card claims that opting out of some procedures is to miss important lessons in professionalism. "By refusing to perform examinations on members of the opposite sex, such students are failing to engage [with] the question of what constitutes a touch that is professional and non-sexual—one that exemplifies a 'cool intimacy' that is still compatible with closeness to a patient" (Card 2012: 604).

It might be argued that a student who has an objection to certain procedures would be wise to choose a career that excluded these practices and, as such, avoid training in the area that gives rise to the objections. The common example in medicine is the student who has a conscientious objection to termination of pregnancy and so chooses not to go into a career in obstetrics and gynaecology and opts out of some aspects of training in this field. However, even if pharmacists could avoid the practice they had an objection to, this would be to miss the point of their programme, which is to educate pharmacists to be fit for practice. Commenting on doctors' training in the UK, the General Medical Council states that "the point of a medical course is to produce a doctor fit for clinical practice. What doctors then choose to do with their career is a matter for them" (General Medical Council 2006). Importantly, opting out of learning about a certain procedure on the grounds that one will object to performing it once qualified is not supportive of the conventional compromise, since this requires pharmacists to make the supply/perform action when a referral is not appropriate.

In the debate on conscientious objections among students, the emphasis is on education: educating students on the full range of skills and competencies but also educating students about moral values. A conscientious objection by a student presents an educational opportunity. The arguments for conscientious objection among qualified healthcare professionals apply to students but with the possibility of being overridden by arguments for educational need. In cases in which the exemption can be upheld without cost to the student's training (e.g. noncore skills), the student could be asked to defend her position with well-articulated arguments that stand up to challenge from peers, thus serving as part of the educational process.

Imperfections and Opportunities for a Better Approach

As is evident from the preceding discussion, justifications for conscience clauses rest on philosophical and practical considerations. The arguments involved are complex and there is no clear answer. Of course, as with many questions in bioethics, there exist real-life pharmacists, patients and governing bodies that need to make decisions and act on them, and they do not always have the luxury of extended periods of deliberation. What is in place, then, is a practical solution that seems to work, albeit with some imperfections. This final section will bring together the discussion so far, synthesising the key concepts and practical points. No attempt will be made to address the inadequacies of the conventional compromise directly. (On the contrary, it is worth re-emphasising some of the problems with the alternatives.) Instead, it is suggested that a different approach be taken to resolving the problem, by paying closer attention to the views and perspectives of the moral actors concerned.

A conscientious refusal in pharmacy can be summed up as a resistance by the pharmacist to perform a procedure, or make a supply, that is otherwise regarded as morally acceptable by the profession because she believes it would be wrong to carry out the procedure or make the supply. As has been shown in this chapter, such a refusal can stand in the way of the patient receiving the healthcare to which she is otherwise entitled. Herein lies the most basic problem: a conflict of rights. It was assumed in this chapter that the rights of the patient to receive healthcare that is ordinarily provided within the healthcare system are relatively uncontroversial. Instead, the attention was on arguments in favour of protecting the pharmacist's integrity. Perhaps the most promising account of integrity is one that articulates integrity as a virtue that keeps in check other virtues. The appeal of this account is that it matches intuitions about integrity being attributed to a person of good character (not merely a description of consistency of values or character). This account is also forgiving of misjudgements, relating integrity not directly to consistent, objectively right moral action, but instead to a commitment to being moral. However, this remains unsatisfactory to some extent; we are still seeking a clear definition of integrity and its relationship to moral action. Integrity is used in everyday language and is certainly not the reserve of philosophers, so it might be useful to look more closely at ordinary usage to understand and define this concept and its role in conscientious refusals.

The conventional compromise offers a practical solution to the conflict between the pharmacist's integrity and the rights of the patient, but, as is the nature of compromises, each side must make sacrifices. For the patient, this is the inconvenience (and perhaps offence) of having to go elsewhere for the pharmacy service. For the pharmacist, this is the obligation to maintain some involvement by providing information and making a referral and the possibility that she will have to make the supply if the patient would be unreasonably burdened by a referral. Philosophically, this is problematic because it is seems the pharmacist is still morally responsible for the very wrongdoing she is attempting to refuse to be complicit in. We also saw how one practical solution of giving advance notification was ethically and practically problematic. The conventional compromise is widely adopted by the pharmacy profession, so it is presumably regarded as a workable, if imperfect, solution. From an academic perspective, the conventional compromise is also arguably the best solution to date. Certainly, its use predates scholarly work on pharmacy ethics. It is also possible that the conventional compromise is interpreted or implemented differently in practice from the way it is discussed in the bioethics literature.

As was shown in this chapter, one of the biggest philosophical problems with the conscience clause is in how to distinguish between refusals that are justified and those that are not. It has been suggested that the justification for refusals could be tested in tribunals or similar (Card 2007: 13; Kantymir and McLeod 2014: 16–17), but this is fraught with problems. It is relatively easy to fix some limits (e.g. the objection must not be based on prejudice and must not be based on claims that are factually incorrect), but other potential limitations present bigger challenges, for example, in defining and testing reasonableness and genuineness. The official nature of such tribunals could all too easily be accompanied by bureaucracy, overregulation and authoritarianism. Further, it would not necessarily make any clearer whether a judgement of refusal was justified, it could tempt pharmacists to shift their responsibility of decision-making onto a panel, and it would risk involving patients as witnesses in disputes (if they are asked to give evidence of the specific nature of the case). What might be more useful is guidance on what the profession regards as justifiable (in broad terms, such as reasonable and genuine). This would come up against the same definitional problems as noted in the context of tribunals but would at the same time clarify the purpose and scope of the conscience clause.[10]

The educational context presents quite a different question. Here, no patient's rights are immediately at stake (though a future patient's welfare could be compromised by the incompetence of an undertrained pharmacist), but there are other reasons not to allow all refusals. The aim of the procedures being objected to is educational and, in many cases, cannot be substituted (so an equivalent of the referral in the conventional compromise would not work). The educational setting

[10] For example, the accepted basis for conscientious refusals by pharmacists in Great Britain is broad: "make sure that if your religious or moral beliefs prevent you from providing a service, you tell the relevant people or authorities and refer patients and the public to other providers" (Royal Pharmaceutical Society of Great Britain 2014: 112).

offers a special opportunity for exploring ideas, values and positions and for training would-be pharmacists in skills such as critical reasoning.

Although, on balance, the arguments presented in this chapter support the conventional compromise, the case is not watertight, and it remains philosophically unsatisfactory. No doubt further progress can be made by continuing to construct and analyse arguments within the existing parameters of the debate, but I suggest we also use methods in empirical ethics to understand how these dilemmas are experienced by the real-life moral actors. A similar suggestion has been made by Cooper et al. (2008b: 441). This would come with its own set of difficulties (including the methodological challenges seen in empirical ethics research), but has the potential to provide insight and further our understanding of this question. For example, this chapter has focused on integrity (because this is a key underlying concept for conscientious objections), but it is possible that conscientious clauses are better supported by other notions, such as liberty, tolerance and political harmony.

Conclusions

The conscience clause exists as an attempt to prevent individual pharmacists from having to act in ways that would go against their conscience and violate their integrity. It allows deviation from the standard accepted practice of the profession. This is a profession that places the patient's interests at its centre. Because of this, we can expect robust arguments about whether and how conscientious objections should be made. There are some convincing arguments in favour of allowing conscientious refusals by way of the conventional compromise, albeit with recognition that there remain some philosophical and practical difficulties with that position.

One possible avenue for further investigation into this area is empirical ethics research. By listening to those with the lived experience of these dilemmas (both pharmacists and patients), we might gain further understanding of the concepts that are considered (in the literature) to be fundamental. By combining philosophical insight and analytic rigour with a richer understanding of what pharmacists actually do in these situations and why, we might understand better how to navigate this practical ethics problem.

References

Ashford, E. 2000. Utilitarianism, Integrity and Partiality. *Journal of Philosophy* 97: 421–439.

Barrett, G., and R. Harper. 2000. Health Professionals' Attitudes to the Deregulation of Emergency Contraception (or the Problem of Female Sexuality). *Sociology of Health & Illness* 22(2): 197–216.

BBC News. 2010, March 10. *Lloyds Pharmacy Probe over Pill Refusal in Sheffield*. http://news.bbc.co.uk/1/hi/england/south_yorkshire/8557816.stm. Accessed Mar 2015.

Benn, P. 2007. Conscience and Health Care Ethics. In *Principles of Healthcare Ethics*, ed. R. Ashcroft et al., 345–350. Chichester: Wiley.

Benson, A., A. Cribb, and N. Barber. 2009. Understanding Pharmacists' Values: A Qualitative Study of Ideals and Dilemmas in UK Pharmacy Practice. *Social Science & Medicine* 68(12): 2223–2230.

British Medical Association. 2015. *Expressions of Doctors' Beliefs*. http://bma.org.uk/practical-support-at-work/ethics/expressions-of-doctors-beliefs. Accessed Apr 2015.

Brock, D.W. 2008. Conscientious Refusal by Physicians and Pharmacists: Who Is Obligated to Do What, and Why? *Theoretical Medicine and Bioethics* 29: 187–200.

Brownlee, K. 2012. *Conscience and Conviction: The Case for Civil Disobedience*. Oxford: Oxford University Press.

Cantor, J., and K. Baum. 2004. The Limits of Conscientious Objection—May Pharmacists Refuse to Fill Prescriptions for Emergency Contraception? *The New England Journal of Medicine* 351(19): 2008–2012.

Card, R.F. 2007. Conscientious Objection and Emergency Contraception. *The American Journal of Bioethics* 7(6): 8–14.

———. 2011. Conscientious Objection, Emergency Contraception, and Public Policy. *The Journal of Medicine and Philosophy* 36: 53–68.

———. 2012. Is There No Alternative? Conscientious Objection by Medical Students. *Journal of Medical Ethics* 38: 602–604.

Chervenak, F., and L.B. McCullough. 2008. The Ethics of Direct and Indirect Referral for Termination of Pregnancy. *American Journal of Obstetrics and Gynecology* 199(3): 232.e1–232.e3.

Childress, J. 1979. Appeals to Conscience. *Ethics* 89(4): 315–335.

Cohen, C. 1968. Conscientious Objection. *Ethics* 78(4): 269–279.

Cooper, R., P. Bissell, and J. Wingfield. 2008a. Ethical, Religious and Factual Beliefs About the Supply of Emergency Hormonal Contraception by UK Community Pharmacists. *The Journal of Family Planning and Reproductive Health Care* 34(1): 47–50.

Cooper R, P. Bissell, and J. Wingfield. 2008b. Ethical Decision-Making, Passivity and Pharmacy. *Journal of Medical Ethics* 34(6): 441–445.

Cox, D., M. La Caze, and M.P. Levine. 2003. *Integrity and the Fragile Self.* Aldershot/Burlington: Ashgate.

Cox D., M. La Caze, and M. Levine. 2013. *Integrity. The Stanford Encyclopedia of Philosophy* (Fall 2013 Edition). http://plato.stanford.edu/archives/fall2013/entries/integrity. Accessed Apr 2015.

Deans, Z. 2013. Conscientious Objections in Pharmacy Practice in Great Britain. *Bioethics* 27(1): 48–57.

Foggo D., and A. Taher. 2007. Muslim Medical Students Get Picky. *The Sunday Times*. http://www.timesonline.co.uk/tol/news/uk/health/article2603966.ece. Accessed Jan 2011.

General Medical Council. 2006. *Core Education Outcomes GMC Education Committee Position Statement*. http://www.gmc-uk.org/Core_Outcomes_1.0.pdf_39260454.pdf. Accessed Mar 2015.

Giubilini, A. 2014. The Paradox of Conscientious Objection and the Anemic Concept of 'Conscience': Downplaying the Role of Moral Integrity in Health Care. *Kennedy Institute of Ethics Journal* 24(2): 59–185.

Harter, T.D. 2015. Toward Accommodating Physicians' Conscientious Objections: An Argument for Public Disclosure. *Journal of Medical Ethics* 41: 224–228.

Hill, T.E. 1998. Four Conceptions of Conscience. In *Integrity and Conscience*, ed. I. Shapiro and R. Adams. New York: New York University Press.

Hope, D.L., M.A. King, and H.L. Hattingh. 2014. Responses of Pharmacy Students to Hypothetical Refusal of Emergency Hormonal Contraception. *The International Journal of Pharmacy Practice* 22: 155–158.

Hunter, J.F.M. 1963. Conscience. *Mind* LXXII(287): 309–334.

Jenkins, I. 1955. The Significance of Conscience. *Ethics* 65(4): 261–270.

Kantymir, L., and C. McLeod. 2014. Justification of Conscience Exemptions in Health Care. *Bioethics* 28(1): 16–23.
Marsh, J. 2014. Conscientious Refusals and Reason-Giving. *Bioethics* 28(6): 313–319.
Mellema, G. 1985. Shared Responsibility and Ethical Dilutionism. *Australasian Journal of Philosophy* 63(2): 177–187.
Mendus, S. 2009. *Politics and Morality*. Cambridge: Polity Press.
Meyers, C., and R.D. Woods. 2007. Conscientious Objection? Yes, but Make Sure It Is Genuine. *The American Journal of Bioethics* 7(6): 19–20.
Millward, M. 2010. Should Pregnant Doctors Work in Termination of Pregnancy Clinics? *British Medical Journal* 340(7743): 425.
Nordstrand, S.J., M.A. Nordstrand, P. Nortvedt, and M. Magelssen. 2014. Medical Students' Attitudes Towards Conscientious Objection: A Survey. *Journal of Medical Ethics* 40: 609–612.
Palanski, M.E., and F.J. Yammarino. 2007. Integrity and Leadership: Clearing the Conceptual Confusion. *European Management Journal* 25(3): 171–184.
Royal Pharmaceutical Society of Great Britain. 2014. *Medicines, Ethics and Practice. The Professional Guide for Pharmacists Edition 38*. London: Pharmaceutical Press.
Savulescu, J. 2006. Conscientious Objection in Medicine. *British Medical Journal* 332: 294–297.
Scherkoske, G. 2013. Whither Integrity I: Recent Faces of Integrity. *Philosophy Compass* 8(1): 28–39.
Stein R. 2005, March 28. Pharmacists' Rights at Front of New Debate. *Washington Post*. http://www.washingtonpost.com/wp-dyn/articles/A5490-2005Mar27.html. Accessed Mar 2015.
Stokes P. 2008, October 3. Mother is Denied Pill by Muslim Pharmacist. *Telegraph*. http://www.telegraph.co.uk/news/uknews/3129625/Mother-is-denied-pill-by-Muslim-pharmacist.html. Accessed Mar 2015.
Strickland, S.L.M. 2012. Conscientious Objection in Medical Students: A Questionnaire Survey. *Journal of Medical Ethics* 38: 22–25.
Sulmasy, D.P. 2008. What Is Conscience and Why Is Respect for It So Important? *Theoretical Medicine and Bioethics* 29: 135–149.
Wicclair, M.R. 2000. Conscientious Objection in Medicine. *Bioethics* 14(3): 205–227.
———. 2006. Pharmacies, Pharmacists and Conscientious Objection. *Kennedy Institute of Ethics Journal* 16(3): 225–250.
———. 2008. Is Conscientious Objection Incompatible with a Physician's Professional Obligations? *Theoretical Medicine and Bioethics* 29: 171–185.
———. 2011. *Conscientious Objection in Health Care: An Ethical Analysis*. Cambridge: Cambridge University Press.
———. 2014. Managing Conscientious Objection in Health Care Institutions. *HEC Forum* 26: 267–283.
Williams, B. 1981. *Moral Luck: Philosophical Papers 1973–1980*. Cambridge: Cambridge University Press.

Part III
Usage

Using Pharmaceuticals to Change Personality: Self-Transformation, Identity, and Authenticity

David DeGrazia

Francis' Request

Francis is not ill, even if occasionally troubled, and he is by most accounts doing rather well. Three years ago he landed a good position as a professor of molecular genetics at a well-respected university. In the interim, he has managed to score a few grants, and the publications are coming along at a pace that will very likely earn him tenure. Outside of work, he has several close friends, who marvel at his success in view of his appearance of being lost, distractible, and tangential. In addition to having a good job and some friends, Francis is physically fit, due in no small part to the long walks he regularly takes in the woods near his home, and he enjoys creative writing despite lacking any clear idea of what he might do with the products of this pastime.

Notwithstanding these positive factors in his life, Francis is dissatisfied with the status quo. In most settings—in fact, whenever he is not in the company of good friends, family members, or a few other people with whom he feels at ease—Francis tends to be withdrawn. His introversion makes things socially difficult for him at professional conferences, committee meetings, and sometimes even in one-on-one interactions with colleagues and students; nor does it help that he tends to be a bit dark in his perceptions of people's behavior and motivations. Although clearly not paranoid in any clinically relevant sense, his friends and family find his social perceptions somewhat cynical, with a tendency to assume the worst rather than giving people the benefit of the doubt. And Francis has become aware of this psychological bias, having noticed that his interpretations are often refuted by follow-up

D. DeGrazia (✉)
Department of Bioethics, National Institutes of Health, Bethesda, MD, USA

Department of Philosophy, The George Washington University, Washington, DC, USA
e-mail: david.degrazia@nih.gov; ddd@gwu.edu

© Springer Science+Business Media B.V. 2017

D. Ho (ed.), *Philosophical Issues in Pharmaceutics*, Philosophy and Medicine 122,
DOI 10.1007/978-94-024-0979-6_11

conversations with people whose intentions he had interpreted darkly or by other sources of corrective information.

Another aspect of his mental life and personality that Francis has begun to question is his enormous "spaciness": his mind's relentless tendency to shift from a particular task at hand to free association mode with long, zigzagging mental forays in the theater of his consciousness even as he works, reads, converses, or engages in some other activity that calls for focusing attention. Although, just like his friends, Francis sometimes marvels at how well he has been able to function in professional life, he wants to focus better on tasks at hand. Greater focus would make his work easier and possibly superior in quality. It would also make it easier to keep up with conversations (without coming across as spacey or tangential) and to perform any number of other tasks such as reading boring instruction manuals, learning how to use new software, and organizing his tax return. With this change, he could enjoy his world of free association—so rich and capacious that his erstwhile therapist reframed his "spaciness" as "spaciousness"—on those occasions, such as when hiking or relaxing, when outward focus was not needed.

So Francis wants change. He wants to have a brighter, more optimistic outlook—especially in social settings—and to be more extraverted. He also wants to be able to focus better when a task calls for sustained attention and to have some control over when his mind wanders around the cosmos of his imagination. In effect, Francis wants to change his personality, where "personality" is understood broadly to include not just social presentation but also inner life or psychological "style." In addition to having this somewhat unusual goal, Francis is clear on the primary means by which he wants to pursue it: pharmaceuticals, specifically a selective serotonin reuptake inhibitor (SSRI) such as Lexapro or Prozac, which has a decent chance of making him sunnier and more socially outgoing, and a stimulant such as Adderall or Focalin, which could help him to maintain focus at will. From his experience with psychotherapy several years ago, Francis knows that many mental health professionals would balk at a patient's goal of changing his personality through the use of medication. They might recommend instead that he return to therapy and maybe work at home on mental exercises (e.g., meditation) to improve focus. But Francis, who has considerable insight and self-knowledge, doubts that these interventions would have much of the desired effect, and he is attracted to the efficiency with which an SSRI and stimulant might take him to his goal.

Francis approaches the therapist whom he saw years ago, makes his request for an SSRI and a stimulant, and explains the basis of his request including the intention to transform his personality in both its social and "inner life" dimensions. His request presents us with several philosophical and ethical issues. With no pretense of offering a comprehensive discussion of this rich conceptual and moral terrain, the present discussion will zero in on three sets of concerns:

1. Is Francis requesting a *medical enhancement*, and, if so, does that make the request morally problematic?
2. Does his plan pose a threat to his *identity* in a morally problematic way?
3. Are Francis' intentions *inauthentic*?

Medical Enhancement

Is Francis' request, in effect, a request for medical enhancement? The ethics of enhancement in biomedical contexts has been a topic of scholarly and clinical interest for several decades.[1] What is enhancement in the relevant sense? The standard way of unpacking this concept is by way of contrast with medical therapy or treatment[2], hence the title of a report on the ethics of enhancement by the President's Council on Bioethics: *Beyond Therapy*.[3] More specifically, enhancements are understood as interventions to improve human form or function that do not respond to genuine medical needs, where the latter are understood either (1) in terms of disease, impairment, illness, or the like or (2) as departures from normal (perhaps species-typical) functioning. Insofar as we are comfortable with the use of medical prowess and resources to the extent that it involves the provision of *therapy* and understand the physician's traditional role as requiring the provision of therapy, the idea of enhancement as something that falls outside this familiar and approved domain invites us to consider the ethical appropriateness of medical enhancement.

More recently, a distinct conception of medical enhancement has emerged, a conception that does not involve a contrast with therapy. Presenting one variant of this distinct conception, Allen Buchanan has helpfully focused on *capacities*: "... a biomedical enhancement is a deliberate intervention, applying biomedical science, which aims to improve an existing capacity that most or all normal human beings typically have, or to create a new capacity, by acting directly on the body or brain" (Buchanan 2011: 23). With an eye toward understanding some cases of embryo selection as involving a sort of enhancement and broadening the definition to include nonmedical as well as medical contexts, I have defined enhancement as "any deliberate intervention that aims to improve an existing capacity, select for a desired capacity, or create a new capacity in a human being" (DeGrazia 2013). Note that these and similar capacity-based understandings of enhancement have the implication that a particular medical enhancement might also be a form of medical therapy. For example, physical therapy following shoulder surgery can enhance a patient's flexibility and functioning while also therapeutically addressing the limited mobility caused by surgery and postoperative sling wearing.

Whether we understand enhancement in the more traditional sense that contrasts it with therapy, or in a capacity-based way with no reference to therapy, it is evident that Francis is requesting a type of medical enhancement from his psychiatrist. First, he is not psychiatrically ill: somewhat dark, yes, but not clinically paranoid; socially withdrawn, yes, but not pathologically so; and spacy, to be sure, but not in a way that impedes functioning enough to warrant a diagnosis of attention deficit

---

[1] See, e.g., Parens 1998, Habermas 2003, President's Council on Bioethics 2003, Elliott and Chambers 2004, Agar 2004, Sandel 2007, Green 2007, Schoene-Seifert and Talbot 2009, Savulescu and Bostrom 2009, Buchanan 2011, and Persson and Savulescu 2012.

[2] See, e.g., Juengst 1998.

[3] See reference in note 1.

disorder.[4] These details suggest that he is not seeking a type of medical therapy but rather something "beyond therapy." Moreover, in view of his goal of improving his capacity to focus on tasks at hand, it is clear that he is requesting a medical enhancement in the capacity-based sense of the term as well.

Does the fact that Francis is seeking an enhancement make his request morally problematic? We may distinguish the question of whether *his making the request* is problematic due to the enhancement factor and the question of whether *his psychiatrist's acceding to it* would be morally problematic for the same reason. Elsewhere I have argued in response to a structurally similar case involving enhancement that the correct answer to both questions is negative (DeGrazia 2000). Rather than rehash the arguments in support of this view, I refer the reader to other works that make what I find to be a compelling case that the mere fact that a particular medical intervention would involve enhancement does not make it morally problematic, even if certain associated issues (e.g., wise use of scarce resources, just distribution of access to the intervention, safety) are relevant to any thorough moral evaluation of the enhancement in question.[5]

A Threat to Identity?

Does Francis' enhancement-seeking request threaten his identity in some significant way? Whatever the precise meaning of "identity," the term carries a connotation of gravitas. This connotation and the sense that seeking a change in personality may pose a threat to one's identity can provoke the feeling that something morally troubling is going on when one makes a request like Francis'. Some authors have clearly had this troubled feeling. For example, Carl Elliott writes:

> What is worrying about so-called "enhancement technologies" may not be the prospect of improvement [one would hope not!] but the more basic fact of altering oneself, of changing capacities and characteristics fundamental to one's identity…. Making him smarter, *giving him a different personality* or even giving him a new face—these things cut much closer to the bone…. They mean, in some sense, transforming him into a new person. (Elliott 2003: 28–29, emphasis mine)

Providing another example, the President's Council on Bioethics states the following: "In seeking by these [technological] means to be better than we are or to like ourselves better than we do, we risk 'turning into someone else,' confounding the identity we have acquired through natural gift cultivated by genuinely lived experiences" (President's Council on Bioethics 2003: 300). The basic idea seems to be that seeking a new personality and/or changing who one is (becoming someone else) poses a morally problematic threat to identity.

To evaluate these and similar claims, we need to be clearer about the relevant sense, or senses, of the term "identity." Then we can ask whether Francis' request or

---

4 See the American Psychiatric Association 2013.

5 See, e.g., Buchanan et al. 2000: Chaps. 4 and 5, Kamm 2009, and Buchanan 2011.

some other types of enhancement using biomedical means would pose a threat to identity. If so, then we can proceed to the question of whether or to what extent the threat to identity is morally significant.[6]

Beginning with "identity"—or, as philosophers often say, "personal identity"—this term has two importantly distinct senses. Identity in the *numerical* sense, the sense that is usually invoked when analytic philosophers discuss personal identity, is a relation that applies not only to persons, or human beings, but to any particular thing, including inanimate objects. Numerical identity is the relation a thing has to itself, both (1) over time and (2) across "possible worlds" (possible states of the world, both actual and counterfactual). Of a wooden desk, for example, we may ask when in the history of its wooden pieces it first came into existence (when fully assembled?) and under what conditions it will go out of existence (when entirely disassembled?). This concerns its numerical identity over time. We may also ask whether this desk would continue to exist if its wooden pieces were replaced, one by one, by plastic pieces of the same shape and size. And would this very desk have come into being if a different wood had been used to create a desk of the same shape and size as the actually existing desk in question? These are questions about the desk's "trans-world" identity.

While perhaps only philosophers can get excited about the criteria of an inanimate object's cross temporal and trans-world identity, others are likely to join in the excitement when the issue is a *human being's* numerical identity. When a person dies, for example, it is plausible to think—at least for those who do not believe in an afterlife—that the individual goes out of existence. Assuming this is correct, then knowing the criteria for numerical identity over time will allow us to determine when someone has died: at the moment when the individual's identity is disrupted so that she no longer exists. According to a biological understanding of human beings' numerical identity, each of us will die when her or his biological life comes to an end (a criterion that requires some interpretation to decide between whole-brain death and irreversible loss of cardiopulmonary functioning).[7] According to the most plausible version of the psychological approach to understanding our identity, each of us will die upon irreversibly losing the capacity for consciousness.[8] If this view is correct, then irreversible comas and irreversible vegetative states—which are compatible with spontaneous respiration and a functioning heart—are actually states of death (even though no jurisdiction treats them as such). Although the criteria for human beings' numerical identity is a topic of vigorous disagreement among philosophers—and, in effect, among physicians, biologists, and laypeople who debate the definition of death and the criteria for a distinct human being's coming into existence—it is a point of conceptual agreement that numerical identity persists as long as one stays in existence. So a genuine threat to identity in this sense would be a genuine threat to one's continued existence.

[6] The reflections on identity and enhancement that follow are developed more fully in DeGrazia 2005a: especially Chap. 6. See also DeGrazia 2005b.

[7] See Olson 1997 and DeGrazia in Luper 2014: 88–99.

[8] See McMahan 2002: Chaps. 1 and 5.

With this much clear, it quickly becomes apparent that Francis' request cannot possibly pose a threat to his identity in the numerical sense, because it is obvious that the desired changes in his personality and inner life would not be tantamount to his dying or literally going out of existence. If he becomes sunnier in outlook, more extraverted, and better able to focus on tasks that require sustained concentration, it will be *he* who has undergone these changes and he who will remember how things were before the changes, and no one who loves him will have reason to grieve his passing. Even if one of his associates finds it apt to say he has become "a new person," the term "new" cannot literally mean "numerically distinct"; it would just mean "qualitatively different." This observation takes us to the second sense of "identity."

Unlike numerical identity, the second sense of identity—*narrative* identity—applies only to persons.[9] That is because only persons have life-narratives from a first-person perspective, that is, implicit (if not also explicit) autobiographies. Narrative identity involves one's self-conception: one's sense of what is most important in one's life and what therefore defines *who one is* or *what sort of person one is* in a qualitative sense. Although a person suffering from amnesia may ask "Who am I?" and have the numerical sense of identity in mind, usually those who in everyday life ask "Who am I?" are trying to figure out what is most important to them as they seek self-given direction in life. Relatedly, it is identity in the narrative sense that falls apart, or threatens to do so, in an identity crisis. In deciding whether to be a teacher or corporate lawyer, whether to be a workaholic or to leave plenty of room for pastimes, whether to marry so-and-so or remain single, whether to become a parent, and what sorts of associates one wants to become close to, a person is defining her narrative identity. Insofar as Francis wants to become a somewhat different sort of person—in the way he relates to and thinks about others and in his capacity to remain focused on particular tasks—he might be described as undertaking to change his narrative identity.

As noted earlier, some commentators have expressed moral concerns about seeking enhancements with an eye toward changing one's identity. But why is this "worrying," to use Elliott's term, and why should we speak of the "*risk*" of "turning into someone else," as the President's Council on Bioethics does? One way to understand such concerns is in terms of a conflation of numerical identity and narrative identity, as in this argument:

1. This use of enhancements threatens one's identity.
2. Threatening one's identity is morally problematic.
3. Therefore, this use of enhancements is morally problematic.

By now, the reader should be able to see the fallacy in this argument. Consider Francis' proposed use of an SSRI and a stimulant. Like other proposals to use enhancements that are likely in the real world within the foreseeable future, this use of enhancements would not threaten his identity in the numerical sense—it would not threaten to end his existence and create a numerically distinct person in his

[9] See Schechtman 1996.

wake—but would at most "threaten" his identity in the narrative sense. That is, it would have the prospect of changing his narrative identity. But premise 2 says that threatening one's identity is morally problematic. This seems true if we are talking about numerical identity: ending someone's existence is a big deal. But to make premise 2 plausible by understanding it in terms of numerical identity is to render the argument fallacious, for by switching the meaning of "identity" from the narrative to the numerical sense, the argument commits the fallacy of equivocation. On this reading of the argument, it is straightforwardly unsound.

It is possible, however, that commentators who are worried about the use of enhancements to alter one's identity are focused entirely on narrative identity. If so, the above argument does not equivocate. But the question then arises, what would be wrong with a threat to one's identity, at least when the "threat" is autonomously chosen? What, that is, is wrong with intending to change one's narrative identity with the use of such enhancements as SSRIs and stimulants? As far as I know, the commentators in question have never furnished a cogent answer to this question.

It is important to realize that we frequently seek to change our narrative identities, by changing ourselves, in ways that do not seem particularly problematic or concerning. A Christian who feels repentant about his sins may want to "turn back to Jesus" and become a better person in his own eyes. If he succeeds, he will regard the change as very important and self-defining—an important change in who he is. Someone else may want to become more philanthropic and less self-centered, and try in this way to become a different person. A dentist may feel that she has had it with her professional trajectory up until the present moment and decide to dedicate herself to her true passion, sculpting. There is nothing morally problematic in these desires for self-change; indeed, they may strike us as admirable. As for Francis' plan to change his personality, it, too, seems unobjectionable. Or, at least, the fact that it seeks to change him in a way that might be important to his self-told inner story, his narrative identity, does not seem inherently problematic.

What About Authenticity and Self-Transformation?

Concerns about authenticity in the context of medical enhancements begin with the ordinary understanding of authenticity as, roughly, *being true to oneself*.[10] The concern is that the use of medical enhancements, or at least some such uses, involves the agent's *not* being true to herself. Beneath this charge is the idea that there is an underlying entity, the self, to which one must maintain a certain relationship of fidelity. This idea, however, suggests that the nature of the self is substantially independent of the agent's will; otherwise, an autonomous decision to transform oneself in a particular way could not involve a failure to be faithful, or true, to oneself. According to the charge of inauthenticity, therefore, the self is in an important sense

[10] The discussion of this section draws several ideas from my 2005a: 108–113 and 2012: 75–78.

"given," an objective entity to which one's life choices, including endeavors to change oneself, should conform.

Without being entirely sure how to conceptualize the self, at least in any depth or detail, I suggest that the model of the self that is assumed in this charge of inauthenticity is misleading. It seems more consistent with current understandings of the self to maintain that we can transform and, in some sense, even invent ourselves—ourselves—at least to some significant extent, an idea that runs counter to the image of the self as given or static. I will argue, further, that self-transformation through pharmaceutical enhancement is not, per se, an inauthentic act that involves a failure to be faithful to oneself (although I will later have to acknowledge some limits to authentic self-transformation). This thesis can draw support from further reflections on authenticity and narrative identity.

At this point, instead of understanding authenticity—as the critique does—simply as being true to oneself, I recommend a slightly more expansive conception: *being true to oneself and presenting oneself to others as one truly is*. Authentic individuals express themselves, without pretense or artifice, through their choices and actions. With little tension between who they are and the personas they present to others, they may strike us as especially natural and at ease in their own skin. Amending the briefer analysis of authenticity by adding the idea of presenting oneself truly to others coheres with the observation that inauthenticity, in some cases, involves presenting oneself *falsely* to others.

Imagine a young couple, vacationing and traveling through Europe, who tell people they meet in their travels that they "met in Oxford." Suppose, also, that they intentionally Anglicize their accents to strengthen the impression they make. In fact, neither of them has ever set foot in England and each of them has exactly one degree, from an undistinguished American state university. The presentation of themselves as Oxonians is motivated by a desire to impress and to hold special status among their new acquaintances.

Obviously, the couple is displaying considerable inauthenticity. The moral wrong in this case involves intentional deception of others. Of course, attempting to deceive others is pro tanto wrong, and the character trait of dishonesty is a moral vice. But inauthenticity sometimes involves *self*-deception. Suppose a college professor tells himself that he is a really effective teacher despite the absence of any significant evidence favoring this assessment and despite an abundance of counterevidence consisting of student evaluations and peer observation reports. He lies to himself, ignoring or discounting the evidence, because he needs to feel that he is a good teacher like both of his parents, whose approval he still seeks. Here inauthenticity takes the form of self-deception, which I understand as a species of failing to be true to oneself. (Whether one judges such self-deception morally wrong is likely to depend on whether one believes that the self-deceiving individual can be expected to catch herself in the distortion and put an end to it.)

Do all cases of inauthenticity involve deception of oneself or others? No, in my view, because some cases apparently involve failures of *autonomy* in the absence of dishonesty. But this complication need not delay our analysis of authenticity in relation to pharmaceutical enhancement. The cases of interest are cases, like that of Francis, in which an agent chooses the enhancement autonomously.

Perhaps we may say that self-enhancement projects that are both honest and sufficiently autonomous are ipso facto authentic. Assuming that this thesis is even close to the truth, it suggests that use of pharmaceuticals for enhancement purposes can be entirely authentic. There is no compelling reason to assume, in advance, that such cases necessarily involve dishonesty, a failure of autonomy, or the violation of any further requirement of authenticity (if I am mistaken that honesty and autonomy are sufficient conditions).

To return to our case, Francis' proposal to change his personality with pharmaceuticals cannot fairly be accused of inauthenticity. Again, there is no reason to doubt that he is acting autonomously in pursuing this change. He is neither deceiving others about how he really is nor caught up in a trap of self-deception. Suppose he carries out his plan successfully. Francis has become more extraverted, less pessimistic in interpreting others' words and deeds, and much more focused when he wants to remain concentrated on a task or activity at hand. Could one accuse him of deceiving others—or even himself—by claiming that he's *just posing* as an outgoing and focused person? Not at all, because he has changed, just as he intended. If one grants this point, but claims that Francis is still in some way being inauthentic, on the grounds that he is not respecting his "natural" personality and cognitive style, we may fairly respond that this claim is rooted in the dubious idea that one is morally bound to be true to oneself by avoiding deliberate self-transformation. I am aware of no good reason to accept this general proscription. Insofar as many people's efforts to transform themselves into better people strike nearly all of us as *admirable*, there is much reason to reject the idea in question.

Some cases of self-transformation, however, are not so easily reconciled with our judgments about authenticity. Suppose, for example, that Jonathan, who is gay, lives in an intolerant community in which he knows that professional and social success are vastly less attainable if one is known to be gay.[11] Rather than hiding his sexual orientation, Jonathan wants to *alter* the aspect of who he is that would need hiding. So he takes a pharmaceutical that reliably has the effect of eliminating homosexual attraction and desire. He has now transformed himself in a way that he considers—in his social context—to be a social enhancement (where what is "improved" is the capacity to avoid certain spontaneous feelings). The challenge here is that Jonathan's self-transformation seems highly inauthentic, yet it is unclear that my account—specifically, the suggestion that self-transformation projects that are honest and sufficiently autonomous are ipso facto authentic—can support this judgment.

One might defend the account as it stands on the grounds that Jonathan's decision seems semi-coerced by his community's homophobia, in which his self-transformation was (arguably) not sufficiently autonomous to meet the conditions for authenticity. But I suspect that this response would miss much of the point of the present challenge. Even if we tailored details of the case so that Jonathan's choice

[11] Thanks to Dien Ho for the example and the challenge to the conception of authenticity defended thus far.

would appear to meet conditions of autonomy, there is something about his deliberate self-transformation—as opposed to the social conditions that motivate it—that seems inauthentic. Yet he is not being dishonest. He knows himself, we may assume, well enough to avoid self-deception. Moreover, following the "enhancement" he is not lying to others: he really does lack homosexual attraction and desire. Yet, one wants to say, there is a sense in which he really *is* gay—and that his "enhancement" involves a failure to be true to himself.

My suggestion is that some acts of self-transformation are inauthentic not because they are dishonest and not because they are insufficiently autonomous but because they are *insufficiently self-respecting*.[12] I think this condition of authenticity is thwarted by Jonathan's self-transformation project. Note that the present qualification to the account of authenticity proposed earlier—the adding of this third condition—has two important implications. First, it sets limits on the extent to which people may transform themselves, whether using pharmaceuticals or other means, authentically. Thus, even if the image of a preexisting self that exists *entirely* independently of one's will is indefensible, the polar-opposite image of a self that is entirely up for voluntary transformation with perfect authenticity is also indefensible. Some acts of deliberate self-transformation fail to show sufficient self-respect and are for this reason inauthentic. Jonathan Glover helpfully discusses "self-creation" as conscious shaping of one's own characteristics "while respecting the constraints of natural shape and grain" (Glover 1988: 136). Although Glover was discussing the extent to which self-transformation is possible, I am appropriating this moderate image as characterizing the extent to which *authentic* self-transformation is possible. We may authentically transform ourselves using pharmaceuticals (and other enhancement technologies) within limits set by the importance of self-respect. Meanwhile, I deny that there is any good reason to think that Francis' self-transformation involves a failure to respect himself and, for this reason, inauthenticity.

So far in this section we have focused on the concept of authenticity and have vindicated Francis' self-transformation project in these terms. We can advance similar points in terms of narrative identity. Assuming again that Francis carries out his self-transformation project successfully, his use of an SSRI and stimulant has had a significant effect on his narrative identity, but there is no reason to think that this effect on his identity is morally problematic. After all, what does narrative identity concern? It concerns a person's self-conception, what she regards as most important to who she is, her self-narrated inner story. And who Francis is includes the sort of person he wanted to become just as much as it includes what he was like beforehand. His inner story is a *continuing* narrative, and, to the extent that circumstances allow him to make autonomous choices and act successfully with regard to who he is, he is the author of that narrative. In this way, it may be helpful to think of deliberate self-transformation as *the autonomous writing of one's self-narrative*. In writing this particular story, Francis expresses his values and priorities. And because he

[12] In earlier work, I gestured at the possibility of including such a self-respect condition without taking a definite stand on whether it was needed (2005a: 241–42).

does so without deception, a failure of autonomy, or insufficient self-respect, Francis' project of self-transformation via use of pharmaceuticals is authentic. Jonathan's self-transformation project, by contrast, involves inauthenticity through insufficient self-respect, which also damages the quality of his self-narrative.

Conclusion

Thus, considered as an across-the-board critique of enhancement via pharmaceutical use, the charge of inauthenticity proves as groundless as the concerns about identity considered earlier (in which numerical and narrative identity appear to be conflated). Francis is the owner of his own life, just as he is the author of his self-narrative. As medical enhancements become increasingly available in the years to come, we had better get used to the phenomenon of deliberate self-transformation by such means. Those means will often include pharmaceuticals. There are important ethical issues in relation to the availability and use of pharmaceuticals, including issues pertaining to distributive justice, the acceptance of risk, threats to autonomy in competitive settings, manipulative marketing, and myriad others. But self-transformation through the use of pharmaceuticals, as the case of Francis has illustrated, should not be judged inauthentic just because it involves significant effects on narrative identity, any more than it should be judged morally problematic in any way related to numerical identity. Again, *some* self-transformation projects involving the use of pharmaceuticals may prove inauthentic due to deficits pertaining to honesty, autonomy, or self-respect, but the possibility of such instances in no way casts doubt on the whole class of enhancement projects under consideration. Once we are reasonably clear on the relationships among personal identity, authenticity, and deliberate self-transformation, we can more profitably devote our energies to the important issues pertaining to justice, safety, manipulation, and the like—issues that already confront us.

References

Agar, N. 2004. *Liberal Eugenics: In Defence of Human Enhancement*. Oxford: Blackwell.

American Psychiatric Association. 2013. *Diagnostic and Statistical Manual of Mental Disorders*. 5th ed. Washington, DC: American Psychiatric Association.

Buchanan, A. 2011. *Beyond Humanity? The Ethics of Human Enhancement*. Oxford: Oxford University Press.

Buchanan, A., D. Brock, N. Daniels, and D. Wikler. 2000. *From Chance to Choice: Genetics and Justice*. Cambridge: Cambridge University Press.

DeGrazia, D. 2000. Prozac, Enhancement, and Self-Creation. *The Hastings Center Report* 30(2): 34–40.

———. 2005a. *Human Identity and Bioethics*. Cambridge: Cambridge University Press.

———. 2005b. Enhancement Technologies and Human Identity. *The Journal of Medicine and Philosophy* 30: 261–283.

———. 2012. *Creation Ethics: Reproduction, Genetics, and Quality of Life*. New York: Oxford University Press.
———. 2013. Moral Improvement, Freedom, and What We (Should) Value in Moral Behavior. *Journal of Medical Ethics* 39. http://jme.bmj.com/content/early/2013/01/24/medethics-2012-101157.full. Accessed 6 Feb 2015.
Elliott, C. 2003. *A philosophical Disease: American Medicine Meets the American Dream*. New York: Norton.
Elliott, C., and T. Chambers, ed. 2004. *Prozac as a Way of Life*. Chapel Hill, NC: University of North Carolina Press.
Glover, J. 1988. *I: The Philosophy and Psychology of Personal Identity*. London: Penguin.
Green, R. 2007. *Babies by Design: The Ethics of Genetic Choice*. New Haven: Yale University Press.
Habermas, J. 2003. *The Future of Human Nature*. Cambridge: Polity Press.
Juengst, E. 1998. What Does *Enhancement* Mean? In *Enhancing Human Traits: Ethical and Social Implications*, ed. E. Parens, 29–47. Washington, DC: Georgetown University Press.
Kamm, F. 2009. What Is and Is Not Wrong with Enhancement? In *Human Enhancement*, ed. J. Savulescu et al., 91–130. Oxford: Oxford University Press.
Luper, S., ed. 2014. *The Cambridge Companion to Life and Death*. Cambridge: Cambridge University Press.
McMahan, J. 2002. *The Ethics of Killing*. Oxford: Oxford University Press.
Olson, E. 1997. *The Human Animal: Personal Identity Without Psychology*. Oxford: Oxford University Press.
Parens, E., ed. 1998. *Enhancing Human Traits: Ethical and Social Implications*. Washington, DC: Georgetown University Press.
Persson, I., and J. Savulescu. 2012. *Unfit for the Future: The Need for Moral Enhancement*. Oxford: Oxford University Press.
President's Council on Bioethics. 2003. *Beyond Therapy: Biotechnology and the Pursuit of Happiness President's Council on Bioethics*. Washington, DC: President's Council on Bioethics.
Sandel, M. 2007. *The Case Against Perfection: Ethics in the Age of Genetic Engineering*. Cambridge, MA: Harvard University Press.
Savulescu, J., and N. Bostrom, ed. 2009. *Human Enhancement*. Oxford: Oxford University Press.
Schechtman, M. 1996. *The Constitution of Selves*. Ithica: Cornell University Press.
Schoene-Seifert, B., and D. Talbot, ed. 2009. *Enhancement: Die Ethische Debatte*. Paderborn: Mentis Verlag.

The Wisdom of Nature: An Evolutionary Heuristic for Human Enhancement

Nick Bostrom and Anders Sandberg

Introduction

The Wisdom of Nature and the Special Problem of Enhancement

We marvel at the complexity of the human organism how its various parts have evolved to solve intricate problems: the eye to collect and preprocess visual information, the immune system to fight infection and cancer, and the lungs to oxygenate the blood. The human brain—the focus of many of the most alluring proposed enhancements—is arguably the most complex thing in the known universe. Given how rudimentary is our understanding of the human organism, particularly the brain, how could we have any realistic hope of *enhancing* such a system?

To enhance even a system like a car or a motorcycle—whose complexity is trivial in comparison to that of the human organism—requires a fair bit of understanding of how the thing works. Isn't the challenge we face in trying to enhance human beings so difficult as to be hopelessly beyond our reach, at least until the biological sciences and the general level of human abilities have advanced vastly beyond their present state?

It is easier to see how *therapeutic* medicine should be feasible. Intuitively, the explanation would go as follows: even a very excellently designed system will occasionally break. We might then be able to figure out what has broken and how to fix it. This seems much less daunting than to take a very excellently designed, unbroken system, and enhance it beyond its normal functioning.

N. Bostrom • A. Sandberg (✉)
Future of Humanity Institute, University of Oxford, Oxford, UK
e-mail: nick.bostrom@philosophy.ox.ac.uk; anders.sandberg@philosophy.ox.ac.uk

© Springer Science+Business Media B.V. 2017

D. Ho (ed.), *Philosophical Issues in Pharmaceutics*, Philosophy and Medicine 122,
DOI 10.1007/978-94-024-0979-6_12

Yet we know that even therapeutic medicine is very difficult. It has been claimed that until circa 1900, medicine did more harm than good (McKeown and Lowe 1974). And various recent studies suggest that even much of contemporary medicine is ineffectual or outright harmful (Newhouse and Group TIE 1993; Frech and Miller 1996; Kirsch et al. 2002). Iatrogenic deaths account for 2–4% of all deaths in the USA (the third leading cause of death according to one accounting (Starfield 2000)) and may correspond to a loss of life expectancy by 6–12 months (Bunker 2001). We are all familiar with nutritional advice, drugs, and therapies that were promoted by health authorities but later found to be damaging to health. In many cases, the initial recommendations were informed by large clinical trials. When even therapeutic medicine, based on fairly good data from large clinical trials, is so hard to get right, it seems that a prudent person has much reason to be wary of purported *enhancements*, especially as the case for such enhancements is often based on much weaker data. Evolution is a process powerful enough to have led to the development of systems—such as human brains—that are far more complex and capable than anything that human scientists or engineers have managed to design.

Surely it would be foolish, absent strong supporting evidence, to suppose that we are currently likely to be able to do *better* than evolution, especially when so far we have not even managed to understand the systems that evolution has designed and when our attempts even just to repair what evolution has built so often misfire!

We believe that these informal considerations contain a grain of truth. Nonetheless, in many particular cases we believe it is practically feasible to improve human nature. The evolution heuristic is our explanation of why this is so. If the evolution heuristic works as we suggest, it shows that there is some validity to the widespread intuition that nature often knows best, especially in relation to proposals for human enhancement. But the heuristic also demonstrates that the validity of this intuition is limited, by revealing important exceptional cases in which we can hope to improve on nature using even our present or near-future science and technology.

The evolution heuristic might be useful for scientists working to develop enhancement technologies. It might also be useful in evaluating beliefs and arguments about the ethics of human enhancement. This is because intuitions about the wisdom of nature appear to play an important role in the cognitive ecology of many anti-enhancement advocates. While sophisticated bioconservatives (aware of the distinction between "is" and "ought") may not *explicitly* base their arguments on the alleged wisdom in nature, we believe that such intuitions influence their evaluation of the plausibility of various empirical assumptions and mid-level moral principles that are invoked in the enhancement discourse, just as the opinions and practical judgments of the pro-enhancement transhumanists look more plausible if one assumes that nature is generally unwise. Addressing such hidden empirical backgrounds, assumptions may therefore help illuminate important questions in applied ethics.[1]

---

[1] On the role of mid-level principles in one area of applied ethics, see Beauchamp and Childress 1979.

The Evolution Heuristic

The basic idea is simple. In order to decide whether we want to modify some aspect of a system, it is helpful to consider why the system has that aspect in the first place. Similarly, if we propose to introduce some new feature, we might ask why the system does not already possess it.

The system of concern here is the human organism. The question why the human organism has a certain property can be answered on at least two different levels, ontogeny and phylogeny. Here the focus is on the phylogeny of the human organism.

We can conceive of a proposed enhancement as an ordered pair (α, A), where α is some specific intervention (e.g., the administration of a drug) and A is the trait that the intervention is intended to realize (e.g., improved memory consolidation). We define an enhancement as an intervention that causes either an improvement in the functioning of some subsystem (e.g., long-term memory) beyond its normal healthy state in some individual or the addition of a new capacity (e.g., magnetic sense).

On this definition, an enhancement is not necessarily desirable, either for the enhanced individual or for society. For instance, we might have no reason to value an enhancement of our sweat glands that increases their ability to produce perspiration in response to heat stimuli. In other instances, we might benefit from increased functionality or a new capacity and yet not benefit from the enhancement because the intervention also causes unacceptable side effects.[2] The evolution heuristic is a tool to help us think through whether some proposed enhancement is likely to yield a net benefit.

The starting point of the heuristic is to pose the *evolutionary optimality challenge*:

> (EOC) If the proposed intervention would result in an enhancement, why have we not already evolved to be that way?

Suppose that we liken evolution to a surpassingly great engineer. (The *limitations* of this metaphor are part of what makes it useful for our purposes.) Using this metaphor, the EOC can be expressed as the question, "How could we realistically hope to improve on evolution's work?" We propose that there are three main categories of possible answers, which can be summarized as follows:

- *Changed tradeoffs.* Evolution "designed" the system for operation in one type of environment, but now we wish to deploy it in a very different type of environment. It is not surprising, then, that we might be able to modify the system better to meet the demands imposed on it by the new environment. Making such modifications need not require engineering skills on a par with those of evolution:

[2] Which side effects are acceptable depends, of course, on the benefits resulting from the enhancement, and these may vary between subjects depending on their goals, life plans, and circumstances.

consider that it is much harder to design and build a car from scratch than it is to fit an existing car with a new set of wheels or make some other tweaks to improve functioning in some particular setting, such as icy roads. Similarly, the human organism, while initially "designed" for operation as a hunter-gatherer on the African savannah, must now function in the modern world. We may well be capable of making some enhancing tweaks and adjustments to the new environment even though our engineering talent does not remotely approach that of evolution.

- *Value discordance.* There is a discrepancy between the standards by which evolution measured the quality of her work and the standards that we wish to apply. Even if evolution had managed to build the finest reproduction-and-survival machine imaginable, we may still have reason to change it because what we value is not primarily to be maximally effective inclusive-fitness optimizers. This discordance in objectives is an important source of answers to the EOC. It is not surprising that we can modify a system better to meet our goals, if these goals differ substantially from the ones that (metaphorically might be seen as having) guided evolution in designing the system the way she did. Again, this explanation does not presuppose that our engineering talent exceeds evolution's. Compare the case to that of a mediocre technician, who would never be able to design a car, let alone a good one, but who may well be capable of converting the latest BMW model into a crude rain-collecting device, thereby *enhancing* the system's functionality as a water-collecting device.
- *Evolutionary restrictions.* We have access to various tools, materials, and techniques that were unavailable to evolution. Even if our engineering talent is far inferior to evolution's, we may nevertheless be able to achieve certain things that stumped evolution, thanks to these novel aids. We should be cautious in invoking this explanation, for evolution often managed to achieve with primitive means what we are unable to do with state-of-the-art technology. But in some cases, one can show that it is practically impossible to create a certain feature without some particular tool—no matter how ingenious the engineer—while the same feature can be achieved by any dimwit given access to the right tool. In these special cases, we might be able to overcome evolutionary restrictions.

In the following three sections, we will explore each of these categories of possible answers to the EOC in more detail and show how they can help us decide whether or not to go ahead with various potential human enhancements.

Our ideas about enhancement in many ways parallel earlier work in evolutionary medicine. Evolutionary medicine is based on using evolutionary considerations to understand aspects of human health (Willams and Nesse 1991; Trevathan et al. 1999). Hosts and parasites have adapted to one another, and analysis of the tradeoffs involved can reveal adaptations that contributed to fitness in the past but are maladaptive today or symptoms that have been misdiagnosed as harmful but may actually aid recovery. Evolutionary medicine also helps explain the incidence of genetic diseases, which can be maintained in the population because of beneficial effects in historically normal environments. Another contribution of evolutionary medicine has been to draw attention to the fact that our modern environment may not always fit a biology designed for Pleistocene conditions and how this mismatch can cause disease. These insights are recycled in our analysis of human enhancement.

Another strand of research relevant to our aims is evolutionary optimization theory, which seeks to determine the abilities and limitations of evolution in terms of producing efficient biological functions (Parker and Smith 1990). While, naively, evolution might be thought to maximize individual fitness (the expected lifetime number of surviving offspring), there are many contexts in which this simplification leads to error. Sometimes it is necessary to focus on the concept of inclusive fitness, which takes into account the effects of a genotype on the fitness of blood relatives other than direct decedents. Sometimes a gene-centric perspective is needed, to account for phenomena such as segregation distortion and junk DNA (Dawkins 1976; Williams 1996/1966). There are also many other ways in which evolution routinely falls short of "optimality," some of which will be covered in later sections.

Changed Tradeoffs

General Remarks on Tradeoffs

Evolutionary adaptation often involves striking a tradeoff between competing design criteria. Evolution has fine-tuned us for life in the ancestral environment, which, for the most part, was a life as a member of a hunter-gatherer tribe roaming the African savannah. Life in contemporary society differs in many ways from life in the environment of evolutionary adaptedness. Modern conditions are too recent for our species to have fully adapted to them, which means that the tradeoffs evolution struck may no longer be optimal today.

In evolutionary biology, the "environment of evolutionary adaptedness" (EEA) refers not to a particular time or place, but to the environment in which a species evolved and to which it is adapted (Hagen 2002). It includes both inanimate and animate aspects of the environment, such as climate, vegetation, prey, predators, pathogens, and the social environment of conspecifics. We can also think of the EEA as the set of all evolutionary pressures faced by the ancestors of the species over recent evolutionary time—in the case of humans, at least 200,000 years (Hagen 2002). Hunting, gathering of fruits and nuts, courtship, parasites, and hand-to-hand combat with wild animals and enemy tribes were elements of the EEA; speeding cars, high levels of trans fats, concrete ghettos, and tax return forms were not.

The import of this for the evolution heuristic is that even if the human organism were a wonderfully well-designed system for life in the EEA, it may not in all respects be well designed for life in contemporary society. If we can identify specific changes to our environment that have shifted the optimal tradeoff point between competing design desiderata in a certain direction, we may be able to find relatively easy interventions that could "retune" the tradeoff to a point that is closer to its present optimum. Such retuning interventions might be among the low-hanging fruits on the enhancement tree, fruits within reach even in the absence of super-advanced medical technology.

Proposed enhancements aiming to retune altered tradeoffs can often meet the EOC. The new trait that the enhancement gives us might have been maladaptive in the EEA even though it would be adaptive now. Alternatively, the new trait might be intrinsically associated with another trait that was maladaptive in the EEA but has become less disadvantageous (or even beneficial) in the modern environment, so that the terms of the tradeoff have shifted. In either case, the enhancement could be adaptive in the current environment without having been so in the EEA, which would explain why we do not have that trait, allowing us to meet the EOC.

We can roughly distinguish two ways in which tradeoffs can change: new *resources* may have become available that were absent, or available only at great cost, in the EEA, or the *demands* placed on one of the subsystems of the human organism may have changed since we left the EEA. Let us consider these two possibilities in turn and look at some examples.

Resources

One of the main differences between human life today (for most people in developed countries) and life in the EEA is the abundant availability of food independently of place and season. In the state of nature, food is relatively scarce much of the time, making energy conservation paramount and forcing difficult energy expenditure tradeoffs between metabolically costly tissues, processes, and behaviors. As we shall see, increased access to nutrients suggests several promising enhancement opportunities. We have also gained access to important new non-dietary resources, including improved protection against physical threats, obstetric assistance, better temperature control, and increased availability of information. Let us examine how these new resources are relevant to potential enhancements of the brain and the immune system.

The Brain

The human brain constitutes only 2% of body mass yet accounts for about 20% of total energy expenditure. Combined, the brain, heart, gastrointestinal tract, kidneys, and liver consume 70% of basal metabolism. This forces tradeoffs between the size and capacity of these organs and between allocation of time and energy to activities other than searching for food in greater quantity or quality (Aiello et al. 2001; Fish and Lockwood 2003).

Unsurprisingly, we find that, in evolutionary lineages where nutritional demands are high and cognitive demands low (such as bats hunting in uncluttered environments), relative brain size is correspondingly smaller (Niven 2005). In humans, brain size correlates positively with cognitive capacity (. ≈ 0.33) (McDaniel 2005).

Holding brain mass constant, a greater level of mental activity might also enable us to apply our brains more effectively to process information and solve problems. The brain, however, requires extra energy when we exert mental effort, reducing the normally tightly regulated blood glucose level by about 5% (0.2 mmol/l) for short (<15 min) efforts and more for longer exertions (Scholey et al. 2001; Fairclough and Houston 2004). Conversely, increasing blood glucose levels has been shown to improve cognitive performance in demanding tasks[3] (Korol and Gold 1998; Manning et al. 1998; Martin and Benton 1999; Winder and Borrill 1998).

The metabolic problem is exacerbated during prenatal and early childhood growth where brain development requires extra energy. Brain metabolism accounts for a staggering 60% of total metabolism in newborns (Holliday 1986), exacerbating the competitive situation between the mother and child for nutritional resources—an unpleasant tradeoff (Martin 1996). Children with greater birth weight have a cognitive advantage (Matte 2001).

Another constraint on prenatal cerebral development is the size of the human birth canal (itself constrained by bipedalism), which historically placed severe constraints on the head size of newborns (Trevathan 1987). These constraints are partly obviated by modern obstetrics and the availability of caesarian section. One way of reducing head size at birth and perinatal energy demands would be to extend the period of postnatal maturation. However, delayed maturation was vastly riskier in the EEA than it is now.

What all this suggests is that cognitive enhancements might be possible if we can find interventions that recalibrate these legacy tradeoffs in ways that are more optimal in the contemporary world. For example, suppose we could discover interventions that moderately increased brain growth during gestation, or slightly prolonged the period of brain growth during infancy, or that triggered an increase in available mental energy. Applying the EOC to these hypothetical interventions, we get a green light. We can see why these enhancements would have been maladaptive in the EEA and why they may nevertheless have become entirely beneficial now that the underlying tradeoffs have changed as a result of the availability of new resources. If the "downside" of getting more mental energy is that we would burn more calories, many of us would pounce at the opportunity.

Not all cognitive enhancement interventions get an immediate green light from the above argument. Stimulants like caffeine and modafinil enable increased wakefulness and control over sleep patterns (Caldwell 2001). But sleep serves various (poorly understood) functions other than to conserve energy (Siegel 2005). If the explanation for why we do not sleep less than we do has to do with these other functions, then reducing sleep might well have more problematic side effects than increasing the amount of calories we need to consume. For any particular intervention,

---

[3] Increasing oxygen levels (another requirement for metabolism) also improves cognition.

such as the administration of some drug, we also of course need to consider the possibility of contingent side effects, i.e., that the drug might have effects on the body other than simply retuning the target tradeoff.

The Immune System

While the immune system serves an essential function by protecting us from infection and cancer, it also consumes significant amounts of energy (McDade 2003). Experiments have found direct energetic costs of immune activation (Demas et al. 1997; Moret and Schmid-Hempel 2000; Ots et al. 2001). In birds immune activation corresponded to a 29% rise of resting metabolic rate (Martin et al. 2003), and in humans the rate increases by 13% per degree centigrade of fever (Elia 1992). In addition, the protein synthesis demands of the immune system are sizeable yet prioritized, as evidenced by a 70% increase in protein turnover in children during infection despite a condition of malnourishment (Waterlow 1984; McDade 2003). One would expect the immune system to have evolved a level of activity that strikes a tradeoff between these and other requirements—a level optimized for life in the EEA but perhaps no longer ideal.

Such a tradeoff has been proposed as part of an explanation of the placebo effect (Humphrey 2002). The placebo effect is puzzling because it apparently involves getting something (accelerated recovery from disease or injury) for nothing (merely having a belief). If the subjective experience of being treated causes a health-promoting response, why are we not always responding that way? Studies have shown that it is possible chemically to modulate the placebo response down (Sauro and Greenberg 2005) or up (Colloca and Benedetti 2005).

One possible explanation is that mobilizing the placebo effect consumes resources, perhaps through activation of the immune system or other forms of physiological health investment. Also, to the extent that the placebo response reduces defensive reactions (such as pain, stiffness, and inflammation), it might increase our vulnerability to future injury and microbial assaults. If so, one might expect that natural selection would have made us such that the placebo response would be triggered by signals indicating that in the near future we will (a) recover from our current injury or disease (in which case there is no need to conserve resources to fight a drawn-out infection and less need to maintain defensive reactions), (b) have good access to nutrients (in which case, again, there is no need to conserve resources), and (c) be protected from external threats (in which case there is less need to keep resources in reserve for immediate action readiness). Consistent with this model, the evidence does indeed show that the healing system is activated not only by the expectation that we will get well soon but also by the impression that external circumstances are generally favorable. For example, social status (Sapolsky 2005), success, having somebody looking after us (House et al. 1988), sunshine, and regular meals might all indicate that we are in circumstances where it is optimal for the body to invest in healing and long-term health, and they do seem to prompt the body to do just that. By contrast, conflict (Kiecolt et al. 1997), stress, anxiety, uncertainty

(McDade 2002), rejection, isolation, and despair appear to shift resources toward immediate readiness to face crises and away from building long-term health.

If this model of the placebo response is correct, several potential avenues of enhancement are worth exploring. One is that since physical safety and reliable access to food are much improved compared to the EEA, it might now be beneficial to invest more in biological processes that build long-term health than was usually optimal in the EEA. We might thus inquire whether the placebo effect and other evolved responses are flexible enough to have adjusted the level of health investment to a level that is optimal under modern conditions. If not, we could benefit from an intervention that triggers a placebo-like response or otherwise increases the body's health investment.

However, while external stresses and resource constraints are reduced in the modern environment, the danger of autoimmune reactions remains. We would therefore have to be careful not to overshoot the target. It is possible that we would benefit from a *lower* baseline immune activity in some parts of the immune system since we are now less at risk of dying from infectious diseases. As an example, the hygiene theory of allergic diseases claims that the reduction in immunological challenge in particular from helminth parasites during the early life increases the risk of allergic disease later in life (Yazdanbakhsh et al. 2002; Maizels 2005). If true, then a downregulation of a particular dendritic cell subpopulation (DC2) sensitive to helminths, but causing allergic reactions, might be desirable. Alternatively, an upregulation of regulatory (DCreg) cells that tend to be lost in unstimulated immune systems might be used to control the DC2 cells.

The evolution heuristic also leads us to consider other potential immune system enhancements. Even if the average activation level of our immune systems were still optimal in the modern era, we now possess more information (a new resource) about the detailed requirements in specific situations. We can use this information to override our bodies' natural response tendencies. For example, recipients of donated organs can benefit from immunosuppressant drugs. Conversely, a patient with early-stage cancer might be better off if her immune system could be induced to mount an immediate all-out assault on the incipient tumor instead of conserving resources for hypothetical future challenges (Boon and van Baren 2003; Dunn et al. 2004).

A more radical enhancement would be to improve DNA repair, which would reduce cancer-causing mutations and improve radiation resistance, at the price of increasing metabolic needs. The modification could be achieved through overexpression of existing DNA repair genes (Wood et al. 2001) or perhaps even by transgenic incorporation of the unique abilities of *Deinococcus radiodurans* (Battista et al. 1999; Venkateswaran et al. 2000). Increased repair would have to be balanced with apoptosis and replacement of irreparably damaged cells (another energy cost). Until recently, increased DNA repair activity might have been too metabolically costly and mutation- prone for evolution to consider it a worthwhile bargain. One of the most well-studied pathways, the PARP-1 pathway, protects the genome from damage but requires so much energy that it can damage cells through energy depletion (de Murcia and Shall 2000; Skaper 2003).

Since the objective of the interventions suggested above is to restore health, one could argue that they should be regarded as therapeutic rather than enhancing. But these classifications are not necessarily incompatible. We could regard the interventions as therapeutic for the subsystems whose functioning has been deteriorated by disease, yet enhancing for the immune system, whose functioning is improved beyond its normal state.[4]

Demands

Just as we have many resources that were denied our hunter-gatherer ancestors, we also face a different set of demands than they did. This suggests further opportunities for enhancement.

Changes in demands on the human organism occur when old demands disappear or are reduced (e.g., less need for long treks to get food, hygienic surroundings reducing demands on the immune system) and when demands grow in strength or new demands arise (e.g., greater need to be able to concentrate on abstract material for long periods, new pathogens spreading in larger societies). The source and nature of a particular demand may also change. For instance, exercise is no longer necessary to gain sustenance, but is instead needed to maintain the body in good shape.

Many "diseases of civilization" are due to these changed demands. For example, our ancestors needed to exert themselves physically to secure adequate nutrition, whereas our easy access to abundant food can lead to obesity. People working indoors do not get the sun exposure that our ancestors had, leading to vitamin D deficiency (Thomas et al. 1998); yet we risk skin cancer when we expose pale skin to the sun during occasional recreational activities. Rapid blood coagulation was beneficial in the past, when there was a high risk of wounding. The increased risk for cardiovascular problems and embolisms was an acceptable tradeoff. Today, the risk of wounding has sharply decreased, making the downsides relatively more important. Reducing coagulation, e.g., by taking low-dose aspirin, can be beneficial given these changed demands (Force 2002), although we risk incidental side effects such as stomach irritation.

While the change in demands can cause or exacerbate problems, it can also alleviate them. The recent emergence of the IT industry appears to have produced a refuge for people with Asperger's syndrome where their preference for structure and detail becomes a virtue and their problems with face-to-face communication less of a disadvantage (Silberman 2001). Deliberate fitting of environments to human evolutionary adaptations and individual idiosyncrasies is a promising adjunct to direct human enhancement for improving human performance and well-being.

---

[4] In like manner, we can view vaccinations both as therapeutic (or, more accurately, prophylactic) and as enhancing.

Literacy and Numeracy

Intellectual capacity, or at least some specific forms of it, seems to have become more rewarded in contemporary society than they were in the EEA. There is a positive correlation in Western society between IQ and income (Neisser et al. 1996; Gottfredson 1997). Higher levels of general cognitive ability are important not just for highly demanding, high status jobs, but also for success in everyday life, such as being able to fill out forms, understand news, and maintain health. As society becomes more complex, these demands increase, placing people of low cognitive ability at a greater disadvantage (Gottfredson 1997, 2004). While general cognitive ability may have been advantageous (and selected for) in our evolutionary past (Gottfredson 2007),[5] numeracy and literacy represent more specific abilities whose utility has increased dramatically in recent times.

Before the invention of writing, the human brain faced no pressure to be literate. In the current age, however, literacy is in very high demand. Failing to meet this demand places an individual at a severe disadvantage in modern society. Since writing is a relatively recent invention (3500 BC) and since it is even more recently that written language has become such a dominant mode of communication, it is plausible that the human brain is not optimized for modern conditions. The fact that the neural machinery needed for writing and reading largely overlaps with that needed to produce and interpret oral communication means that the mismatch between evolved capacity and present demands is not as great as it might have been. Nevertheless, as the phenomenon of dyslexia demonstrates, it is possible to have deficits in language processing that are relatively specific to written language, possibly arising from minor variations in phonological processing (Goulandris et al. 2000). Dyslexia also appears to be linked to enhanced or atypical visuospatial abilities (von Karolyi et al. 2003; Brunswick et al. 2007). These abilities might have been useful in the EEA, but today literacy is usually more important for achieving life goals. If our species had been using written language for a couple of million years and reproductive fitness had depended on literacy, dyslexia might have been much rarer than it is.

Modern society also places much greater demands on advanced numerical skills than we faced in the EEA. In hunter-gatherer societies, numeracy demands appear to have been limited to being able to count to five or ten (Pica et al. 2004). In the modern world, one is at a major disadvantage if one cannot understand at least basic arithmetic. Many occupations require a grasp of statistics, calculus, geometry, or higher mathematics. Programming skills open up additional employment possibilities. Good logical and analytical skills create further opportunities in our information-dense, technology- mediated, and generally formalized modern society. These skills were much less useful in the Pleistocene.

---

[5] It should be noted that IQ correlates negatively with fertility in many modern societies (Udry 1978; Vancourt and Bean 1985; Vining et al. 1988). This might be an example of value discordance between human values and evolutionary fitness.

The altered nature of the demands we face suggests opportunities for enhancement by readjusting tradeoffs that are no longer optimal. For example, number relations appear to be handled by brain circuits closely linked to spatial cognition of external objects and affected by spatial attention abilities (McCord 2000; Hubbard et al. 2005). Hence, enhancement of this type of spatial attention (Green and Bavelier 2006), possibly at the expense of remote or peripheral attention, could be a useful enhancement. Similarly, enhancements in reading ability at the expense of the dyslexia-related visuospatial abilities might gain support from the EOC.

Concentration

The importance of being able to concentrate on abstract thinking and tasks with little sensory feedback has increased significantly in modern times relative to the importance of peripheral awareness. In the EEA, peripheral awareness was crucial for detecting predators and enemies, while an ability to exclude other stimuli had few applications. We may hence have evolved attention systems with a tendency to be too easily distracted in a modern setting. It has been suggested that ADHD is a form of "response-readiness" that was more adaptive in past environments (Jensen et al. 1997). Concentration enhancers may therefore be feasible and promising in modern settings, enabling users to meet high demands for sustained attention. Drugs such as methylphenidate (Ritalin) are already used to treat ADHD and occasionally also for enhancement purposes (Farah et al. 2004).

Dietary Preferences and Fat Storage

One tradeoff involving food availability relates to the question of how much nutrition the body should store in fatty deposits. If high-calorie foods are scarce and food availability highly variable, it is optimal for an individual to crave high-calorie foods and to store lots of energy in fat deposits as insurance against lean times. We still need an appetite today, and we still need fat deposits, but—at least in the developed world—they are much less important now than in the past. Many people's natural set points of appetite and body fat are higher than optimal, leading to increased morbidity. In wealthy modern societies, where a Mars bar is never far away, the risks of obesity and diabetes outweigh the risk of undernutrition (Fontaine et al. 2003), and a sweet tooth is maladaptive.

This suggests that it might be possible to enhance human health by finding effective ways to downregulate our cravings for fat and sugar or by reducing the absorption and storage of these calories in fatty tissues. Such an enhancement might take various forms: nutritional advice, diet pills, artificial sweeteners, indigestible substances that taste like fat, weight-loss clubs, hypnotherapy, and, in the future, gene therapy. The evolution heuristic suggests that our natural proclivities to consume and store nutrients might be a case where we could benefit from going against the wisdom of nature. Independent considerations and possibly further research would be needed to determine the most effective way of doing this, given that weight loss itself is a longevity risk factor (Gaesser 1999) and that those who are mildly overweight have lower mortality than those who are underweight or obese (Flegal et al. 2005). Possibly an aversion to unhealthy foods and eating habits would be more effective and safer than a general downregulation of appetite. The heuristic tells us only that there are no general "wisdom of nature" reasons to retain our current bodyweight set points; it does not by itself tell us which approaches to changing them would be safest.

The Interplay Between Resources and Demands

The picture is complicated by the fact that some phenomena zigzag across the two subcategories of changed tradeoffs (resources and demands). Transport vehicles and machinery are new resources that reduce the demand for physical exertion. The effect is that most of us get less exercise in the course of our daily routines. Yet our bodies appear to be designed for physical activity, so a sedentary life causes a variety of health problems. New resources (gyms, exercise equipment, parks, and jogging clubs) have been developed to help us overcome the problems of a sedentary lifestyle. But now a new demand arises: we need the energy and self-motivation to make use of these resources—a demand that many find it difficult to meet.

In a case like this, there are multiple potential intervention points where a change could result in an improvement of our lives. One approach would be to design our environment in such a way as to force us to be more physically active. Elevators could be removed, motor vehicles banned from certain areas, and so forth. Another approach would be to attempt to redesign our bodies so that they would not be dependent on frequent physical exertion to remain healthy. On this approach, we might try to develop pharmaceuticals that trigger effects in the body similar to those normally caused by exercise (such as the IGF-1/MGF signaling pathways, which are stimulated by exercise or muscle damage) (Baldwin and Haddad 2002; Goldspink 2005). Yet another approach would be to attempt interventions that increase our energy and self-motivation, thereby making it easier for us to exercise on our own initiative. For instance, there might be pharmaceuticals that would give us more energy or strengthen our willpower, or perhaps a habit of regular workouts instilled in childhood would carry over into adult life.

Whether any of these interventions will work, and, if so, which one would be the most effective and have the best balance of benefits over burdens, cannot be determined a priori. This is an empirical question, whose answer may depend on changing social circumstances, levels of technology, personal preferences, and other factors. One should note that it is not only biological interventions which can have undesirable side effects. Removing elevators might cause some health benefits for people forced to climb the stairs, but it may also deny access for people with mobility impairments and cause unnecessary inconvenience to others. Encouraging high levels of physical activity in children might have overall health benefits, but it might also lead to more injuries, more worn-out knees and hip joints later in life, and less time for nonphysical activities.

Another illustration of the complex interplay between new resources and new demands is offered by the case of addictive drugs. Alcohol, heroin, and crack cocaine are comparatively novel resources. The availability of these resources creates a new demand on the human organism: the ability to avoid becoming addicted to harmful drugs that hijack the brain's reward system. Individuals vary in how they metabolize these drugs and how their brains react to exposure. Again, the solution might be to develop new resources (e.g., detox clinics), temporary pharmacological interventions (methadone), permanent biological modifications (vaccines), educational initiatives (drug awareness programs), or social policies (criminalization). Alternatively, one might attempt to develop safer, nonaddictive substitutes for harmful drugs (Nutt 2006). There are many possible ways to defy or to work around the wisdom of nature.

Value Discordance

General Remarks on Value Discordance

We have discussed opportunities for enhancement arising from the changed tradeoffs we face in the modern world compared to those of the EEA. (A great engineer built a system for use in a certain environment; we adapt it for use in a different environment.) In this section, we discuss another source of enhancement opportunities: the discordance between evolutionary fitness and human values. (A great engineer built a system that efficiently serves one purpose; we tinker with it to make it serve a different purpose.)

While our goals are not identical to those of evolution, there is considerable overlap. We value health, and health increases inclusive fitness. We value good eyesight, and good eyesight is useful for survival. We value musicality and artistic creativity, and these talents helped to attract mates in the EEA. If we are hoping to enhance some attribute for which the concordance in objectives is perfect, the present category will not give any help in meeting the EOC. We then either have to find an answer from one of the other categories or else suspect that what appears to be an easy enhancement will in fact come at a large hidden cost.

While some of our traits are both valuable to us and conducive to fitness, many attributes that we value would either not have promoted inclusive fitness in our natural environment or else would not have been fitness promoting to a sufficient extent to result in a profile of traits that is optimal from the perspective of our own values. There is a plethora of capacities or characteristics to which we assign a value that exceeds the contribution these characteristics made to survival and reproduction.

One obvious example is contraceptive technology. Vasectomy, birth control pills, and other contraceptive methods enhance our control over our reproductive systems, severing the link between sex and reproduction. We may value such enhancements

Table 1 Some traits that may promote individual well-being

Emotional well-being
Freedom from severe or chronic pain
Friendship and love
Long-term memory
Mathematical ability
Awareness and consciousness
Musicality
Artistic appreciation and creativity
Literary appreciation
Confidence and self-esteem
Healthy pleasures
Mental energy
Ability to concentrate
Abstract thinking
Longevity
Social skills

because they make family planning easier and increase choice. But evolution would frown on these practices. The great engineer would not regard the absence of an easy reproductive off-switch as a defect. When our goals differ from hers, it is unsurprising that we are able to modify her design in ways that make it better (by our lights) even if our design skills fall far short of hers.[6]

We can distinguish (at least) two distinct sources of such value discordance. The first is that the characteristics that would maximize an individual's inclusive fitness are not always identical to the characteristics that would be best for her. The other is that the characteristics that would maximize an individual's inclusive fitness are not always identical to those that would be best for society, or impersonally best. If our goal is to identify potential interventions that individuals would have prudential reasons for wanting, then we may perhaps set aside the second source of value discordance. If, however, we are interested in addressing ethical and public policy matters, then it is relevant to consider value discordance arising from either of these two sources. Let us consider each in turn.

Good for the Individual

What characteristics promote individual well-being? There is a vast ethical and empirical literature on this question, which we shall not attempt to review here. For our purposes, it will suffice to list (Table 1) some candidate characteristics, ones

[6] Evolution might still have the last laugh if in the long run she redesigns our species to directly desire to have as many children as possible or to have an aversion against contraceptives. Cultural "evolution" might beat biological evolution to the punch.

which may with some plausibility be taken to be among those that contribute to individual well-being in a wide range of circumstances. This list is for illustration only. Other lists could be substituted without affecting the structure of our argument.[7]

To illustrate the idea, take mathematical ability. Suppose that we believe that having greater mathematical ability would tend to make our lives go better—perhaps because it would give us competitive advantages in the job market, perhaps because appreciating mathematical beauty is a value in itself, or perhaps because we believe that mathematical ability is linked to other abilities that would increase our well-being. We then pose the EOC: why has evolution not already endowed us with more mathematical ability than we have?

It is possible that answers to this EOC may be found in the other categories we discuss in this chapter (changed tradeoffs or evolutionary restrictions). Yet suppose that is not so. We may then appeal to an answer in the value discordance category. Even if greater mathematical capacity would have been maladaptive in the EEA and even if it would still be maladaptive today, it may nevertheless be good for us, because the good for humans is different from what maximizes our fitness.

But we are not yet done. What the evolution heuristic teaches us in this case is that we must expect that the intervention will have some effect that reduces fitness. If we cannot form any plausible idea of what sort of effect the intervention might produce that would reduce fitness, then we must suspect that the intervention will have important effects that we have not understood. That should give us pause. A fitness-reducing effect that we have not anticipated might be something very bad, such as a serious medical side effect. The EOC hoists a warning flag. If, however, we can give a plausible account of why the proposed intervention to increase mathematical ability would reduce fitness, *and yet we judge this fitness-reducing effect as desirable or at least worth enduring for the sake of the benefit*, then we have met the EOC.

This does not guarantee that the enhancement will succeed. It is still possible that the intervention will fail to produce the desired result or that it would have some unforeseen side effect. There might be more than one sufficient reason why evolution did not already make this intervention to enhance mathematical ability. But once we have identified at least one sufficient reason, the warning flag raised by the EOC comes down. We have shown that one potential reason for thinking that the enhancement will fail (the "wisdom of nature" reason) does not apply to the present case.

As an example, evolution has not optimized us for happiness and has instead led to a number of adaptations that cause psychological distress and frustration (Buss 2000). The "hedonic treadmill" causes us quickly to adapt to positive experiences and to seek more, as goods we have gained become taken for granted as a new

[7] The items in the list need not be final goods. Characteristics that are mere *means* to more fundamental goods can be included. For example, even if one thinks that musicality or musical appreciation is not intrinsically good, one can still include them in the list if one believes that they tend—as a matter of empirical fact—to promote well-being (e.g., by creating opportunities for enjoyment).

Table 2 Some traits that may promote the social good

Extended altruism
Conscientiousness and honesty
Modesty and self-deprecation
Originality, inventiveness, and independent thinking
Civil courage
Knowledge and good judgment about public affairs
Empathy and compassion
Nurturing emotions and caring behavior
Just admiration and appreciation
Self-control, ability to control violent impulses
Strong sense of fairness
Lack of racial prejudice
Lack of tendency to abuse drugs
Taking joy in others' successes and flourishing
Useful forms of economic productivity
Healthy longevity

status quo (Diener et al. 1999). Sexual jealousy, romantic heartaches, status envy, competitiveness, anxiety, boredom, sadness, and despair may have been essential for survival and reproductive success in the EEA, but they take a toll in terms of human suffering and may substantially reduce our well-being. An intervention that caused an upward shift in hedonic set point, or that downregulated some of these negative emotions, would hence meet the EOC: we can see why the effect would have been maladaptive in the EEA and yet believe that we would benefit from these effects because of a discordance between inclusive fitness and individual well-being.

Good for Society

Many characteristics that promote individual well-being also promote the social good, but the two lists are unlikely to be identical. Table 2 lists some candidate traits that might contribute to the good of society.

As with the list for individual well-being, this one is for illustration only. One could create alternative lists for various related questions, such as traits that are good for humanity as a whole, or for sentient life, or for a particular community or traits that specifically help us become better moral agents. While the lists may overlap, they will likely disagree about some characteristics or their relative importance. The evolution heuristic can be applied using any such list as input.

To use such a list with the EOC, we proceed in the same way as with the "good for the individual" source of value discordance. For example, we might have a drug that appears to make those who take it more compassionate. This might seem like a good thing, but why has evolution not already made us more compassionate? Presumably, evolution could easily have produced an endogenous substance with similar effects to the drug; so the likely explanation is that a higher level of compassionateness would not have increased inclusive fitness in the EEA. We may press on and ask *why* it is that greater compassionateness would have been maladaptive in the EEA. One may surmise that such a trait would have been associated with evolutionary downsides—such as reduced ability credibly to threaten savage retaliation, a tendency to spare the lives of enemies allowing them to come back another day and reverse their defeat, an increased propensity to offer help to those in need beyond what is useful for reciprocity and social acceptance, and so forth. But these very effects, which would have made heightened compassionateness maladaptive for an individual in the EEA, are precisely the kinds of effects which we might believe would be beneficial for the common good today. We do not have to assume that the relevant tradeoffs have changed since the EEA. Even in the EEA, it might have had net good effects for a local population of hunter-gatherers if one person was born with a mutation causing an unusually high level of compassionateness, even though that individual himself might have suffered a fitness penalty. If we accept these premises, then the hypothetical drug that increases compassionateness would pass the EOC. It would be a case where we have reason to think that the wisdom of nature has not achieved what would be best for society and that we could feasibly do better.

Evolutionary Restrictions

General Remarks on Evolutionary Restrictions

The final category of answers to the EOC focuses on the fact that there are certain limitations in what evolution can do. Using the "great engineer" metaphor, we may say that we can hope to achieve certain things with our ham-handed tinkering that stumped evolution, because we have access to tools, materials, and techniques that the great ingenious engineer lacked.

Metaphors aside, we can identify several restrictions of evolution's ability to achieve fitness-maximizing phenotypes even in the EEA. These are important, because in some cases they will indicate clear limitations in the "wisdom of nature" and a fortiori cases where there is room for potentially easy improvements. At a high level of abstraction, we can divide these restrictions into three classes:

- *Fundamental inability*: evolution is fundamentally incapable of producing a trait A.
- *Entrapment in local optimum*: evolution is stuck in a local optimum that excludes trait A.
- *Evolutionary lag*: evolution of trait A takes so many generations that there has not yet been enough time for it to develop.

These three classes, which are discussed in more detail in the following three subsections, are not sharply separate. For example, one reason why a trait may take a vast number of generations to develop is that it requires escaping from one or more local optima. And given truly astronomical timescales, even some traits that we shall regard as fundamentally beyond evolution's reach might conceivably have evolved. However, the three classes are distinct enough to deserve individualized attention.

Fundamental Inability

Biology is limited in what it can build. DNA can only code for proteins, which have to act on moieties in a water-based cellular environment using the relatively weak chemical forces that a protein can muster. This makes it very unlikely that any terrestrial organism could produce diamond, for instance, since the synthesis of diamondoid structures requires significant energy.[8] And while bacteria can produce microscopic metal crystals (Klaus et al. 1999), there is no way to unite them into contiguous metal. Hence, evolution cannot achieve diamond tooth enamel or a titanium skeleton, even if these traits would have improved fitness.

Examples can be multiplied. It is unlikely that evolution could have evolved high-performance silicon chips to augment neural computation, even though such augmentations might have provided important benefits. A theoretical design of artificial red blood cells has been published, calculating the performance of a potentially feasible physical structure for transporting oxygen and carbon dioxide in the blood (Freitas 1998). This design, which is not limited by the materials and pressures that can be achieved using biology, would enable performance far outside the range of natural red blood cells.

Radical departures from nature are apt to raise a host of separate questions regarding biocompatibility and functional integration with evolved systems. But at least there is no mystery as to why we would not already have evolved these enhancements even if they would have increased inclusive fitness in the EEA.

Enhancements that evolution is fundamentally incapable of producing can therefore meet the EOC. When invoking "fundamental inability," it is important to determine that the inability does not pertain merely to the specific means one intends to use to effect the enhancement. If evolution would have been able to employ some different means to achieve the same effect, the challenge would remain to explain why evolution has not achieved the enhancement using that alternative route.

[8] Adding a carbon dimer to a diamond surface using a nanotechnological tool would take more than 6.1 eV (Merkle and Freitas 2003), about 20 times more energy than is released by the ATP hydrolysis that powers most enzymatic actions.

Entrapment in Local Optimum

Evolution sometimes gets stuck on solutions that are locally but not globally optimal. A locally optimal solution is one where any small change would make the solution worse, even if some big changes might make it better.

Being trapped in a local optimum is especially likely to account for failure to evolve polygenic traits that are adaptive only once fully developed and incur a fitness penalty in their intermediary stages of evolution. In some cases, the evolution of such traits may require an improbable coincidence of several simultaneous mutations that might simply not have occurred among our finite number of ancestors. A crafty genetic engineer may be able to solve some of the problems that were intractable to blind evolution. A human engineer can think backward, starting with a goal in mind, working out what genetic modifications are necessary for its attainment.

The human appendix, a vestigial remnant of the cecum in other mammals, while having some limited immunological function (Fisher 2000), easily becomes infected. In the natural state, appendicitis is a life-threatening condition and is especially likely to occur at a young age. There is also evidence that surgical removal of the appendix reduces the risk of ulcerative colitis (Koutroubakis and Vlachonikolis 2000; Andersson et al. 2001). It appears that removal of the appendix would have increased fitness in the EEA. However, a *smaller* appendix increases the risk of appendicitis. Carriers of genes predisposing for small appendices have higher risks of appendicitis than noncarriers and, presumably, lower fitness (Nesse and Williams 1998). Therefore, unless evolution could find a way of doing away with the appendix entirely in one fell swoop, it might be unable to get rid of the organ; whence, it remains. An intervention that safely and conveniently removed it might be an enhancement, increasing both fitness and quality of life.

Another source of evolutionary lock-in is antagonistic pleiotropy, referring to a situation in which a gene affects multiple traits in both beneficial and harmful ways. If one trait is strongly fitness-increasing and the other mildly fitness-decreasing, the overall effect is positive selection for the gene (Leroi et al. 2005). One example is the ε4 allele of apolipoprotein E. Having one or two copies of the allele increases the risk of Alzheimer disease in middle age but lowers the incidence of childhood diarrhea and may protect cognitive development (Oria et al. 2005). Antagonistic pleiotropy has also been discussed in relation to theories of aging. The local optimum here is to retain the genes in question, but the global optimum would be to eliminate the antagonistic pleiotropy by evolving genes that specifically produced the beneficial traits without detrimental effects on other traits. Over longer timescales, evolution usually gets around antagonistic pleiotropy, for example, by evolving modifier genes that counteract the negative effects (Hammerstein 1996), but such developments can take a long time, and in the meanwhile, a species remains trapped in a local optimum.

Yet another way in which evolution can get locked into a suboptimal state is exemplified by the phenomenon of heterozygote advantage. This refers to the common situation where individuals who are heterozygous for a particular gene (i.e., have two different alleles of that gene) have an advantage over homozygote individuals (who have two identical copies of the gene). Heterozygote advantage is responsible for many cases of potentially harmful genes being maintained at a finite frequency in a population.

The classic example of heterozygote advantage is sickle-cell gene, where homozygote individuals suffer anemia while heterozygote individuals benefit from improved malaria resistance (Allison 1954; Cavalli-Sforza and Bodmer 1999). Heterozygotes have greater fitness than both types of homozygote (those lacking the sickle-cell allele and those having two copies of it). Balancing selection preserves the sickle-cell gene in populations (at a frequency that varies geographically with the prevalence of malaria). The "optimum" that evolution selects is one in which, by chance, some individuals will be born homozygous for the gene, resulting in sickle-cell anemia, a potentially fatal blood disease. The "ideal optimum"—everybody being heterozygous for the gene—is unattainable by natural selection because of Mendelian inheritance, which gives each child born to heterozygote parents a 25% chance of being born homozygous for the sickle-cell allele.

Heterozygote advantage suggests an obvious enhancement opportunity. If possible, the variant allele could be removed and its gene product administered as medication. Alternatively, genetic screening could be used to guarantee heterozygosity, enabling us to reach the ideal optimum that eluded natural selection.

The phenomenon of heterozygote advantage points to potential enhancements beyond reducing susceptibility to diseases such as malaria and sickle-cell anemia. For instance, there is some indirect evidence that at least Type I Gaucher's disease (and possibly other sphingolipid storage diseases) is linked to improved cognition, given the significantly higher proportion of sufferers in occupations correlated with high IQ (Cochran et al. 2006). This, and other circumstantial evidence, is used by the authors of the cited study to argue that heterozygote advantage can explain the high IQ test scores and the high prevalence of Type I Gaucher's disease among Ashkenazi Jews. Should this prediction be borne out by finding an IQ advantage for heterozygote carriers of the diseases, it would suggest that screening to promote heterozygosity, or genetic interventions to induce it, would be viable forms of cognition enhancement that meet the EOC.

One other kind of evolutionary entrapment is worth noting here, that of an evolutionarily stable strategy (ESS), "a strategy such that, if all the members of a population adopt it, no mutant strategy can invade" (Smith 1982). One way in which a species can become trapped in an ESS is through sexual selection. In order to be successful at wooing peahens, peacocks have to produce extravagant tails which serve to advertise the male's genetic quality. Only healthy peacocks can afford to produce and carry top notch tails. It is adaptive for peahens to prefer to mate with peacocks that sport an impressive tail, and given this fact, it is also adaptive for peacocks to invest heavily in their plumage. It is likely that the species would have been better off if it had evolved some less costly way for males to signal fitness. Yet no individual peacock or peahen is able to defect from the ESS without thereby removing themselves from the gene pool. If there had been a United Nations of the peafowl, through which the birds could have adopted a coordinated millennium plan to overcome their species' vanity, the peacocks would surely soon be wearing a more casual outfit.

The concept of an ESS can be generalized to that of an evolutionarily stable state. A population is said to be in an evolutionarily stable state if its genetic composition is restored by selection after a disturbance, provided the disturbance is not too large (ibid). Such a population can be genetically monomorphic or polymorphic. Thus, while ESS refers to a specific strategy that is stable if everybody adopts it, an evolutionary stable state can encompass a set of strategies whose distribution is stable under small perturbations. It has been suggested that the human population has been in a stable state in the EEA with regard to sociopathy, which can be seen as a defector strategy which can prosper when it is rare but becomes maladaptive when it is more common (Mealey 1995).

Another way in which evolution can fail to produce solutions that are fitness-maximizing for organisms is intragenomic conflict, in which phenomena such as meiotic drive, transposons, homing endonuclease genes, B-chromosomes, and plasmids result from natural selection among lower-level units such as individual genes (Burt and Trivers 2006). In cases where we can identify intragenomic conflict as responsible for a suboptimal outcome, there is an opportunity for enhancement that can meet the EOC (provided we have the technological means to make the requisite interventions). Genes or traits that would not have evolved, or which would not have been stable against intragenomic competition, could be inserted, possibly, supported by interventions removing some of the competing genetic elements.

Evolutionary Lag

Evolution takes time—often, a long time. If conditions change rapidly, the genome will lag. Given that conditions for humanoid ancestors were quite variable—due to migration into new regions, climate change, social dynamics, advances in tool use, and adaptation in pathogens, parasites, predators, and prey—our species has never been perfectly adapted to its environment. Evolution is running up fitness slopes, but when the fitness landscape keeps changing under its feet, it may never reach a peak. Even if beneficial alleles or allele combinations exist, they may not have had the time to diffuse across human populations. For some proposed enhancements, evolutionary lag can therefore provide an answer to the EOC.

This source of answers to the EOC is related to the changed tradeoff category, but with the difference that here we are focusing on ways in which even during the EEA we were not perfectly adapted to our environment. Even if we set aside the dramatic ways in which resources and demands have changed since the introduction of agriculture, there may still be instances of earlier evolutionary lags that have not yet been truncated and which may point to opportunities for enhancement.

There are many factors limiting the speed of evolution (Barton and Partridge 2000). Some are inherent in the process itself, such as the mutation rate, the need for sufficient genetic diversity, and the constraint that selection can only encode a few bits into the genome per generation (Worden 1995). A recessive beneficial mutation will spread to an appreciable fraction of a fixed well-mixed population in time inversely proportional to its selective advantage. For example, if the mutation gives a 0.1% increase in fitness, it will take 9200 generations (230,000 years assuming 25 years per generation) to reach 50% of the population from a starting level of 0.01%. For a 10% fitness advantage, just 92 generations (2300 years) are needed (Cavalli-Sforza and Bodmer 1999). Population structure and especially low-population bottlenecks can accelerate the spread significantly.

In nature, the strength of selection for a trait is generally quite weak. A review of published studies (Hoekstra et al. 2001) found the distribution of selection strengths across species to be exponential, with a small median magnitude: for most traits and in most systems, directional selection is fairly weak. Selection via survival appears to be weaker than selection through mating success, making sexual selection a big factor. Quadratic selection gradients, indicating the "sharpness" of fitness peaks, were also found to be exponentially distributed and with small median. This implies that stabilizing selection (reducing genetic diversity once a population has reached a local fitness peak) is often fairly weak. Indirect selection (where trait fitness depends on another correlated trait) also appears to be playing only a minor role (ibid). These results suggest that beneficial new traits are likely to spread slowly.

A population living in a heterogeneous or changeable environment may not be able to converge on a single fitness peak but will be spread out around it. This might reduce extinction risks for the lineage, since there will always be some individuals that are well adapted if the conditions change, and the lineage will survive more easily than if a less dispersed population had to ascend the current gradient toward the top through a region of low survivability.

It is possible to detect empirically the presence of genetic variations under positive fitness pressure through their signatures (Sabeti et al. 2006). These signatures range from multimillion-year timescale changes in gene sequence (mostly useful to point out ongoing or recurrent selection) to changes in genetic diversity caused by the rapid spread of a beneficial mutation in the past 250,000 years to the differences between human populations which can indicate genetic selection over the last 50,000–75,000 years. Such long-term selection evidence is mainly useful for understanding the selection pressures in the EEA.

There is evidence for recent positive selection in humans (Voight et al. 2006). Some of it may be in response to climate variations, producing a wide range of variation in salt-regulating genes in populations far from the equator (Thompson et al. 2004). Genes involved in brain development have also been shown to have been under strong positive selection with new variants emerging over the last 37,000 years (Evans et al. 2005) and 5800 years (Mekel Bobrov et al. 2005).

There is evidence that genes related to the brain have evolved more quickly in the human lineage than in other primates and rodents (Dorus et al. 2004). The rapid growth of the brain in the human lineage also suggests that its size must be controlled by relatively simple genetic mechanisms (Roth and Dicke 2005). Despite this, it should be noted that the selection differential per generation for human brain weight during the Pleistocene was only 0.0004 per generation (Cavalli-Sforza and Bodmer 1999: 692): even under fast evolution, brain size was limited by tradeoffs.

If we find a gene that has a desirable effect and that evolved recently and has not yet spread far despite showing evidence of positive selection, interventions that insert it into the genome or mimic its effects would likely meet the EOC. A simple example would be lactose tolerance. While development of lactose intolerance is adaptive for mammals since it makes weaning easier, dairy products have stimulated selection for lactase in humans over the last 5000–10,000 years (Bersaglieri et al. 2004; Tishkoff et al. 2007). This is so recent that there has not been time for the trait to diffuse to all human populations. (Populations that have domesticated cattle but do not have lactose tolerance instead make use of fermented milk or cheese.) Taking lactase pills enables lactose-intolerant people to digest lactose, widening the range of food they can enjoy. This enhancement clearly passes the EOC.

Discussion

The evolution heuristic instructs us to consider, for an apparently attractive enhancement, why we have not already evolved the intended trait if it is really such a good idea. We called this question the evolutionary optimality challenge, and we have described three broad categories of possible answers and given some examples of particular enhancements for which it is possible to meet the EOC and which, therefore, seem comparatively promising as intervention targets that may be feasible in the relatively near term and which may have on balance beneficial effects.

In general, when we pose the EOC for some particular proposed enhancement, we might discover one of several things:

1. Current ignorance prevents us from forming any plausible idea about the evolutionary factors at play.
2. We come up with a plausible idea about the relevant evolutionary factors, and this reveals that the proposed modification would likely not be a net benefit.
3. We come up with a plausible idea about the relevant evolutionary factors, and this reveals why we would not already have evolved to have the enhanced capacity even if it would be a net benefit.
4. We come up with several plausible but mutually inconsistent ideas about the relevant evolutionary factors.

The first possibility means that we have no clear idea about why, from a phylogenetic perspective, the trait that is the target of the proposed enhancement is the way it is. This should give us pause. If we do not understand why a very complex evolved system has a certain property, there is a considerable risk that something will go wrong if we try to modify it. The case might be one of those where nature does know best. Like an overambitious tinkerer with merely superficial understanding of what he is doing while he is making changes to the design of a master craftsman, the potential for damage is considerable, and the chances of producing an all-things-considered improvement are small.

We are not claiming that it is always inadvisable to try an intervention when we have no adequate understanding of the subsystem we intend to enhance. We might have other sources of evidence that afford us sufficient assurance that the intervention will work and will not cause unacceptable side effects, even without understanding the evolutionary functions involved. For example, we might have used the intervention many times before and found that it works well. Alternatively, we might have evidence from a closely analogous subsystem, such as an animal model, that suggests that the intervention should work in humans too. In such cases, the evolution heuristic delivers only a weak recommendation: that absent any good answer to the EOC, we should proceed only with great caution. In particular, we should be alert to the possibility that the proposed intervention will turn out to have significant (but perhaps subtle) side effects.

The second possibility is that we succeed in developing a plausible understanding of the pertinent evolutionary factors, and, having done so, we find our initial hopes about the proposed modification undermined. None of the three categories we have described yields a satisfactory answer to the EOC: the relevant tradeoffs have not changed since the EOC, there is no relevant value discordance, and no evolutionary restriction would have prevented the modification from already having evolved by now. In this case, we have strong reason for thinking that the enhancement intervention will fail or backfire. If we proceed, the wisdom of nature will bite us.

The fourth possibility is that we come up with two or more plausible but incompatible evolutionary accounts of the evolutionary factors at play. In this case, we can consider the implications of each of the different evolutionary accounts separately according to the above criteria. If all yield green lights, we are encouraged to proceed. If some of the evolutionary accounts yield green lights but others yield red lights, then we face a situation of uncertainty. We can use standard decision theory to determine how to proceed—we can take a gamble if we feel that the balance of probabilities sufficiently favors the green lights; if not, we can attempt to acquire more information in order to reduce the uncertainty or forgo the potential enhancement and try something else.

The evolution heuristic is not a rival method to the more obvious way of determining whether some enhancement intervention works: testing it in well-designed clinical trials. Instead, the heuristic is complementary. It helps us ask some useful questions. By posing the EOC and carefully searching for and evaluating possible answers in each of the three categories we described, we can (a) identify promising candidate enhancement interventions, to be explored further in laboratory and clinical studies, and (b) better evaluate the likelihood that some intervention which has shown seemingly positive results in clinical studies will actually work as advertised and will not have unacceptable side effects of a hidden, subtle, or long-term nature.

Conclusion

There is a widespread belief in some kind of "wisdom of nature." Many people prefer "natural" remedies, "natural" food supplements, and "natural" ways of improving human capacities such as training, diet, and grooming. "Unnatural" interventions are often viewed with suspicion, and this attitude seems to be especially pronounced in relation to unnatural ways of enhancing human capacities, which are viewed as unwise, shortsighted, and hubristic. We believe that such attitudes also exert an influence on beliefs about the kind of matters that arise in bioethical discussions of human enhancement.

While it is tempting to dismiss intuitions about the wisdom of nature as vulgar prejudice, we have suggested that these intuitions contain a grain of truth, especially as they pertain to human enhancement. We have attempted to explicate this grain of truth as the evolutionary optimality challenge.

After posing this challenge, the evolution heuristic instructs us to examine three broad categories of potential ways of meeting the challenge: changed tradeoffs, value discordance, and evolutionary restrictions. These categories correspond to systematic limitations in the wisdom of nature idea. For some potential enhancement interventions, the challenge can be met with an answer from one of these categories; for other potential interventions, the challenge cannot be met. The latter interventions merit suspicion, and attempting them may indeed be unwise, shortsighted, and hubristic. The former interventions, in contrast, do not defy the wisdom of nature and have a better chance of working.

By understanding both the sense in which there is validity in the idea that nature is wise and the limits beyond which the idea ceases to be valid, we are in a better position to identify promising human enhancements and to evaluate the risk-benefit ratio of extant enhancements. If we are right in supposing that intuitions about the wisdom of nature exert an inarticulate influence on opinion in contemporary bioethics of human enhancement, then the evolution heuristic—while primarily a method for addressing empirical questions—may also help to inform our assessments of more normatively loaded items of dispute.[9]

References

Aiello, L.C., N. Bates, and T. Joffe. 2001. In Defense of the Expensive Tissue Hypothesis. In *Evolutionary Anatomy of the Primate Cerebral Cortex*, ed. D. Falk and K.R. Gibson, 57–78. Cambridge: Cambridge University Press.

Allison, A.C. 1954. Protection Afforded by Sickle Cell Trait against Subtertian Malarial Infection. *British Medical Journal* 1: 290–294.

[9] For their comments, we are grateful to Rebecca Roache for helpful comments on an earlier draft of this paper and to the audience at the *TransVision 2006* conference in Helsinki, Finland, for useful questions.

Andersson, R.E., G. Olaison, C. Tysk, and A. Ekbom. 2001. Appendectomy and Protection against Ulcerative Colitis. *New Engl J Med* 344(11): 808–814.

Baldwin, K.M., and F. Haddad. 2002. Skeletal Muscle Plasticity—Cellular and Molecular Responses to Altered Physical Activity Paradigms. *American Journal of Physical Medicine & Rehabilitation* 81(11): S40–S51.

Barton, N., and L. Partridge. 2000. Limits to Natural Selection. *Bioessays* 22(12): 1075–1084.

Battista, J.R., A.M. Earl, and M.J. Park. 1999. Why is Deinococcus Radiodurans so Resistant to Ionizing Radiation? *Trends in Microbiology* 7(9): 362–365.

Beauchamp, T.L., and J.F. Childress. 1979. *Principles of Biomedical Ethics*. Oxford/New York: University Press.

Bersaglieri, T., P.C. Sabeti, N. Patterson, T. Vanderploeg, S.F. Schaffner, J.A. Drake, M. Rhodes, D.E. Reich, and J.N. Hirschhorn. 2004. Genetic Signatures of Strong Recent Positive Selection at the Lactase Gene. *American Journal of Human Genetics* 74(6): 1111–1120.

Boon, T., and N. van Baren. 2003. Immunosurveillance against Cancer and Immunotherapy—Synergy or Antagonism? *New Engl J Med* 348(3): 252–254.

Brunswick, N., G.N. Martin, L. Marzano, and N. Savill. 2007. *Visuo-Spatial Ability, Handedness and Developmental Dyslexia: Just how Sinister was Andy Warhol?* Presentation to the 25th European Workshop on Cognitive Neuropsychology, Italy 2007.

Bunker, J.P. 2001. The Role of Medical Care in Contributing to Health Improvements within Societies. *International Journal of Epidemiology* 30(6): 1260–1263.

Burt, A., and R.L. Trivers. 2006. *Genes in Conflict: The Biology of Selfish Genetic Elements*. Harvard: Belknap Press.

Buss, D.M. 2000. The Evolution of Happiness. *The American Psychologist* 55(1): 15–23.

Caldwell, J.A. 2001. Efficacy of Stimulants for Fatigue Management: The Effects of Provigil and Dexedrine on Sleep-Deprived Aviators. *Fatigue in Transportation (Part F)* 4: 19–37.

Cavalli-Sforza, L.L., and W.F. Bodmer. 1999. *The Genetics of Human Populations*. New York: Dover Publications.

Cochran, G., J. Hardy, and H. Harpending. 2006. Natural History of Ashkenazi Intelligence. *Journal of Biosocial Science* 38(5): 659–693.

Colloca, L., and F. Benedetti. 2005. Placebos and Painkillers: Is Mind as Real as Matter? *Nature Reviews. Neuroscience* 6(7): 545–552.

Dawkins, R. 1976. *The Selfish Gene*. Oxford: Oxford University Press.

de Murcia, G., and S. Shall, ed. 2000. *From DNA Damage and Stress Signaling to Cell Death: Poly ADP-Ribosylation Reactions*. Oxford: Oxford University Press.

Demas, G.E., V. Chefer, M.I. Talan, and R.J. Nelson. 1997. Metabolic Costs of Mounting an Antigen-Stimulated Immune Response in Adult and Aged C57BL/6J Mice. *American Journal of Physiology—Reg I* 42(5): R1631–R1637.

Diener, E., E.M. Suh, R.E. Lucas, and H.L. Smith. 1999. Subjective Well-Being: Three Decades of Progress. *Psychological Bulletin* 125(2): 276–302.

Dorus, S., E.J. Vallender, P.D. Evans, J.R. Anderson, S.L. Gilbert, M. Mahowald, G.J. Wyckoff, C.M. Malcom, and B.T. Lahn. 2004. Accelerated Evolution of Nervous System Genes in the Origin of *Homo sapiens*. *Cell* 119(7): 1027–1040.

Dunn, G.P., L.J. Old, and R.D. Schreiber. 2004. The Immunobiology of Cancer Immunosurveillance and Immunoediting. *Immunity* 21(2): 137–148.

Elia, M. 1992. Organ and Tissue Contribution to Metabolic Rate. In *Energy Metabolism: Tissue Determinants and Cellular Corollaries*, ed. J.M. McKinney and H.N. Tucker, 61–79. New York: Raven.

Evans, P.D., S.L. Gilbert, N. Mekel-Bobrov, E.J. Vallender, J.R. Anderson, L.M. Vaez-Azizi, S.A. Tishkoff, R.R. Hudson, and B.T. Lahn. 2005. Microcephalin, a Gene Regulating Brain Size, Continues to Evolve Adaptively in Humans. *Science* 309(5741): 1717–1720.

Fairclough, S.H., and K. Houston. 2004. A Metabolic Measure of Mental Effort. *Biological Psychology* 66(2): 177–190.

Falk, D., and K.R. Gibson, ed. 2001. *Evolutionary Anatomy of the Primate Cerebral Cortex*. Cambridge: Cambridge University Press.

Falkner, F., and J.M. Tanner, ed. 1986. *Human Growth: A Comprehensive Treatise*. New York: Plenum Press.

Farah, M.J., J. Illes, R. Cook-Deegan, H. Gardner, E. Kandel, P. King, E. Parens, B. Sahakian, and P.R. Wolpe. 2004. Neurocognitive Enhancement: What Can We Do and What Should We Do? *Nature Reviews. Neuroscience* 5(5): 421.

Fish, J.L., and C.A. Lockwood. 2003. Dietary Constraints on Encephalization in Primates. *American Journal of Physical Anthropology* 120(2): 171–181.

Fisher, R.E. 2000. The Primate Appendix: A Reassessment. *The Anatomical Record* 261(6): 228–236.

Flegal, K.A., B.I. Graubard, D.F. Williamson, and M.H.H. Gail. 2005. Excess Deaths Associated with Underweight, Overweight, and Obesity. *Journal of the American Medical Association* 293(15): 1861–1867.

Fontaine, K.R., D.T. Redden, C.X. Wang, A.O. Westfall, and D.B. Allison. 2003. Years of Life Lost Due to Obesity. *Journal of the American Medical Association* 289(2): 187–193.

Force, U.S.P.S.T. 2002. Aspirin for the Primary Prevention of Cardiovascular Events: Recommendation and Rationale. *Annals of Internal Medicine* 136(2): 157–160.

Frech, H.E., and R.D. Miller. 1996. *The Productivity of Health Care and Pharmaceuticals: An International Comparison*. UCLA Research Program in Pharmaceutical Economics and Policy 97–1. http://repositories.cdlib.org/pep/97-1/. Accessed 11 Aug 2016.

Freitas, R.A. Jr. 1998. Exploratory Design in Medical Nanotechnology: A Mechanical Artificial Red Cell. *Artificial Cells, Blood Substitutes, and Immobilization Biotechnology* 26: 411–430.

Gaesser, G.A. 1999. Thinness and Weight Loss: Beneficial or Detrimental to Longevity? *Medicine and Science in Sports and Exercise* 31(8): 1118–1128.

Goldspink, G. 2005. Mechanical Signals, IGF-I Gene Splicing, and Muscle Adaptation. *Physiology* 20: 232–238.

Gottfredson, L.S. 1997. Why G Matters: The Complexity of Everyday Life. *Intelligence* 24(1): 79–132.

———. 2004. Life, Death, and Intelligence. *Journal of Cognitive Education and Psychology* 4(1): 23–46.

———. 2007. Innovation, Fatal Accidents, and the Evolution of General Intelligence. In *Integrating the Mind: Domain General versus Domain Specific Process in Higher Cognition*, ed. M.J. Roberts, 387–425. Hove: UK Psychology Press.

Goulandris, N.K., M.J. Snowling, and I. Walker. 2000. Is Dyslexia a Form of Specific Language Impairment? A Comparison of Dyslexic and Language Impaired Children as Adolescents. *Annals of DysleExia L* 50: 103–120.

Green, C.S., and D. Bavelier. 2006. Enumeration versus Multiple Object Tracking: The Case of Action Video Game Players. *Cognition* 101(1): 217–245.

Hagen, E.H. 2002. *What is the EEA? (detailed answer) Evolutionary Psychology FAQ*. http://www.anth.ucsb.edu/projects/human/epfaq/eea2.html. Accessed 2 July 2006.

Hammerstein, P. 1996. Darwinian Adaptation, Population Genetics and the Streetcar Theory of Evolution. *Journal of Mathematical Biology* 34(5–6): 511–532.

Hoekstra, H.E., J.M. Hoekstra, D. Berrigan, S.N. Vignieri, A. Hoang, C.E. Hill, P. Beerli, and J.G. Kingsolver. 2001. Strength and Tempo of Directional Selection in the Wild. *Proceedings of the National Academy of Sciences of the United States of America* 98(16): 9157–9160.

Hofsten, C., and L. Backman, ed. 2002. *Social, Developmental, and Clinical Perspectives*, Psychology at the Turn of Millennium. Vol. vol. 2. Hove: Psychology Press.

Holliday, M.A. 1986. Body Composition and Energy Needs During Growth. In *Human Growth: A Comprehensive Treatise*, ed. F. Falkner and J.M. Tanner, 101–117. New York: Plenum Press.

House, J.S., K.R. Landis, and D. Umberson. 1988. Social Relationships and Health. *Science* 241(4865): 540–545.

Hubbard, E.M., M. Piazza, P. Pinel, and S. Dehaene. 2005. Interactions between Number and Space in Parietal Cortex. *Nature Reviews. Neuroscience* 6(6): 435–448.

Humphrey, N. 2002. Great Expectations: The Evolutionary Psychology of Faith-Healing and the Placebo Response. In *Social, Developmental, and Clinical Perspectives*, Psychology at the Turn of Millennium, ed. C. Hofsten and L. Bäckman, Vol. 2, 225–246. Hove: Psychology Press.

Jensen, P.S., D. Mrazek, P.K. Knapp, L. Steinberg, C. Pfeffer, J. Schowalter, and T. Shapiro. 1997. Evolution and Revolution in Child Psychiatry: ADHD as a Disorder of Adaptation. *Journal of the American Academy of Child Psychiatry* 36(12): 1672–1679.

Kiecolt, J.K., R. Glaser, J.T. Cacioppo, R.C. MacCallum, M. Snydersmith, C. Kim, and W.B. Malarkey. 1997. Marital Conflict in Older Adults: Endocrinological and Immunological Correlates. *Psychosomatic Medicine* 59(4): 339–349.

Kirsch, I., T.J. Moore, A. Scoboria, and S.S. Nicholls. 2002. The Emperor's New Drugs: An Analysis of Antidepressant Medication Data Submitted to the US Food and Drug Administration. *Prevention & Treatment* 5: 23.

Klaus, T., R. Joerger, E. Olsson, and C.G. Granqvist. 1999. Silver-Based Crystalline Nanoparticles, Microbially Fabricated. *Proceedings of the National Academy of Sciences of the United States of America* 96(24): 13611–13614.

Korol, D.L., and P.E. Gold. 1998. Glucose, Memory, and Aging. *The American Journal of Clinical Nutrition* 67(4): 764S–771S.

Koutroubakis, I.E., and I.G. Vlachonikolis. 2000. Appendectomy and the Development of Ulcerative Colitis: Results of a Metaanalysis of Published Case-Control Studies. *The American Journal of Gastroenterology* 95(1): 171–176.

Leroi, A.M., A. Bartke, G. De Benedictis, C. Franceschi, A. Gartner, E.S. Gonos, M.E. Fedei, T. Kivisild, S. Lee, N. Kartaf-Ozer, M. Schumacher, E. Sikora, E. Slagboom, M. Tatar, A.I. Yashin, J. Vijg, and B. Zwaan. 2005. What Evidence is there for the Existence of Individual Genes with Antagonistic Pleiotropic Effects? *Mechanisms of Ageing and Development* 126(8): 421–429.

Maizels, R.M. 2005. Infections and Allergy—Helminths, Hygiene and Host Immune Regulation. *Current Opinion in Immunology* 17(6): 656–661.

Manning, C.A., W.S. Stone, D.L. Korol, and P.E. Gold. 1998. Glucose Enhancement of 24-H Memory Retrieval in Healthy Elderly Humans. *Behavioural Brain Research* 93(1–2): 71–76.

Martin, R.D. 1996. Scaling of the Mammalian Brain: The Maternal Energy Hypothesis. *News in Physiological Sciences* 11: 149–156.

Martin, P.Y., and D. Benton. 1999. The Influence of a Glucose Drink on a Demanding Working Memory Task. *Physiology & Behavior* 67(1): 69–74.

Martin, L.B. 2nd, A. Scheuerlein, and M. Wikelski. 2003. Immune Activity Elevates Energy Expenditure of House Sparrows: A Link between Direct and Indirect Costs? *Proceedings of the Biological Sciences* 270(1511): 153–158.

Matte, T.D. 2001. Influence of Variation in Birth Weight within Normal Range and within Sibships on IQ at Age 7 Years: Cohort Study. *British Medical Journal* 323(7314): 310–314.

McCord, J.M. 2000. The Evolution of Free Radicals and Oxidative Stress. *The American Journal of Medicine* 108(8): 652–659.

McDade, T.W. 2002. Status Incongruity in Samoan Youth: A Biocultural Analysis of Culture Change, Stress, and Immune Function. *Medical Anthropology Quarterly* 16(2): 123–150.

———. 2003. Life History Theory and the Immune System: Steps toward a Human Ecological Immunology. *Yearbook of Physical Anthropology* 46: 100–125.

McDaniel, M.A. 2005. Big-Brained People are Smarter: A Meta-Analysis of the Relationship between in Vivo brain Volume and Intelligence. *Intelligence* 33(4): 337–346.

McKeown, T., and C.R. Lowe. 1974. *An Introduction to Social Medicine*. Oxford: Blackwell Scientific.

McKinney, J.M., and H.N. Tucker, ed. 1992. *Organ and Tissue Contribution to Metabolic Rate, Energy Metabolism: Tissue Determinants and Cellular Corollaries*. New York: Raven.

Mealey, L. 1995. The Sociobiology of Sociopathy—An Integrated Evolutionary Model. *The Behavioral and Brain Sciences* 18(3): 523–541.

Meikle, A., L.M. Riby, and B. Stollery. 2005. Memory Processing and the Glucose Facilitation Effect: The Effects of Stimulus Difficulty and Memory Load. *Nutritional Neuroscience* 8(4): 227–232.

Mekel-Bobrov, N., S.L. Gilbert, P.D. Evans, E.J. Vallender, J.R. Anderson, R.R. Hudson, S.A. Tishkoff, and B.T. Lahn. 2005. Ongoing Adaptive Evolution of ASPM, a Brain Size Determinant in *Homo sapiens*. *Science* 309(5741): 1720–1722.

Merkle, R.C., and R.A. Freitas. 2003. Theoretical Analysis of a Carbon-Carbon Dimer Placement Tool for Diamond Mechanosynthesis. *Journal of Nanoscience and Nanotechnology* 3(4): 319–324.

Moret, Y., and P. Schmid-Hempel. 2000. Survival for Immunity: The Price of Immune System Activation for Bumblebee Workers. *Science* 290(5494): 1166–1168.

Neisser, U., G. Boodoo, T.J. Bouchard, A.W. Boykin, N. Brody, S.J. Ceci, D.F. Halpern, J.C. Loehlin, R. Perloff, R.J. Sternberg, and S. Urbina. 1996. Intelligence: Knowns and Unknowns. *The American Psychologist* 51(2): 77–101.

Nesse, R.M., and G.C. Williams. 1998. Evolution and the Origins of Disease. *Scientific American* 279(5): 86–93.

Newhouse, J.P., and Group TIE. 1993. *Free for All? Lessons from the RAND Health Insurance Experiment*. Cambridge, MA: Harvard University Press.

Niven, J.E. 2005. Brain Evolution: Getting Better all the time? *Current Biology* 15(16): R624–R626.

Nutt, D.J. 2006. Alcohol Alternatives—A Goal for Psychopharmacology? *Journal of Psychopharmacology* 20(3): 318–320.

Oria, R.B., P.D. Patrick, H. Zhang, B. Lorntz, C.M.D. Costa, G.A.C. Brito, L.J. Barrett, A.A.M. Lima, and R.L. Guerrant. 2005. APOE4 Protects the Cognitive Development in Children with Heavy Diarrhea Burdens in Northeast Brazil. *Pediatric Research* 57(2): 310–316.

Ots, I., A.B. Kerimov, E.V. Ivankina, T.A. Ilyina, and P. Horak. 2001. Immune Challenge Affects Basal Metabolic Activity in Wintering Great Tits. *Proceedings of the Biological Sciences* 268(1472): 1175–1181.

Parker, G.A., and J.M. Smith. 1990. Optimality Theory in Evolutionary Biology. *Nature* 348(6296): 27–33.

Pica, P., C. Lemer, V. Izard, and S. Dehaene. 2004. Exact and Approximate Arithmetic in an Amazonian Indigene Group. *Science* 306(5695): 499–503.

Roberts, M.J., ed. 2007. *Integrating the Mind: Domain General versus Domain Specific Process in Higher Cognition*. Hove: UK Psychology Press.

Roth, G., and U. Dicke. 2005. Evolution of the Brain and Intelligence. *Trends in Cognitive Sciences* 9(5): 250–257.

Sabeti, P.C., S.F. Schaffner, B. Fry, J. Lohmueller, P. Varilly, O. Shamovsky, A. Palma, T.S. Mikkelsen, D. Altshuler, and E.S. Lander. 2006. Positive Natural Selection in the Human Lineage. *Science* 312(5780): 1614–1620.

Sapolsky, R.M. 2005. The Influence of Social Hierarchy on Primate Health. *Science* 308(5722): 648–652.

Sauro, M.D., and R.P. Greenberg. 2005. Endogenous Opiates and the Placebo Effect: A Meta-Analytic Review. *Journal of Psychosomatic Research* 58(2): 115–120.

Scholey, A.G., S. Harper, and D.O. Kennedy. 2001. Cognitive Demand and Blood Glucose. *Physiology & Behavior* 73(4): 585–592.

Siegel, J.M. 2005. Clues to the Functions of Mammalian Sleep. *Nature* 437(7063): 1264–1271.

Silberman, S. 2001. The Geek Syndrome. *Wired* 9(12): 174.

Skaper, S.D. 2003. Poly(ADP-ribosyl)ation Enzyme-1 as a Target for Neuroprotection in Acute Central Nervous System Injury. Curr Drug Targets. *CNS & Neurological Disorders* 2(5): 279–291.

Smith, J.M. 1982. *Evolution and the Theory of Games*. Cambridge: Cambridge University Press.
Starfield, B. 2000. Is US Health Really the Best in the World? *Journal of the American Medical Association* 284(4): 483–485.
Thomas, M.K., D.M. Lloyd-Jones, R.I. Thadhani, A.C. Shaw, D.J. Deraska, B.T. Kitch, E.C. Vamvakas, I.M. Dick, R.L. Prince, and J.S. Finkelstein. 1998. Hypovitaminosis D in Medical Inpatients. *The New England Journal of Medicine* 338(12): 777–783.
Thompson, E.E., H. Kuttab-Boulos, D. Witonsky, L. Yang, B.A. Roe, and A. Di Rienzo. 2004. CYP3A Variation and the Evolution of Salt-Sensitivity Variants. *American Journal of Human Genetics* 75(6): 1059–1069.
Tishkoff, S.A., F.A. Reed, A. Ranciaro, B.F. Voight, C.C. Babbitt, J.S. Silverman, K. Powell, H.M. Mortensen, J.B. Hirbo, M. Osman, M. Ibrahim, S.A. Omar, G. Lema, T.B. Nyambo, J. Ghori, S. Bumpstead, J.K. Pritchard, G.A. Wray, and P. Deloukas. 2007. Convergent Adaptation of Human Lactase Persistence in Africa and Europe. *Nature Genetics* 39(1): 31–40.
Trevathan, W. 1987. *Human Birth: An Evolutionary Perspective*. New York: Aldine de Gruyter.
Trevathan, W., E.O. Smith, and J.J. McKenna, ed. 1999. *Evolutionary Medicine*. Oxford: Oxford University Press.
Udry, J.R. 1978. Differential Fertility by Intelligence—Role of Birth Planning. *Social Biology* 25(1): 10–14.
Vancourt, M., and F.D. Bean. 1985. Intelligence and Fertility in the United States—1912–1982. *Intelligence* 9(1): 23–32.
Venkateswaran, A., S.C. McFarlan, D. Ghosal, K.W. Minton, A. Vasilenko, K. Makarova, L.P. Wackett, and M.J. Daly. 2000. Physiologic Determinants of Radiation Resistance in Deinococcus Radiodurans. *Applied and Environmental Microbiology* 66(6): 2620–2626.
Vining, D.R., L. Bygren, K. Hattori, S. Nystrom, and S. Tamura. 1988. IQ/Fertility Relationships in Japan and Sweden. *Personality and Individual Differences* 9(5): 931–932.
Voight, B.F., S. Kudaravalli, X.Q. Wen, and J.K. Pritchard. 2006. A Map of Recent Positive Selection in the Human Genome. *PLoS Biology* 4(3): 446–458.
von Karolyi, C., E. Winner, W. Gray, and G.F. Sherman. 2003. Dyslexia Linked to Talent: Global Visual-Spatial Ability. *Brain and Language* 85(3): 427–431.
Waterlow, J.C. 1984. Protein Turnover with Special Reference to Man. *Quarterly Journal of Experimental Physiology* 69(3): 409–438.
Williams, G.C. 1996/1966. *Adaptation and Natural Selection*. Princeton: Princeton University Press.
Willams, G.C., and R.M. Nesse. 1991. The Dawn of Darwinian Medicine. *The Quarterly Review of Biology* 66(1): 1–22.
Winder, R., and J. Borrill. 1998. Fuels for Memory: The Role of Oxygen and Glucose in Memory Enhancement. *Psychopharmacology* 136(4): 349–356.
Wood, R.D., M. Mitchell, J. Sgouros, and T. Lindahl. 2001. Human DNA Repair Genes. *Science* 291(5507): 1284–1289.
Worden, R.P. 1995. A Speed Limit for Evolution. *Journal of Theoretical Biology* 176(1): 137–152.
Yazdanbakhsh, M., P.G. Kremsner, and R. van Ree. 2002. Immunology—Allergy, Parasites, and the Hygiene Hypothesis. *Science* 296(5567): 490–494.

Against the Drug Cure Model: Addiction, Identity, and Pharmaceuticals

Şerife Tekin, Owen Flanagan, and George Graham

Addiction affects the person as an embodied agent in the world, not just as a brain on a body's shoulders. Addictive behavior patterns involve not only brain chemistry, but an addict's interpersonal relationships, social and physical environment, and personal identity, all of which contribute to habits that cause or perpetuate addiction. No doubt, drugs may be useful in treating some features of addiction. But unfettered optimism about pharmaceuticals in the treatment of addiction is scientifically unwarranted and therapeutically imprudent.

Introduction

Suppose you are an alcoholic and you walk into an outpatient addiction clinic. Financially, interpersonally, and occupationally, your drinking is having harmful consequences. Your life is becoming undone.

Your physician has recommended the clinic and you have an appointment with the medical director. Here is what the director, a medical doctor, says when she meets you:

> Alcoholism in particular and addiction in general is a disease of the brain reward system. In your case, your substance abuse and heavy drinking have been caused by activation and

Ş. Tekin (✉)
Department of Philosophy and Religious Studies, Daemen College, Amherst, NY, USA
e-mail: stekin@daemen.edu

O. Flanagan
Department of Philosophy, Duke University, Durham, NC, USA
e-mail: ojf@duke.edu

G. Graham
Department of Philosophy, Georgia State University, Atlanta, GA, USA
e-mail: ggraham@gsu.edu

© Springer Science+Business Media B.V. 2017
D. Ho (ed.), *Philosophical Issues in Pharmaceutics*, Philosophy and Medicine 122,
DOI 10.1007/978-94-024-0979-6_13

> dysregulation of endogenous midbrain processes, which hijack brain circuits that normally insure more prudent consumption. Mood stabilizers like lithium and Depakote are not drugs for you because you do not suffer from bipolar depression and, besides, mood stabilizers do not act on mesolimbic reward systems. Your consumption behavior often is impulsive, true, but you are not grandiose in your thinking and you do not suffer from comorbid excesses of mania or anxiety. Recent research with Parkinson's disease, a disease that involves the limbic areas of the brain, suggests that a dopamine agonist is likely to be effective with your particular brain reward system pathology. We don't want to induce motivational apathy in you, but certainly some dopaminergic pharmacology, even if not as specific as, say, a D3 reuptake inhibitor, will prove promising.

The doctor talks. You "fog out." You are not opposed to taking medication, but somehow this is not what you had expected to hear at the clinic. The doctor's terms are too technical and esoteric for you. You also don't like to think of yourself as diseased, although you are not sure just what a disease is. You wish for help that looks at your life as a whole, its stresses and circumstances, and not just at choices of medicine that target parts of your brain.

Sitting in her office, you feel lost. You should have known this would happen. A poster on the waiting room wall, which, alas, you had not bothered to read, says:

> Recent advances in brain imaging methods, as well as increased sophistication in modeling the brain's reward systems, have facilitated the study of neural mechanisms associated with addiction, for example, the processes involved in motivation, decision-making, pleasure seeking, and inhibitory control. As a result, scientists now can delineate the neurological causes of addiction. Pharmaceuticals that target and change these neural mechanisms can facilitate early intervention, leading to full and lasting recovery. Effective treatment of an addictive behavior requires some form of pharmaceutical intervention. Speak to your doctor now about which drug or drugs are best suited for you.

You should have read it before you spoke to the director, of course. You failed to do that. You may also have helped yourself by reading more about the roles of drugs in treatment. There are good general audience books on the subject by respected psychiatrists—some quite sobering in their cautions (Frances 2013). But we three philosophers don't blame you for feeling lost or failing to do your homework. We propose that it is misguided to construe addiction as, or just as, a brain chemistry disorder, something which can be adequately treated by pharmaceutical intervention. Addiction affects the whole person in his or her complexity as an embodied agent in a social world, not simply as a brain on a body's shoulders. And while you do, of course, have one of those on your shoulders, your own addictive behavior has, to your mind, a narrative, a story, a self behind it that this particular clinic is not prepared to hear.

The grip of the chemical disease or neurological disorder model within addiction science reflects a kind of scientific myopia blind to the fact that, in the words of Richard Bentall, "recovery involves the development of new meaning and purpose in one's life," as a complex person "grows beyond the catastrophic effects" of illness (Bentall 2009: 266).

Our goals in this paper may be immodest in their boldness, but they are moderate in their exercise or scope. First, we outline a model of addiction that applies to a case like yours. This model construes addiction as person-level phenomenon, as opposed to a sub-personal brain chemistry phenomenon. The model conceives of the addict as a person and addiction as a comprehensive, destructive way of being. If addicts are persons and addiction is a lifestyle (with all sorts of distinctive phenomenological and behavioral aspects), then addiction is not a brain disease. Addictions are not in the head.

Can we treat parts of the brain to help addicts overcome addiction? Almost certainly. But it would be a big, costly mistake to conceive of addiction as a brain disorder. Let us explain.

Our model of addiction has two parts: first, persons are historical beings whose lives are best understood by the self and others in terms of narrative. A person or a self is a "center of narrative gravity" (Dennett 1991; see also Flanagan 1991; Tekin 2011, 2013a). Especially for people in modern worlds, the narratives have great complexity. Often, individuals are what we call "Person, as multiplex," involving many roles, multiple relations and ways of being, as well as aspirations for being a better or more successful person:

> [Most of us] have for ourselves multifarious projects and plans nested together in various, possibly ever-adjusting, relations of priority and expansiveness. For many, most, perhaps all of us persons, we develop a narrative self-interpretation of ourselves as persons and perpetually evaluate how well we are doing in becoming who we aim to be and in accomplishing what we aim to accomplish. A basic way in which to understand the inter-relationships between our past, present, and future is to conceive of the lives we lead as an unfolding story. (Flanagan 2013a: 2)

In the addict's case, one aspect of her complexity, really of her multiplexity, is that she experiences herself as living out some sort of contradiction. This is the second aspect of our model: normally the sort of person who seeks help for her addiction suffers *twin normative failure*s (see Flanagan 2013a; Graham in press).

Addiction is partly constituted by the addict's recognition that (1) her immoderate use of alcohol is connected to a failure to "live up to the hopes, expectations, standards, and ideals she has for a good life for herself" (Flanagan 2013a: 1, see also Graham 2013: 178–179) and (2) that she can't successfully moderate or quit on her own; she fails to "execute normal powers of effective rational agency" (Flanagan 2013a: 1). The addict comes to understand herself correctly as unable to live up to her standards, ideals, or aspirations, as well as to norms for the multiple roles she occupies and wishes to be successful at—as mother, wife, professor, prelate, pediatrician, or politician—the successful negotiation of which is part of what it will mean for her to be the person she aims to be. The addict realizes that the meaning, worth, and success of her life (possibly her life itself) depend on her not using, but she uses. Given the two failures combined, addiction undermines the goodness and the rationality of the addict's life. The addict is not the person she should be and, in properly reflective moments, wants to be.

Our proposal is that this picture of the addict as a person and historical being, and of addiction as a lifestyle, not a disease of the brain, is true—even obviously so—and that it s*hould* be used as a scientific treatment tool that helps to understand and intervene in addiction. These are not therapeutically idle concepts. Pharmaceuticals may and sometimes must play a role in the complex causal-explanatory matrix of treating addiction, but they are not the only helpful instruments. Unfettered optimism about the effectiveness of pharmaceuticals in the treatment of addiction is scientifically unwarranted and therapeutically imprudent. One extremely important reason is that the varieties of neural activity that can be relevant to an addictive behavior pattern are multiform. They are the physical-biochemical substrates of such psychological phenomena as the failure to reflect before acting, rapid onset of boredom, inability to cope with stressful or anxious situations or unforeseen obstacles, habitual preference reversals, fragile self-esteem and self-respect, and a host of other impairments and challenges to an addict's rational and normatively interpreted reason-responsive agency.[1] Our goal is not to resist pharmaceutical assistance altogether in addressing addiction. Selective use of drugs may help with aspects of addiction, but they can't conceivably cure addiction by re-equilibrating damaged parts of the brain. Why? Because addiction is not in the brain nor is it caused by the brain.

We focus on a prototypical addictive pattern where a particular substance, such as alcohol, becomes the object of an individual's attention and periodically repeated activity, despite knowledge of its harmful consequences and repeated attempts to moderate or quit the behavior (Graham 2013: 178–179). This means we focus both on what is sometimes called the "unwilling addict" and on cases of alcohol abuse. These are addicts who may actively seek or at least willingly accept help from other persons in order to alleviate their harmful condition. Not all addicts, of course, are unwilling. Some are resigned. Some are indifferent. Some are sometimes one, then the other. Nor is the unwillingness necessarily easy to define or recognize empirically. But the intuitive idea behind an addict's unwillingness is this: he is addicted; he knows he is addicted, recognizes the harmful consequences of being addicted, wants to avoid the consequences, and appreciates that doing so requires breaking the pattern (quitting, refraining, avoiding relapse). Such addicts may be able to quit on their own; some need help; some know that they need help; some seek it. You ended up in the wrong office. You now want help from a more person-centered source.

The Nature of Addiction

Addictions are heterogeneous. The patterns picked out as addictive are only more or less generalizable, and there may be different aspects of a pattern that help to assign a person to the category of being addicted.

[1]A sound theory of pharmaceuticals for addiction's substrates must resist falling prey to what may be called the pharmaceutical temptation of the one over the many, which is a variety of monocausal thinking. The neural-causal etiology of addiction is not some one single personal subsystem; for example, it is not a dictatorship of the neural reward system. The best hope for an addict may sometimes be to combine psychotherapies with complements of drugs. But treatment, drugs used or not used, is enfeebled without appreciating that addicts are persons and addiction is a way of being in the world.

Consider the example we are using—a case of addiction to alcohol (a drug, a substance). There are many varieties of drug-related addictive behavior, depending on the drug of choice (DoC), the individual person, the social and cultural contexts within which the addiction progresses, and the person and social practices involving the consumption of the DoC. Often personal preoccupation with the "release" and relief that comes after using the DoC or with craving the DoC when the drug is absent are parts of addictive behaviors—parts but not essential or necessary parts. Addiction can occur without craving or felt release from craving (West 2006). But for purposes of this paper, let's assume craving and release/relief to be elements of addiction (again, of the unwilling addict). Based on characteristics such as craving and release and noting the recent advances in understanding the neural correlates of motivation, decision-making, pleasure seeking, and inhibitory control, some theorists argue that addiction is "a disorder of the brain's reward system" (Gastfriend 2005: 1514). That is not our belief.

No addiction bypasses the brain, to be sure, but whether the brain is not functioning as it should or is ill, diseased or disordered is not to be judged by whether it supports personally harmful or unwanted behavior. Nature did not design the human nervous system to guarantee personal prudence. Learning on a variable ratio schedule of reinforcement (as the brain does) is certainly fitness-enhancing. It kept our ancestors pursuing prey, despite numerous failures and attempts to capture them with unpredictable outcomes. But subjecting oneself to the very same ratio of reinforcement at a racetrack or casino can lead to significant imprudent harm. Drinking liquids is also fitness-enhancing to be sure. Without doing so we would shrivel up and die. But no respected evolutionary biologist who thinks that brains were designed, in part, to control hydration also thinks that it was designed to support drinking too much scotch too often in a bar in Brooklyn or the Bronx. Alcoholism comes like a booby prize for drinking the wrong stuff too often. True, eventually alcohol consumption may harm the brain, when, say, damage to the dopaminergic system or akinesia sets in. But the brain of someone with alcoholism is not disordered or diseased just because it is the brain of someone with alcoholism. Neither is it relevant for whether the brain is disordered that an alcoholic may feel that they have lost control over their drinking. The brain also is designed to entrain motor movements and actions that can proceed with minimal deliberate control after learning. Failing to pay full attention to one's heavy bias to drink is often a prelude to relapse on the part of an otherwise unwilling addict, but these behaviors can be explained without assuming that the brain is broken.

Addictive patterns are formed by an inability to resist impulses that lead to significant harm. A fuller account of the "cognitive-motivational dynamics of addiction" requires an explanation of the "impairment in taking evaluative stock of oneself and of exerting reason-responsive control" (Graham 2013: 181). Developing this fuller and normatively infused account requires thinking of persons as reason-responsive, normative, and self-interpreting agents. Neither brains nor genes, for that matter, are the basic, root or sole cause of addiction; it is not brains or genes that become addicted, it is persons who get addicted (Flanagan 2013a, b; Graham 2013; Tekin 2013b).

An addict (of the unwilling sort) fails to execute normal powers of effective rational agency, according to the twin normative failure model of addiction. She decides not to use the DoC but uses anyway. She assesses her addictive behavior as a rational agent and judges that she does not want to continue, yet her self-assessment does not produce the outcome of effectively and permanently ending her use of the DoC. In a normative sense, the individual "fails to live up to the hopes, expectations, standards, and ideals she has for a good life for herself because of her addiction" (Flanagan 2013a: 1). She believes that life without addiction is a better life, a life in which she can flourish. Still she fails to act in accordance with her belief.

The model that asks us to conceive of the person as a narrative being who comes to exhibit, and often to recognize, that she suffers twin normative failures enriches our understanding of the depth, clinical presentation, and phenomenology of addiction. Nonetheless, some might argue that the model is not scientifically useful, as it makes the concept of the self central to understanding addiction by engaging with the reason responsiveness of the self, self-interpretation, self-assessment, and other self-related feelings and attitudes. This self, it may be objected, is neither empirically tractable nor easily describable; thus, it is not a useful scientific focus if we want to understand and address addiction. In fact, the concept of the self was omitted from the Diagnostic and Statistical Manual of Mental Disorders (DSM), the primary manual of mental disorders used for research and clinical purposes, starting with its third edition (1980), precisely because it was considered an unscientific concept reminiscent of Freudian psychoanalytic approaches to mental disorders (for more on this topic, see Tekin 2014a, b, c, 2015; Parnas et al. in press; Schaffner and Tabb in press).

But let's not get entangled in Freudianism. Yes, the word "self" is ambiguous. Sometimes it just refers to each and every person him or herself. I am myself. You are yourself. My self is none other than me. Your self is none other than you. Sometimes it refers to an internal subject of awareness, a homunculus, separate from the body. What follows gives a detailed explanation of what we refer to when we write of the self.

The Multitudinous Self

We need to say more about the concept of the person as a narrative self-interpreting being. The specific narrative conception we advocate is not a simple linear one, but what we call the picture of the self as multiplex (Flanagan 1991), even multitudinous (Tekin 2014b). We'll explain what it means to say that a person, the self, is multiplex and multitudinous, and how this conception is an empirically and philosophically plausible conception of personhood that captures the complexities of "real people," (Wilkes 1988) including those with psychopathologies.[2] Much like

[2] It is worth noting the distinction between multiplex self and multitudinous self. The multiplex self, a concept introduced by Flanagan, explains what the personal identity of those with typical

other scientific models of other sorts of phenomena, the model of the multitudinous self represents a complex real-world system and can be used for many different purposes by scientists and clinicians. Model builders in science use abstractions and idealizations to make a complex real-world system subject to manipulation, so that particular questions about complex phenomena can be answered. For example, in weather forecasting, scientists create models based on their interests, i.e., the weather forecast for tomorrow vs. the weather forecast for next summer. In a similar vein, the multitudinous self-model sorts out various dimensions of selves or persons, which scientists and clinicians can use as abstractions and idealizations as they make the resulting model fit their interests and purposes. So, we can understand, scientifically study, and intervene on addiction through the multitudinous self.

Here briefly is how the model may be described.

Assume that self, the person, is dynamic, complex, relational, and multi-aspectual, a more or less integrated configuration of capacities, processes, states, and traits supporting a degree of agential capacity subjected to various psychopathologies (Neisser 1988; Jopling 2000; Tekin 2014a, b, c). The multitudinous self-model is built around Ulric Neisser's (1988) account of the self as a complex configuration specified by various kinds of information originating from the subject and her social and physical environment. Neisser argues that the various forms of information individuating the self are quite different, making it plausible to suggest that each establishes a different "self." He tracks five distinguishable if ultimately inseparable selves:

The *ecological self*, or the embodied self in the physical world, which perceives, acts, and interacts with the physical environment

---

cognition consists in. It refers to persons playing multiple roles simultaneously; i.e., the self is multiplex because the person represents and exhibits different parts of themselves to different audiences in different environments. For instance, "my philosopher self, my baseball-coach self, my religious self, my parental self—are played for different audiences. Different audiences see who we are differently" (Flanagan 1996: 71). Suggesting that the self is multiplex means that identity exists despite chronic transformations and synchronic conflicts between these different facets of a person, since there is a narrative connectedness between them. This narrative connectedness is due to the authorial work of the agent, who tells the story of her life and thus holds different strains together. If this unifying authorship falls apart, we end up with atypical cognition: we no longer deal with a multiplex self, but rather with multiple selves. The concept of multitudinous self, introduced by Tekin, on the other hand, considers the self to be complex matrix, individuated and constituted by five dimensions: ecological, interpersonal, temporal, private, and conceptual (following the Neisserian selves). Unlike the multiplex self, multitudinous self considers psychopathology or atypical cognition to be a possible feature of the self. Multiplex self, in so far as it is a conceptual representation of the self to the self, and others, can be situated within the conceptual dimension of the multitudinous self. The model of multitudinous self aims to (i) get at the complexity of "real people," (ii) provide opportunities for scientists to use abstractions and idealizations and study it scientifically, and (iii) encourage a wholesome approach to psychopathology without sidestepping the complexity of persons. The inspiration for the name of this model is the poem "Song of Myself" by Walt Whitman, in which he proclaims, "Do I contradict myself? Very well, then, I contradict myself; (I am large—I contain multitudes)." Special thanks to Flanagan, who steered Tekin in the direction of these lines, hence the word "multitudinous."

The *interpersonal self*, or the self embedded in the social world, which constitutes and is constituted by intersubjective relationships with others

The *temporally extended self*, or the self in time, which is grounded in memories of the past and anticipation of the future

The *private self* which is exposed to experiences available only to the first person and not to others

The *conceptual self*, which represents the self to that individual by drawing on the properties or characteristics of the person and the social and cultural context to which she belongs

Neisser's individual and collaborative work empirically tracks these five selves or aspects of multiplexity in cognitive sciences, including developmental psychology, social psychology, cognitive psychology, and neuroscience, making it a methodology which we believe to be useful for research on psychopathology.

The multitudinous self represents all five dimensions as features of a single person, instead of construing them as separate aspects. This self-organizing and self-interpreting system or person is a locus of agency that remains more or less integrated through time. Each aspect is experienced from the first-person point of view, but can be empirically traced from the second- and third-person points of view, making them instrumental in connecting the subject to herself and to the physical, social, and cultural environment in which she is situated.

The ecological dimension of the multitudinous self represents embodiment in the physical world: brain mechanisms, genetic makeup, and the body's shape and structure enable an individual to perceive and be shaped by the immediate physical environment, to act on and be acted on by it, and to respond to it. Through the ecological dimension, a person enters the physical world and participates in and manipulates it. In the process, she molds and is molded by the physical world. In this sense, the ecological aspect of the self is specified by the body, the physical conditions of a particular environment, and the active perceptual exploration of and response to these conditions. It is present from birth and continues over time, across varying physical and social conditions (Jopling 1997, 2000).

The ecological aspect of the multitudinous self might help track addictive behavior in a number of ways. For instance, there is something going on in the body—in the brain cells, brain's reward system, hormones, genes, etc.—of an individual when he becomes addicted to a certain DoC. Evidence suggests that addiction involves an impaired psychobiology, for example, imprudent desire for the DoC and "eventually a malfunctioning 'off-switch,' such that once one starts drinking there is no telling when (days, weeks, years) one stops" (Flanagan 2013b). The physical basis for addiction is manifested in a series of engagements observable through the ecological dimension of the self. Consider the following examples. During intense craving in the absence of a DoC, the addict's hands may shake. She may get restless and irritable. She may experience sleep disturbances and anxiety, waking up in the middle of the night with nausea or with a desire to consume the DoC. The focus is such that in some cases, it may be "the only tune or story in the addict's head, and nothing else drives it out" (Graham 2013: 178). At times during craving, individuals may even experience life-threatening symptoms such as delirium and hallucinations. This continues until the individual gives in and takes the DoC.

The alcoholic is restless, irritable, and discontented if her required and planned behavior is delayed. This mood lifts, or, perhaps, the awfulness settles, with the first drink. The first drink leads normally to other drinks and drunkenness. The next day she hates himself again. The cycle repeats (Flanagan 2013a, b).

The manifestation of addictive behavior in the ecological dimension of the self is not only experienced by the addict herself but observed by her loved ones; scientists who study her brain, hormones, genes, etc.; and her medical practitioners. The addict experiences the craving and her body's transformation. People around her may observe her restlessness, anger, frustration, and perhaps even hostility toward her loved ones. They may see her neglecting her responsibilities. From the third-person perspective, the doctor may notice she is more tense and quieter than normal, while scientists may gather evidence about the level of the DoC in her blood.

Another possible way of individuating addictive behavior in terms of the ecological dimension of the self comes from animal studies. One neural basis for addiction, according to certain animal studies, lies in the mesolimbic dopamine and brain reward system. It may be possible to investigate whether this area is equally compromised in human addicts. Although scientists may find that a compromised mesolimbic reward system is a necessary condition for human addiction, it does not follow that it is the only necessary condition (see Graham 2013: 191–193). Remember, we are self-interpreting normative beings. Nonhuman animals are not, at least not in the way we are.

Both humans and animals harbor short-term brain pleasure centers, but we, full-on multiplex persons, know what life spans are and engage in long-term planning and reason-responsiveness. Nonhuman animals do not. Our impulses can get us into trouble, and we will know it and regret it and wish to do something about it. A rat will lack any such self-interpretative spin.

The intersubjective aspect of the self is individuated by "species-specific signals of emotional rapport and communication" between the self and other people (Neisser 1988: 387). From earliest infancy, a person enters a social world through interaction with her caregivers (See Trevarthen 1980; Neisser 1988; Fogel 1993; Bowlby 1969; Stern 1993). Through the intersubjective dimension of the self, the person begins the "I-Thou" or interpersonal relationships of care and concern, through which her identity is formed, enriched, or (when care and concern is absent) sometimes impoverished and through which she contributes to others' own identity formation, enrichment, and impoverishment.

Most humans have natural desires for companionship, and most of us recognize, even if only inchoately, that we cannot survive, develop ourselves as persons, or live good lives, that is, lives which are happy as well as meaningful and fulfilling, without situating ourselves in complex socio-moral relationships with others. And despite wide cultural variation in the exact norms governing social practices, we all typically engage in normatively governed practices of "lending and borrowing, promising and consenting, buying and selling, making friends, entering into marriage, establishing a family, offering and accepting aid, and so forth" (Flanagan 2013a: 2 within the single quotes, Deigh 2010: 20).

The intersubjective dimension of the self also tracks or follows addictive behavior patterns in multiple ways. First, forms of addictive behavior and practices in the consumption of DoC progress in a particular kind of social environment. Consider one familiar kind of alcoholic lifestyle, i.e., "the male life of public and gregarious heavy drinking" (Flanagan 2013b: 870). Flanagan writes:

> Because social drinking, especially among males, is widely endorsed, involves a host of well-known social scripts, and because alcoholism takes time to develop, it is uncommon for there to be male alcoholic individuals who become addicted. (Flanagan 2013b: 871)

Here, DoC use becomes the context through which individuals socialize in their professional lives. They go for a drink after a conference and talk about business using DoC. Intoxication seems secondary in these socializations, but for an addict, it eventually becomes the first goal.

Next, addictive behavior not only results in harm to the self, the person, but it also harms others, as the addict engages within a social world (Graham 2013). Recurrent use of the DoC may result in a failure to fulfill major role obligations at work, school, or home. Recurrent social or interpersonal problems may be exacerbated by the effects of the DoC. Typically, significant problems start or get worse in the family. For instance, under the influence of the DoC, the addict may be violent toward her loved ones or give up important social, occupational, or recreational activities.

The temporally extended aspect of the self consists of the person in time and memories of the individual's past and anticipation for the future. It relies on autobiographical memory and other stored information. What the individual recalls depends on what she now believes, as well as what she stored. Addictive behavior is manifest in the temporally extended self in a variety of ways. Consider the common situation of an addict's preoccupation with planning the next DoC intake. In her memoir *Drinking: A Love Story*, Caroline Knapp says while she was doing her job at the office, she would be planning that moment in the afternoon when she could go for a drink with her colleagues (Knapp 1997). That moment of reward would be in the back of her mind, giving structure to the rest of her actions. The minute the clock hits 4, she would ask if anyone wanted a "quick" drink at the bar. It had to be called a "quick" drink, she writes, as she sought to create the illusion of a busy life to which she would shortly return. In reality, her life was planned around changing locations to continue drinking; for example, she might be planning to go home to open a bottle of cognac. If getting to the "happy hour" was delayed, Knapp writes, she would feel restless and uncomfortable overcome with the desire to drink. Addicts often talk about how their perception of the world is tainted by planning the next substance intake or that they cannot focus on anything but the anticipation of the next "high."

The private aspect of the multitudinous self traces the individual's felt experiences that are not phenomenologically available to anyone else (such as feelings of pain or disappointment). This first developmentally appears when children notice that some of their experiences are unique to them. Addiction is manifest in the private aspect of the self. Caroline Knapp describes drinking alone (addicts increasingly do this to hide their drinking from others) as entering into a room of one's own and closing the blinds, turning inward. The DoC becomes the only company she enjoys.

Finally, let us turn to the conceptual aspect of the multitudinous self. Self-concepts selectively represent the self to the self, to the person. They are the products of the dynamic interaction between the four aspects of the self and the features of the social and cultural environment. In turn, self-concepts inform and shape the aspects of the self, as well as some features of the social and cultural environment. They are thus informed by the features of the four aspects of the multitudinous self and by the individual's embodied experiences in the world, for example, illness (Neisser 1988; Jopling 1997; Tekin 2011). Consider each in turn.

Self-concepts include ideas about and evaluations of our physical bodies (ecological aspect), interpersonal experiences (intersubjective aspect), the kinds of things we have done in the past and are likely to do in the future (temporally extended aspect), and the quality and meaning of our thoughts and feelings (private aspect) (see Jopling 1997, 2000; Neisser 1988). For instance, an individual's self-concept as a "friendly person" is the product of the intersubjective aspect of her selfhood and also of the norms of friendliness in the culture of which she is a part. Self-regarding feelings and attitudes, such as self-confidence, security, self-esteem, self-respect, and social trust, emerge as we develop self-concepts and as the different dimensions of the self interact with the social and cultural world, through an exchange between the self and others.

Self-concepts are also informed by pathologies to which the individual is subjected. This influence is mediated by the changes that occur in the ecological, intersubjective, temporally extended, and private aspects of the self owing to pathology, by the scientifically based or folk-psychological knowledge available to the individual about her illness, and by her self-narratives in making sense of her condition (Tekin 2010, Tekin 2011, 2014a, b, c, 2015). For example, alcohol addiction affects an individual's intersubjective dimension by, say, making it difficult for her to keep promises, such as picking up her son from school at the time she promised or meeting with her client at the specified time. Failing to follow through on promises and breaching others' trust over time may alter her self-image as a reliable person. She may develop feelings of frustration or even hatred toward herself. She may decide to stop making promises or even taking on responsibilities that require keeping promises. In addition, the addict may develop "self-regarding reactive attitudes of bewilderment, disappointment, and shame" about her addiction (Flanagan 2013a: 6).

Consider relapse as another example. The addict refrains from the addictive behavior during certain periods; however, cessation ultimately proves unsuccessful:

> They "fall back" into the detrimental behavior after a period of temporary stoppage. After the relapse, the individual self-interprets himself as a failure, relapse becomes a source of shame, regret, self-blame, and embarrassment or as grounds for diminished self-confidence or self-esteem. (Graham 2013: 179)

These experiences influence one's self-concept: for example, the individual may think he lacks self-discipline. What an addict learns about the course of his condition from various scientific and popular media may also lead him to alter his self-concepts. For example, memoirs may offer sources of better self-understanding and motivation. Or learning more about the scientific research on the link between the anomalies in brain's reward system and addiction may cause the addict to consider himself a victim, inhibiting his motivation to improve his condition.

Self-concepts are not only representations of the self to the self; they are also action-guiding (Tekin 2014a, b, c, 2015). Our ideas about ourselves inform how we behave. Generally speaking, a person's self-concept of her physical strength affects her physical activities. She may or may not try to lift a suitcase, depending on how strong she feels and how heavy she perceives the suitcase to be. Similarly, her concept of her intelligence and ability to learn new philosophical material will influence what she can actually learn or how well she does in a job interview. In the context of addiction, the self-concepts formed or altered in this vein influence future actions. Hopelessness in the face of repeated relapses and self-concepts such as being weak-willed may diminish an addict's ability to quit the addictive behavior. Alternatively, after a few months' success at staying away from the DoC, the individual may be heartened and continue to try hard. In addition, perceiving herself as someone who needs help, an addict may reach out to the communities of other individuals with addiction who have experienced a similar condition. The success of Alcoholics Anonymous programs partially owe to this.

In sum, self-concepts motivate the person to think, act, and behave in certain ways, restricting or expanding his or her possibilities for action (Tekin 2011, 2014a, b, c, 2015; Jopling 1997).

Limitations of Pharmaceutical Remedies for Addiction

The model of addiction and the addict we have proposed is simple. Addicts are multiplex persons, and addiction is a lifestyle or for the addict an integral part of her overall life. The alcoholic pattern normally comes to be experienced as harmful by the addict herself. It is not a way of being that meshes well with the set of goals and aspirations she has for a life well lived. In fact, the alcoholic lifestyle may be killing her physically, psychologically, and spiritually. Most addicts come to experience themselves as suffering from a way of being that can't coexist with the other strands of the dynamic multiplex equilibrium we think makes for a good life. Multiplex persons are used to juggling many roles, commitments, and relations. For the addict, this strand, this part of her being, is intruding on all the rest. She can no longer succeed as a multiplex. The DoC comes to win all competitions among aspects of herself. The alcoholic comes to see that she is not living well, that using is defeating that project, but she cannot regain control of herself.

The concern that addicts viewed as persons suffering twin normative failures is not scientifically grounded or useful is misguided. By comparison, both the brain reward system failure (or chemical disorder or disease model of addiction) and proposed pharmaceutical solutions are scientifically incomplete because they step around the complexity of the multitudinous self and fail to embrace the multifaceted aspects of and heterogeneous character of addiction. According to the twin normative failure model of addiction, the individual fails to fulfill her promise to herself by not quitting despite resolving to do so. She fails to "execute normal powers of effective rational agency" (Flanagan 2013a). The individual also fails to live up to her own standards of a good life—a life lived in moral-social space; her expectations are not fulfilled because of her addiction (Flanagan 2013a). Both can be mapped onto the five dimensions of the self, which, as shown, are empirically trackable and phenomenologically responsive.

An addict can assess her practices along five dimensions according to her consumption of a particular DoC. For example, an alcoholic can trace her addictive behavior through her ecological dimension. She knows she is drinking too much, too often. She can observe the symptoms of withdrawal when she has not had a drink for a while. During cravings, her hands shake, she gets restless, and she feels nauseated. In addition, she can also trace her drinking behavior through her intersubjective dimension. She does not remember the last time she was intimate with someone without drinking. She becomes aggressive when she drinks, so much so that she has alienated her loved ones. Her relationships with her partner and children have significantly deteriorated. She can also make sense of the shifts in the temporally extended dimension of herself. All she is planning is the next time she will drink; other future commitments are secondary. Her life narrative is impoverished because she no longer imagines her child's graduation or other things that otherwise would matter a great deal. As for her private dimension, she is aware that she is not happy with her life and her only relief is consuming more alcohol. Based on these observed experiences of the five dimensions of the multitudinous self, she resolves to quit drinking. Her assessment is that her experiences in each of these dimensions will improve. She promises herself to quit drinking, start exercising, reinvigorate her relationships, etc. However, she cannot execute her decisions and, thus, fails to deliver on her promises. And she is aware at a conceptual level that her rational executive functions have deteriorated.

The second normative failure is the individual's failure to living up to her personal standards of what constitutes a good life—in social space. The alcoholic woman mentioned above will correctly assess that she would have a good life if the dimensions of her self-experience were different. Ecologically, she knows her quality of life would improve if she did not crave alcohol, waking up in the middle of the night to reach for the next glass. She knows that she wants to improve her intersubjective relationships; she wants to reconnect to her family. She wants to change so that her temporally extended dimension revolves around larger life plans—not preoccupied with alcohol and drinking. At a private level, she does not want to continue hiding her drinking from others or to experience deep feelings of shame and guilt. All these are necessary for the good life she imagines herself having. However, by failing to execute rational control over her drinking, she also fails to meet her own standards of a good life.

The picture of addicts as multiplex persons who suffer twin normative failures is empirically credible. Therapeutic interventions can be developed to target normative failures by conceptualizing them from the different angles recommended by the multitudinous self-model. Furthermore, this model is more powerful than the competition. Consider again the claim that addiction is just (or primarily) a brain chemistry problem, adequately treated by pharmaceutical interventions alone. The brain chemistry failure model of addiction only embraces the ecological dimension of the self by individuating the physical dimension of addictive behavior and encouraging treatments limited to pharmaceuticals. In contrast, by engaging with the complexity of the selves or multiplexity of persons, the twin normative failure model of addiction facilitates effective interventions through multiple methods by focusing on each dimension. The strength of the twin normative failure model is that it accomplishes everything that the brain chemistry failure model would accomplish. The twin normative failure model would embrace, for instance, the invention of a miracle drug that completely and permanently removes the obsession with the DoC, by suggesting that the improvement in the ecological dimension of the self will gradually improve all the other dimensions of the self. The brain chemistry model, on the other hand, would fail to engage with the interpersonal or temporal aspects of addiction, in the absence of a miracle drug.

Following this analysis, one may ask whether it is plausible to imagine a cocktail of drugs to address all the dimensions of the self compromised by addiction. Thus, a drug regimen that treats alcoholism might include components that treat lack of self-esteem and shyness, curb the desire to drink, and help the addict sleep better. Our answer is yes, this is conceptually or imaginatively possible, but only insofar as the person continues to work on addressing her twin failures through her work on herself as well as psychotherapeutic and interpersonal support. In short, we remain unconvinced that drugs can replace the work one does on one's self *qua* self or *qua* others in addressing addiction; they can only support it.

You've left the doctor's office and not committed yourself to treatment at the clinic, not out of disrespect for her or because you feel empowered to do it by yourself or because you dismiss the use of drugs to control your condition. You have the feeling that whatever pharmaceuticals you take, if you can be assisted by them or may even need them, should be highly dependent upon the context of your relationship with the doctor who prescribes them and the psychotherapist who will discuss them with you. You want therapy to connect with you as a person, not as a receptacle. You are a multiplex person. You contain multitudes. Your addiction isn't in your head, or in your brain, or in a subsystem of your brain. You are an addict, specifically an alcoholic. Addiction is a way of describing a certain destructive way you find yourself being in the world. You need to change the way you live. Luckily there are multiple ways of entry to leverage a life and get it back on track. We recommend whole person cures.

References

Bentall, R. 2009. *Doctoring the Mind: Is Current Treatment of Mental Illness Really Any Good?* New York: New York University Press.

Bowlby, J. 1969. *Attachment and Loss*. London: Hogarth Press.

Deigh, J. 2010. *An Introduction to Ethics*. Cambridge: Cambridge University Press.

Dennett, D. 1991. *Consciousness Explained*. Boston: Little, Brown, and Company.

Flanagan, O. 1991. *Varieties of Moral Personality*. Cambridge, MA: Harvard University Press.

———. 1996. *Self Expressions: Mind, Morals, and the Meaning of Life (Philosophy of Mind)*. New York: Oxford University Press.

———. 2013a. The Shame of Addiction. *Frontiers in Psychiatry* 4(120): 1–11.

———. 2013b. Identity and Addiction: What Alcoholic Memoirs Teach. In *The Oxford Handbook of Philosophy and Psychiatry*, ed. K.W.M. Fulford et al., 865–888. New York: Oxford University Press.

Fogel, A. 1993. *Developing Through Relationships: Origins of Communication, Self, and Culture*. London: Harvester Wheatsheaf.

Frances, A. 2013. *Saving Normal*. New York: Harper Collins.

Gastfriend, D.R. 2005. Physician Substance Abuse and Recovery: What Does It Mean for Physicians and Everyone Else. *Journal of the American Medical Association* 293: 1513–1515.

Graham, G. 2013. *The Disordered Mind: Philosophy of Mind and Mental Illness*. 2nd ed. London: Routledge.

———. in press. *The Abraham Dilemma: A Divine Delusion*. Oxford: Oxford University Press.

Jopling, D. 1997. A 'Self' of Selves. In *The Conceptual Self in Context*, ed. U. Neisser et al., 249–268. New York: Cambridge University Press.

———. 2000. *Self-Knowledge and the Self*. New York: Routledge.

Knapp, C. 1997. *Drinking: A Love Story*. New York: Dial Press.

Neisser, U. 1988. Five Kinds of Self-Knowledge. *Philosophical Psychology* 1: 35–39.

Schaffner, K., and K. Tabb. in press. Hempel as a Critic of Bridgman's Operationalism: Lessons for Psychiatry from the History of Science (a Response to Bovet and Parnas). In *Philosophical Issues in Psychiatry III*, eds. Kendler, K.S., et al. Oxford: Oxford University Press. Chapter 24.

Stern, D. 1993. The Role of Feelings for an Interpersonal Self. In *The Perceived Self: Ecological and Interpersonal Sources of Self-Knowledge*, ed. U. Neisser, 205–215. New York: Cambridge University Press.

Tekin, Ş. 2011. Self-concept through a Diagnostic Looking Glass: Narrative and Mental Disorder. *Philosophical Psychology* 24(3): 357–380.

———. 2013a. How Does the Self Adjudicate Narratives? *Philosophy, Psychiatry, and Psychology* 20(1): 25–28.

———. 2013b. "Will I be Pretty, Will I be Rich?" The Missing Self in Antidepressant 682 Commercials. *The American Journal of Bioethics* 13(05): 19–21.

———. 2014a. Self-Insight in the Time of Mood Disorders: After the Diagnosis, Beyond the Treatment. *Philosophy, Psychiatry, Psychology* 21(2): 139–155.

———. 2014b. The Missing Self in Hacking's Looping Effects. In *Classifying Psychopathology: Mental Kinds and Natural Kinds*, ed. Kincaid et al., 227–256. Cambridge, MA: MIT University Press.

———. 2014c. Psychiatric Taxonomy: At the Crossroads of Science and Ethics. *Journal of Medical Ethics* 40: 513–514.

———. 2015. Against Hyponarrating Grief: Incompatible Research and Treatment Interests in the DSM-5. In *The DSM-5 in Perspective, History, Philosophy and Theory of the Life Sciences*, ed. S. Singy et al. Dordrecht: Springer Press.

Trevarthen, C. 1980. The Foundations of Intersubjectivity: Development of Interpersonal and Cooperative Understanding in Infants. In *The Social Foundations of Language and Thought: Essays in Honor of J. S. Bruner*, ed. D. Olson, 316–342. New York: W.W. Norton.
West, R. 2006. *Theory of Addiction*. Oxford: Blackwell.
Wilkes, K. 1988. *Real People*. Oxford: Oxford University Press.

Preventive Self-Help and the Six Nonnaturals: Remedies from Burton's *Anatomy of Melancholy*

Jennifer Radden

Introduction

Burton's *Anatomy of Melancholy* (1621) provides us with a compendium of remedies against melancholy moods and dispositions, some, such as the herbs hellebore and borage, so popular since ancient times that they might be said to correspond to fluoxetine in today's pharmacopeia. Several ideas and themes in the *Anatomy* with relevance for the understanding and treatment of depression in our own times are introduced in the following chapter. Most central for our discussion, these herbal and other remedies were to be taken as part of a broader regimen of which no single part could be omitted. The regulation of exercise, fresh air, sleep, diet, evacuation, and feelings, believed to together keep the bodily humors in healthy balance, demanded habits and practices that were essential accompaniments to one another and to other measures. This was eclectic and holistic healing, *only effective* when combined. Adhering to this regimen was, for the most part, the individual's own responsibility: an extensive self-help program. It was also preventive medicine, thought to anticipate, monitor, and ward off the symptoms of melancholy before they became entrenched and difficult to treat. Underlying and reinforcing these features and the recommendations they implied, I want to point out, were general ideas about the causes of, and remedies for, melancholy. And these ideas are consonant with some recent challenges to the application of mainstream disease conceptions when it comes to depression. Inasmuch as Burton seems to eschew "common cause" etiological models, as he does "magic bullet," or single-remedy, assumptions, this account corresponds surprisingly closely to the alternative, network-based models proposed for mood disorder.

Thus, because much remains unsettled in the way affective disorders, and particularly depression, are understood and treated, Burton's prescriptions for eclectic,

J. Radden (✉)
Philosophy Department, The University of Massachusetts Boston, Boston, MA, USA
e-mail: Jennifer.Radden@umb.edu

© Springer Science+Business Media B.V. 2017

D. Ho (ed.), *Philosophical Issues in Pharmaceutics*, Philosophy and Medicine 122,
DOI 10.1007/978-94-024-0979-6_14

preventative, and largely self-directed healing, together with his underlying assumptions about the nature of causes and effects, deserve our careful attention. In the chapter that follows, I first provide some preliminary qualifications, background about Burton, his book, and herbal remedies for melancholy in his time, and a summary of the relevant ideas from the *Anatomy*. I then turn to the contrasts and similarities between ideas about remedying melancholy and about understanding and treating depression in our own day.

Preliminary Qualifications

The setting where hellebore and borage were recommended for melancholy differs vastly from today's, of course, as historical scholarship has illustrated.[1] The exact relationship between the disorders of melancholy and today's depression is complex and contested, moreover, and we invite anachronism by *equating* the earlier-described condition with present-day complaints (Radden 2003, 2009, 2013; Varga 2013; Foucault 2006; Berrios 2011). Yet strong analogies unite the sad and fearful moods characterizing the melancholy depicted by Burton with the suffering of those diagnosed as depressed today. What Burton calls "an habit of Melancholy" possesses characteristics that are held and defended in relation to depression: in the diversity of its symptoms; its link to anxious, apprehensive, and fearful states; its association with objectless and pervasive moods as well as more directed feelings; its manifestation in both severe, even delusional, states and subclinical conditions and traits; and its ruminative and solitary symptom profile. Parallels such as these are sufficient warrant for the present exploration.

The Book

Burton lived between 1577 and 1640; most of his life was spent quietly, at Oxford. His famous book on melancholy includes medicine and psychology from ancient sources as well as his own times. A 900-page first edition (1621) was followed by four ever-larger versions, together with one published posthumously (1651).[2] The book is divided into three parts (partitions), with the causes and cures of melancholy the topics of the first and second partitions, respectively, and those are our focus. The *Anatomy* is more a compendium of earlier writing about melancholy than an original, or systematic, work. And it gives little evidence of the scientific observation and method associated with the emerging thinking of the early seventeenth

[1] Recent historical works include Arikha (2007), Gowland (2006), Lund (2010), and Schmidt (2007).

[2] All references are to the 1632 edition of the *Anatomy of Melancholy* (Burton 1989–2000, Thomas Faulkner, Nicolas Kiessling, and Rhonda Blair eds).

century. Rather, it combines the humoral theories of the Greeks and Romans, Stoic ideas about the regulation of the passions, the Aristotelianism of medieval Christianity, the astrology of Renaissance humanism, and late Renaissance theories and observations about melancholy (such as that of the French doctor André du Laurens), with the homely style of self-help manuals.

Humors, Faculties, and Mental Processes

Following ancient humoral lore codified in Galenic medicine, constantly fluctuating mixtures make up individual humoral states ("complexions" or "temperatures"), and the humoral imbalances indicating melancholy are depicted as varying unceasingly within each person.[3] With their origins in Hippocratic and Aristotelian writing, these theories had been faithfully employed until Burton's day. Melancholy symptoms included disordered moods, particularly unwarranted sadness and fear. Imbalances between the bodily humors (black and yellow bile, phlegm, and blood) explained temperamental variation among different people as well, serving to link abnormal melancholy moods with normal temperaments.

The assumptions about the mind and mental processes in the *Anatomy* reflect Aristotelian faculty psychology where, within the sensible soul, the apprehensive faculty comprises the five outer senses as well as the three inward ones of commonsense, memory, and phantasy or imagination. The imagination and passions continually affect one another, and embodied, multi-directional, causal interactions result: "...the body works upon the mind by his bad humours, troubling the spirits, sending gross fumes into the brain, and so...disturbing the soul and all the faculties of it, with fear, sorrow, etc...: so, on the other side, the mind most effectually works upon the body, producing by his passions and perturbations miraculous alterations..." (I,2,3,1: 247). Mental states will be reflected in the particular, momentary, humoral temperature of any given person, and sometimes, but not always, humoral states are causally responsible for changes within the mind. Imaginings affect bodily states and feelings, and bodily states similarly affect psychic functioning, these different states "treading a ring," in Burton's description of their unceasing multi-directional interactions or feedback loops.

Melancholy is associated with an overactive or disordered imagination that "misinforms" the heart, introducing misleading or false ideas that prompt feelings inappropriate to their object.[4] (The imagination is so restless, operative, and quick in those inclined to melancholy, Burton says, that it is likely to "... work upon itself, melancholize, and be carried away instantly, with some fear, jealousy, discontent, suspicion, some vain conceit or other" (II,2,6,2: 107).) The resonant phenomenology

[3] For this aspect of humoral lore, see Arikha (2007) and Gowland (2006).

[4] The imagination, in his words "misinforming the heart, causeth all these distemperatures, alteration and confusion of spirits and humours" (I,2,3,1: 249). On the role of imagination, see Haskell (2011).

of melancholy distress is typified by moods of sadness and apprehension without, or without sufficient, cause or warrant: "fear and sadness without cause" are melancholy's "most assured signs, inseparable companions, and characters" (I,3,1,2: 384). Reason should, and usually can, rule the passions and the imagination (II,2,6,1: 103). In more severe and entrenched cases of the disorder, however, judgment is more thoroughly affected. Failing to curb the dangerous excesses of the imagination, reason loses its own grip on logic, perpetuating error and confusion.

Hellebore and Borage

Of all the remedies for preventing, averting, and healing melancholy, hellebore and borage arguably correspond most closely to fluoxetine in today's pharmacopeia. Helleborus or Hellebore names a herbaceous perennial flowering plant from the botanical family of *Ranunculaceae*. White and black hellebore were each included in ancient and medieval medicine, although the white hellebore is in our day recognized to have been a different plant. It was a "strong purger upward" (II,4,2,1: 228), while "that most renowned plant," black hellebore, is described as a "famous purger of melancholy, which all antiquity so much used and admired" (II,4,2,2: 231).[5] Borage (*Borago officinalis*) is a member of the comfrey family and has similarly ancient origins as a remedy, known for its cooling properties and its use with mood regulation and mental disorder. During the years Burton worked on the *Anatomy*, astrologer and herbalist Nicholas Culpeper was writing his own comprehensive and authoritative volume on herbal medicine (first published in 1653). There, describing hellebore as the "herb of Saturn" to emphasize the astrological connection with melancholy, Culpeper notes that "The roots are very effectual against all melancholy diseases, such as are of long standing, as quartan agues, and madness…" (154).[6] Borage is identified as an herb of Jupiter (another of the planets linked to melancholy); all or any of its leaves, flower, and seeds, Culpeper says, "are good to expel melancholy" (53). That hellebore and borage were of particular importance in the treatment and prevention of melancholy is suggested by their place in the engraved frontispiece Burton had designed for the 1632 edition.[7]

These two plants were part of a much larger number used in Burton's time, known to him, and recommended in the *Anatomy*. And herbal medicine not only employed a range of plants, these treatments used every part of each one, Burton notes: "substance, juice, roots, seeds, flowers, leaves, deconcoctions, distilled waters, extracts, oils, etc…," in addition to conserves and syrups derived from them (II,4,1,3: 216). These substances were simple and compound, gentle and violent,

[5] The use of black hellebore as a purgative by no less of an authority than Hippocrates is taken to confirm that, employed carefully, this plant is a strong and effective remedy against melancholy.

[6] Culpeper's work was first published as *The English Physician, or Herball.* (Culpeper 1985).

[7] Borage and Hellebore are depicted in two of the eleven scenes and described as "The best medicines that ere God made" (Argument of the Frontispiece (lxii)).

upward (to produce vomiting) and downward (as enemas) in their purgative effects, and are all carefully explained in the *Anatomy*.[8]

Evidence-based support for any treatment was limited in Burton's era, and he was anyway more inclined to put his trust in medical treatises than direct observation. Yet we cannot dismiss the confidence he shows here, as we shall see, since these remedies were supposed effective only in conjunction with an array of other interventions. Moreover, like many of his contemporaries (including Montaigne), Burton recognized the power of expectation effects and would have in part credited to them whatever healing occurred. The particular power of the imagination to bring about such effects in those tending toward melancholy is one of the most distinctive themes in the *Anatomy*, and Burton may be said to anticipate the link subsequently confirmed between placebo effects and unipolar depression.

Remedies: The Six Nonnaturals

Burton's recommendations about the six nonnaturals of Galenic medicine resemble other popular English writing on health from the Renaissance, which provided simplified humoral ideas to explain prescribed regimens.[9] Exercise, fresh air, sleep, diet, evacuation, and perturbations provoked by the passions were "nonnatural" properties in not themselves being constituents of the body, while affecting it. By regulating each of these aspects of our daily regimen, we should for the most part avoid the humoral imbalance associated with melancholy.

The passions were intermediary states, located in the interstices between the wholly mental and the solely corporeal. Because melancholy was primarily a disorder of affect, the perturbations of the mind, or passions, were particularly implicated in it. The effects of the passions on the imagination, and its on them, each engendered melancholy. If we give reins to such feelings (lust, anger, ambition, and pride are those named) and "follow our own ways," Burton explains, " we...heap upon us this of Melancholy, and all kinds of incurable diseases..." (Burton I,6,3,2: 128). That the impressive yet alarming power of the imagination is finally much more implicated in melancholy than any humoral arrangements is a central tenet of the *Anatomy*. ("...much more ought the cause of melancholy be ascribed to this [imagination] alone than to the distemperature of the body" (I,2,3,1: 249).) These ideas are derived from classical philosophy as well as medieval Christianity. In Hellenistic thought and the neo-Stoicism of the Renaissance, the passions combine with false and distorted beliefs to prevent contentment and tranquility; they invite temptation and sin, in Christian doctrine.

[8] The purgative function of these herbs, it has been emphasized, was symbolic as much as actual; purging purified: "...the replacement of the thick, overloaded blood of melancholics, heavy with bitter humours, with the light, clear blood whose fresh movement would dissipate delirium" (Foucault 2006: 310).

[9] On the prevalence of such self-help texts in that era, see Lund (2010).

Explaining how the risks of entrenched melancholy are to be avoided, Burton emphasizes that melancholy has an infinite number of causes: not only behavioral and mental habits, and momentary humoral fluctuations springing from individual temperament, tendencies, and bodily changes, but all manner of experiences, even astrological signs and, arguably, demonic influences.[10] Remedies and preventive measures and cures reflected an equivalent array.

Eclectic, Preventive, Self-Help

Attention to the six nonnaturals was primarily preventive medicine, and it was the responsibility, to a great degree, of the individual. Melancholy must be avoided and averted before it has become habituated. Of the good rules and precepts we learn from physicians, it is noted, the first is to "withstand the beginning." For "he that will but resist at first may easily bee a conquerer at the last" (III,2,5,2: 207).

Insisting that the person can and should become his own physician, Burton follows the ideas laid down by his Stoic guides. Since it entailed mental and behavioral habits, attention to the six nonnaturals by which more severe melancholy conditions were to be averted lays within a person's grasp. These habits could be acquired and exercised through nothing more (or less) than regular practice, close attention, and self-control. In addition, such autonomy and self-mastery were always to be encouraged: with confidence in the power of reason to avoid moral failing, Burton strongly approved of self-help.

Themes about avoiding the perturbations of the soul in the *Anatomy* also show the influence of Stoicism.[11] The practices ascribed to Chrysippus and described by Epictetus and Seneca seek truth through clarity, coherence, and logical validity. They involve careful and rigorous assessment of evidence and sound reasoning. (When "discontent…sad and heavy," Burton says, we must ask *why* and *upon what ground*? Consider of it, he admonishes, "examine it thoroughly" (II,2,6,1: 103, emphasis added).) Attentively employed, these practices will serve to avert, dispel, and heal melancholy symptoms. Daily cognitive exercises aimed to eliminate unwarranted belief and ill-formed judgment must particularly be directed toward flights of the imagination; moreover, appealing as it is, the unfettered play of the imagination brings danger.[12] Imaginings and feelings are an endless goad to one another, our imaginings and errant beliefs prompt feelings, and our feelings in turn

[10] Scholars differ somewhat in the extent to which they interpret Burton's seemingly cautious remarks on demons. See Gowland (2006).

[11] The influence of Epicureanism in Europe came later, during the second half of the seventeenth century (Long 1986: 242).

[12] These "phantastical and bewitching thoughts… so urgently, so continually set upon, creep in, insinuate, possess, overcome, distract and detain…[some people] cannot go about their more necessary business, stave off or extricate themselves, but are ever musing, melancholizing, and carried along…" (I,2,2,6).

foster imaginings. To curb unruly and overweening passions, we must monitor and correct the fantasies and willful blindnesses that result from the solitary, careless, self-indulgent ruminating he calls "melancholizing." (Distraction was also essential in avoiding and averting melancholy: regular conversation with others, for example, particularly friends (II,2,6,2: 106). Solitary pursuits and idle habits must be avoided (I,2,2,6: 238). Work, business, reading, travel, social activity—such "behavioral" means must be sought to turn us away from melancholy thoughts and moods and dangerous imaginings that would unsettle our feelings.)

The primary emphasis within these efforts was early prevention. Melancholy may be treated or at least mitigated, as he says, by "... some contrary passion, good counsell and perswasion, if it be withstood in the beginning, maturely resisted, and as those ancients hold, *the nayles of it be pared befor they grow too long*" (III,3,4,1: 306).[13]

Prevention and Self-Help for Depression

Turning now to depression: the Institute of Medicine Report (Mrazek and Haggerty 1994; O'Connell et al. 2009) has recommended that prevention be defined as those interventions that occur prior to the onset of a clinically diagnosed disorder, i.e., the initial onset of what is taken to be a single, chronic disease.[14] Conformity with this usage is uneven across clinical and research settings, however. Among assessments of preventive efforts, some blur or disregard the distinction between first and subsequent depressive episodes, while others have focused on second and subsequent, rather than initial, ones.[15] And measures of prevention have included qualitative issues of severity, as well as various quantitative ones relating to chronicity (e.g., the number and length of episodes and the length between reoccurrences).[16]

Although well established for some other chronic medical complaints, preventive efforts to avoid or avert initial episodes of depression have received little systematic and sustained research attention (McLaughlin 2011). This is now changing.[17] In the 1990s, projections drew widespread attention to the magnitude of depression as a serious and growing public health problem, measuring and predicting the cost of untreated depression in terms of mortality and morbidity (Murray and Lopez

[13] Burton uses the words "cure" and "cured" in describing these treatments but confusingly to our modern ears, since cure has come to mean completed and successful remedy.

[14] From the National Research Council and Institute of Medicine Committee on the Prevention of Mental Disorders and Substance Abuse Among Children, Youth, and Young Adults are recommendations for "interventions *before the disorder occurs* [that offer] the greatest opportunity to avoid the substantial costs to individuals, families, and society that these disorders entail" (O'Connell et al. 2009: 1, emphasis added).

[15] For some perspective, one estimate gives the relapse rate of symptoms during remission and recurrence after an episode of untreated depression at between 50 and 80% (Biegler 2011: 153).

[16] See Kupfer et al. (1989).

[17] See Cuijpers (1997).

1997).[18] A public health perspective has brought recognition that the young must be the target of preventive interventions: "behavioral vaccines" are proposed for mental, emotional, and behavioral (MEB) disorders in young people (Embry 2011, Merry et al. 2012). At the same time, cognitive forms of therapy, with the subject an active learner capable of not only monitoring but adjusting and regulating affective states, have become popular. Closer to the lines Burton proposed for melancholy, depression has been associated with efforts to alter cognitive habits, a didactic approach with a ready aid in the Internet that brings new models for the self-directed patient (Hollandare et al. 2011; Van Voorhees et al. 2011; Van Straten et al. 2008).

Burton acknowledges a degree of severity which, when reached, leaves the sufferer beyond the help provided by preventive measures and incapable of self-help. And similarly, some data seems to suggest that only milder forms of depression can be effectively treated with adjustments to some combination of healthy habits and CBT, the more severe, intractable cases requiring a different approach.[19]

There is little consensus today over what is in our power to change directly by adopting healthy habits, what we must accept, and that for which it is appropriate, or even imperative, to seek expert treatment. Added to that, self-help approaches contain assumptions about the patient's agency and autonomy that are contrary to the passive role often attributed to the depression patient. Acknowledgment that the psychiatric patient is powerless in the face of a debilitating disease -often advocated to reduce negative attitudes -has been particularly emphasized for depression, where depleted agency, a diminished sense of autonomy, and reduced motivation are characteristic aspects of the symptom picture. The admonition to recognize and seek help for one's depression is frequently heard. Yet how, and how much, we can help ourselves remains a nagging, unresolved issue at the heart of depression treatment. And since those who suffer depression vary within the course of their own disorder in the extent to which agency and autonomy are compromised, as well as varying from one another, it is not one that seems likely to find easy resolution.

That self-help is an ethically preferable response if it can be achieved and is effective is rarely doubted. The long-term, personal benefits of taking an active role in one's treatment (here placed in contrast to using antidepressant drugs) have been noted by Joanna Moncrieff: "If people believe that …they have only improved because a drug helped to rectify a chemical defect or imbalance, then they are likely to fear the recurrence of depression with every difficult period of their lives…doctors and health professionals want to help people to help themselves over depression... What they fail to realize is that every prescription they issue conveys a message of hopelessness and powerlessness" (Moncrieff 2009: 172–3).

---

[18] The Global Burden of Disease Study's predictions that depressive disorder would rise to second among the causes of disability were confirmed for some countries by 2002, by which time it ranked first in others (Mathers et al. 2008).

[19] Recent research has distinguished "true" melancholic from non-melancholic depression employing criteria including severity and treatment responsiveness (Taylor and Fink 2006, Shorter and Fink 2010, Shorter 2011, and Parker and Hadzi-Pavlovic 1996).

It seems entirely plausible that a person's self-conception and agency will be affected as Moncrieff describes and that altering the sense of self and efficacy in these far-reaching ways is especially important with episodic disorders such as depression. Yet this desirable state of self-efficiency is approached by degrees. Even following a drug regimen, which may involve tolerating uncomfortable side effects, or enduring a sequence of medications before finding the right dosage and type, offers some experience of agency (and at some stages of her depression, it might also be all the patient is capable of). Certainly most doctors and health professionals want to help people help themselves, as Moncrieff notes. The controversy here lies not over the matter of *valuing* self-help so much as over whether self-help can be effectively achieved.

Other Similarities and Differences

Burton's eclectic approach has superficial echoes in the plethora of alternatives and a seemingly numberless range of traditional and nontraditional remedies offered for depression and anxiety today. But there remain differences. Consensus about the treatment of depression is lacking over at least three features endorsed by Burton for melancholy. First, Burton sees each of his remedies and preventives as at best complementary. All the nonnaturals and behavioral and cognitive habits must be adhered to together. Alone, they would be ineffective. In treatment for depression today, by contrast, differing remedies are often portrayed as alternatives that are individually sufficient, or even necessary and sufficient, for healing. The second difference, and Burton's ace in the hole here, is his firm confidence in the healing power of the imagination. He recognizes and applauds what we would today deem placebo effects; for him, they were important ingredients in effective treatment. (Sometimes, he observes, nothing more than "a strong conceit or apprehension…will take away Diseases" (I,2,3,2: 253).) By contrast, although placebo effects have consistently shown themselves to affect, and reduce, depression, within today's research protocols, these effects are designated as inert; and the clinical use of placebos is prohibited on ethical grounds. And finally, Burton's account is multi-causal, while many contemporary models are not. Every possible factor of personal natural history is named a source of melancholy. Each of these features of Burton's account is applicable to depression, we'll now see.

Integrated Complementary Remedies

Only together can the various remedies for melancholy be effective, Burton insists. And since the last decade of the twentieth century, a worldwide increase in what are known as complementary and alternative medicine approaches (CAMs) has been observed (Frass et al. 2012). Even with growing endorsement of CAMs, the parallels between the approaches and practices prescribed in the *Anatomy*, and today's depression treatments, are less than complete, however. In particular, complementary and alternative approaches are as often placed *in contrast* to the approach and remedies offered in mainstream medical settings as integrated within them (Rhead 2014; Eardley et al. 2012; Frass et al. 2012). Belying its original meaning, complementary medicine is construed as a replacement or substitute for more traditional approaches. Closer to Burton's ideas are the "integrated" and "holistic" medicine that emphasize the synergistic effects arising out of combined approaches.[20]

CAMs and integrated and holistic medicine are widely found in many parts of medicine. The treatment of depression has been comparatively slow to adopt this model (although research initiatives are now beginning to assess its effectiveness) (Grimaldi-Bensouda et al. 2012).[21] And data on lifestyle and mental health have identified and apparently confirmed the treatment effectiveness of what are called therapeutic lifestyle changes (TLCs) along the lines Burton proposes for averting melancholy (Walsh 2011).[22] Holistic and integrated treatments still remain outside the treatment repertoire of modern psychiatric medicine, however, where, to repeat, the depression sufferer especially is more often depicted as powerless in the face of her disorder and, when the disorder is severe at least, too disabled to help herself.[23]

Burton's Ace in the Hole, Placebo

All experience is mediated by expectations, Burton recognized, and these expectations were evidence of the power of the imagination. (Such power was exemplified in a reiterated warning to the melancholy man, issued in the third edition (and thus, presumably, reflecting the audience reception of earlier ones), that reading about melancholy might invoke it (II,4,2,3: 236).) In today's psychology, the phenomena variously known as context, psychosocial, positive care, or placebo effects or meaning responses are acknowledged (Nayak and Patel 2014). These effects form an integral part of experimental design in the study of treatment effectiveness. That certain symptoms are particularly receptive to influence through expectation in

[20] "Integrated medicine" has been defined as combining mainstream biomedicine with complementary and alternative medicine, so that such effects can be attained (Dobos and Tao 2011: 11).

[21] Research on patients seeking treatment for anxiety and depressive disorders (SADD), designed to measure the effectiveness of more holistic approaches, found them comparable to the effects of conventional (and homeopathic) medicine (Grimaldi-Bensouda et al. 2012).

[22] Lifestyle includes "exercise, nutrition and diet, time in nature, relationships, recreation, relaxation and stress management, religious or spiritual involvement, and service to others" (Walsh 2011: 579). For a review of the effect of combined treatments on depression, see Pampallona et al. (2004). See also Stahl et al. (2014).

[23] Aside from mainstream psychiatry, several consumer-driven movements including the recovery model have more strongly emphasized these approaches, it must be added, although these will not be discussed further here. See, for example, Jacobson and Greenley (2001).

ways akin to those Burton describes has been widely demonstrated (Finnis et al. 2010). Thus, those with mild to moderate depression especially show measurable improvement on the inert pills provided as placebos that almost matches the improvement shown by those receiving more active interventions such as antidepressants (Brown 1998; Kirsch 1999; Kirsch et al. 2002; Turner et al. 2008). There are also nocebo effects, when expectations of sickness, disability, or deficiency and the affective and imagined states associated with such expectations actually induce or augment them (Mora et al. 2011). Although ethical constraints prevent the full study of nocebo effects, they have been demonstrated in contemporary studies (Hahn 1999).

Researchers have become wary of the kind of broad definition of placebo (and nocebo) effects relied on in earlier times.[24] The mechanisms by which these effects occur have proven to be various. So it is more accurate to speak of placebo effects in the plural, even in the limited instance of the kinds of symptoms Burton was concerned with. But matters around when and how placebo (and nocebo) effects occur remain somewhat contentious, due to ethical and methodological impediments to preventing their thorough study.[25] For example, confounding factors are recognized. The natural course of a disease is one (many depressions are self-limiting); there are the statistical phenomenon of regression to the mean, as well as response bias (in the patient who wants to please, for instance), and the effects of concurrent treatments (Finnis et al. 2010). Because of ethical and methodological limitations like these, efforts to understand, define, and explain placebo effects remain, today, without an authoritative formulation. In its general features, however, Burton's account of these effects finds affirmation. They are made possible by what has been described as a complex interplay between "belief, hope, expectation, and emotions" (Jopling 2008: 146).[26]

[24] Placebo responses are complex, multiform, and multimodal; it has been observed, involving many interacting neuronal systems that orchestrate changes in pain, motor control, mood, anxiety, memory, and motivation; and both conscious expectations and unconscious conditioning "move the molecules of change along the distributed biological systems, subserving cognition, emotions, pain control, reward and learning" (Nayak and Patel 2014: 74).

[25] For a review of these differences, see Harrington (1997) and Jopling (2008) and Finnis et al. (2010). The anthropological perspective is to be found in Moerman (2002). The Declaration of Helsinki (2000) prohibits as unethical placebo-controlled trials for life-threatening conditions (or when proven safe and effective treatments are available) (World Health Organization 2001). Moreover, there is a strong ethical prohibition on deceiving patients, which complicates the design of placebo-controlled studies. For the data, see Jopling (2008: 117–40); for the explanations, see Harrington (1999); and on definitional issues, see Moncrieff et al. (2004) and Jopling (2008).

[26] Recent efforts to define placebo effects, together with a revised definition, are provided by Jopling (2008: 132–47).

Multiple Cause and Remedy Models of Depression

In its underlying ideas and assumptions, the *Anatomy* departs from standard contemporary analyses of depression, suggesting instead a model of disease causation with parallels to the network models very recently proposed as alternatives to more traditional assumptions. Two of these traditional assumptions are distinguishable. Most commonly accepted and familiar is an "etiological" model of disease whereby depressive disorder names and depressive symptoms are attributed to, an underlying organic state of disorder or dysfunction. This way of thinking has been associated with medical psychiatric orthodoxy, including the "harmful dysfunction" definition of mental disorder found in the 4th and recent 5th editions of the DSM (APA 2013). The following, typical definition of depression provides another example: "Depressive disorder is a long term, relapsing condition associated with high levels of disability and mortality. *It has a neurobiological basis* and is associated with *functional and structural brain abnormalities*" (Palazidou 2012: 1, emphasis added). Beneath the potentially misleading appearances of mere symptoms, it is here presupposed, real disorders involve objective dysfunctions that account for why the symptoms occur, as well as the way they cluster as they do to produce a characteristic profile. A related assumption, known as the common cause hypothesis, is that a single cause is so implicated (Cramer et al. 2011). On this model, all symptoms arise from, and are explained by, the same underlying dysfunction, imbalance, disorder, or diathesis condition (Kramer 2005; APA 2013). Depression symptoms are the product of underlying, endogenous causes, albeit that those causes remain, as yet, incompletely identified. The etiological and common cause hypotheses are separable, but in descriptions of depressive disorder, they are customarily combined. Contemporary conceptions of prevention, noted earlier, also reflect the common cause model, as the language of "relapse" and "remission" indicates. Although not continuous, episodes of depression are united by their common source: they are one chronic condition, rather than several.

Alternatives to the common cause etiological model hypothesize that depression symptoms are not the result of any single or underlying, endogenous disease process or dysfunction, but instead comprise a causally interrelated cluster, or network, of features, attributable to diverse causes, including the stressful life events with which depression has been observed to be correlated (Kendler et al. 2011; Borsboom and Cramer 2013; Wichers 2014; Zachar 2014; Cramer et al. 2011).[27] While they may each be *realized* in biological states, the symptoms within any network are not each traceable to underlying biological dysfunction or disease. The totality of symptoms making up a disorder may not result from *underlying* causes at all, single or multiple. It is rather that, together, and over time, symptoms will arise and trigger one another to form a circuitry of causal relationships. Feedback loops result within

[27] Empirical research on the networks model of depression is thus far limited. But one study of symptoms caused by stressful life events has shown the network model to have significant predictive advantages over common cause hypotheses (Cramer et al. 2011).

these symptom clusters, and particular features will enter into a large number of mutual relationships with other symptoms in the network. Over time, a "vicious circle of experiences" will result; this is a severe disorder and a pattern that an individual is increasingly unable to escape (Wichers 2014: 1353).

The strongest argument in support of a network model, it has been asserted, is that it seems implausible to assume that symptoms like insomnia and fatigue, for example, are only correlated because both are the result of an underlying liability to develop depression (Cramer et al. 2011). It seems obvious that insomnia might *directly* cause fatigue and that a pattern of symptoms might emerge that goes from *fatigue to concentration problems to self-reproach to depressed mood to insomnia* and so on – in a loop that, with repetition, and through time, brings about a depressive episode.

A working presupposition of the Hypothesis of depression, network model is the presence of a finite set of symptoms forming a cluster that is relatively cohesive and stable. Such syndromes will vary in their degree of coherence, stability, and boundary fuzziness, leaving them dissimilar to the categorical kinds often associated with etiological models. And their status as distinct entities may shift with changes in the networks involved. The interrelated symptoms making up the network need not share any commonality, either. Some symptoms and symptom clusters in the domain overlap with others, but no shared feature need unite all of them (Zachar 2014: 238).

Burton's account of the causes and symptoms of melancholy has much in common with the network model. First, it acknowledges that melancholy's initiating causes include many that are the sort of stressful life events identified as causes in network models of depression. Thus, as we saw, the interaction Burton describes cannot be entirely captured in any set of uni-directional, body-to-mind, causal sequences. The rapid and incessant shifts and changes making up the course of melancholy in any individual anticipate the network model in the chaotic plurality of its multilevel, multicausal explanations, and also in the incessant feedback loops, where elements are transformed from cause to symptom to cause and back. And, contrary to much that has been asserted or assumed about causation in humoral medicine, underlying humoral states in no consistent way cause symptoms, as they might be supposed to do following the etiological model of disease. The arrangement of the humors is sometimes assigned a causal role; but no regular pattern of causation can be traced from bodily "temperature" to melancholy symptoms.

Also conforming to the network model is the temporally extended natural history of melancholy states at the center of Burton's analysis. The examples used earlier, where symptoms (sadness, solitariness) become in their turn causes of further symptoms, illustrate a *temporal sequence*. By following the course of the disorder this way, it is possible to attribute its presence and extent. Severe melancholy is identified with the "chronic," habituated condition.[28] The transformation from brief

[28] The relation between more and less severe forms of depression is disputed, those distinguishing melancholic and non-melancholic depression asserting a categorical separation. Moreover antidepressant medicines have long been known to outperform placebo more robustly when severe disorder is involved (Klein 1974).

and passing states of melancholy into the intractable condition that resists all efforts at amelioration is a process. Because entrenched and more severe forms of the disorder result from repetition, this is a final parallel between Burton's melancholy and depression understood on a network model. The model relies on a cumulative effect. Although it will reflect other factors as well (e.g., temperamental tendencies and environmental triggers), severity is partly a function of the number of feedback repetitions linking symptoms together.

Just as an illimitable number of causes for melancholy are permitted in the *Anatomy* account, so it seems to indicate that an almost infinite range of treatments, preventatives, and remedies will also likely be needed in any given case of melancholy symptoms. No one of these might alone be sufficient, Burton indicates, but *together they may succeed.* "If not alone, yet *certainely conjoyned* [they] may do much" (III,2,5,2: 207). Associated with today's common cause, etiological disease models, where a single causal factor is assigned to account for the symptom cluster relied on for the diagnosis of depression, there is a tendency to think in terms of a single, isolated treatment response: with a sufficiently targeted treatment, a single cause can be expected to require only a single remedy. This "magic bullet" attitude has been most evident in the claims made on behalf of antidepressant drugs (Healy 1997: 21).[29] The misleading specificity associated with magic bullet thinking about depression was central to David Healy's critique of these assumptions. Until the last decades of the nineteenth century, he asserts, the idea that there might be a specific remedy for a specific ill was "tantamount to quackery" (Healy 1997: 10). On the version of the common cause, etiological disease model, Healy refers to as the "bacteriological model" of depression, a specific, underlying vulnerability or dysfunction is necessary, if not sufficient, to explain all symptoms. That is the "disease specificity," to which "therapeutic specificity" was an obvious response (Healy 1997: 10). The mistaken idea of therapeutic specificity, Healy thus attributes to the misapplication of models from one disease (the infections successfully treated by antibiotics) to another (depression).

In the case of depression, Healy has argued, neither disease nor therapeutic specificities are appropriate. Depression is itself nonspecific: the category is better seen to refer to a generalized neurasthenia that, when it responds to antidepressant drugs, does so because their *effect is also nonspecific.* Appealing to humoral medicine, with its emphasis on illness as imbalance, he asserts that today's antidepressant drugs probably work as "humoral" remedies, in the manner of old-fashioned tonics affecting sleep and appetite (Healy 1997: 259). In spite of these allusions to humoral medicine, and the holism common to each account, Healy's analysis apparently differs from Burton's inasmuch as for Burton expectation may in some considerable part contribute to healing. If placebo effects are necessary though not sufficient for therapeutic effectiveness, then they will interfere with the intended effects. No specific remedy can be effective for all melancholy symptoms because of the vagaries of placebo effects that, as Burton says, might allow one individual to be entirely healed merely by a strongly held idea (I,2,3,2). What is importantly shared between

[29] Healy traces the initial magic bullet to 1930s work resulting in the first antibiotics (sulfa drugs).

the *Anatomy* account and Healy's, however, is the presupposition that neither the causes nor the treatments of these disorders are specific in their effects. No single cause engenders melancholy; moreover, no single remedy will alone be likely to be sufficient to treat it. And, if Healy is correct, the same is true for depression.[30]

Should a multicausal account of depression require a multi-remedial approach, we need to understand why. So how is Healy's allegiance to nonspecific remedies to be understood? Part of his discussion, it must be pointed out, is directed at the incomplete neuroscience that attributed spurious specificity to antidepressant substances whose effects are, or were at the time of his writing, incompletely understood. The regulations instituted to protect consumers are the source of some of this problem. By the middle of the twentieth century in the USA, Healy emphasizes, the FDA designated certain drugs as prescription only, insisted on randomized control trials, and encouraged the pharmaceutical industry to develop specific treatments for what were judged to be specific diseases (Healy 1997).

In light of network suppositions, two further explanations for this link between multiple causes and multiple treatments deserve attention. Integrated and holistic medical approaches emphasize the synergistic effects arising out of combining remedies, as we saw. Such enhanced effects might occur and if so might serve to diminish, disrupt, or sever intra-network bonds. On a different (although compatible) hypothesis, it can be argued that no specific remedy will be effective, since the networks comprising the totality of depression symptoms are insufficiently coherent and/or stable. This is a position that we might fairly attribute to Burton with regard to melancholy, but it can also be found in the work of sociologist Alain Ehrenberg (2010), who similarly rejects any common cause or essentialist account of depression. There has been a transformation of the meaning of depression, Ehrenberg asserts, prompted in part by changes in the goal, refinements, and effects of the newer antidepressant medicines. Stimulants to action replaced the older idea of a cure for pathology. And no longer a curable illness, depression had become, and was by our contemporary psychopharmacology treated as, a chronic, incurable personality trait of inhibited action. Insufficient or excess serotonin is not the cause of the several disorders SSRIs treat, but rather the "neurochemical vector of a person's equilibrium," systemically regulating the balance between the opposing dimensions of inhibition and impulsiveness (Ehrenberg 2010: 169). Whether depression symptoms have so wholeheartedly changed their nature, and in this particular way, we might debate. But such changes seem entirely compatible with the network model and with Healy's concerns over magic bullet reasoning. Many remedies are required to address the range and variability of symptoms included in the loose category of depression.

[30] That a multifactorial etiology for depression argues for a multipronged approach to intervention is voiced in some recent analyses. See Jacob (2012).

Conclusion

Much of today's depression treatment, geared as it is to common cause, etiological models of depression and their correlates in the magic bullet remedies presupposed by medical psychiatry and psychopharmacology, seems to assume the answer to depressive suffering lies in a single treatment or remedy such as fluoxetine. Reading Burton about melancholy would encourage us to think otherwise, however. His remedies are notable not only for his themes of prevention and self-help but also for his emphasis on the necessity of their combination; if not alone, as he says, "yet *certainly conjoined*," these remedies could succeed. Purgative doses of hellebore may do little by themselves perhaps, but as truly complementary medicine, they may be powerful.

While Burton's conclusions have bearing on milder states of disorder, it may be supposed they are inapplicable to more debilitating and severe depressions. Such a suggestion overlooks Burton's emphasis on the difference between passing and habituated or chronic forms of melancholy. For him, the entrenched, more severe forms of the disorder are those resulting from repeated feedback loops linking symptoms to causes and back. Unchecked repetition is in this respect an indicator of severity. On Burton's practical criterion, more severe and chronic forms of melancholy are defined by reference to remediation. When we can no longer adhere to the required behavioral and cognitive regimen and need to seek outside help, our melancholy has become a recalcitrant and dangerous disease. Rather than the network model being applicable only to milder forms of melancholy, Burton offers a criterion of severity accommodated by that model. And so, perhaps, might depression.

For all the differences, present-day responses to depression can be informed by Burton's *Anatomy.* His emphasis on prevention in the strongest sense of averting or avoiding initial episodes as well as reducing subsequent ones, on self-help through the practice of healthy "lifestyle" habits, *and on the combined effect of these treatments*, deserves our attention as a corollary to the present-day network models of depression within which they are so nicely accommodated.

References

American Psychiatric Association. 2013. *Diagnostic and Statistical Manual of Mental Disorders.* 5th ed. Washington, DC: American Psychiatric Association.

Berrios, H. 2011. Preface to Diseases of the Imagination and Imaginary Disease in the Early Modern Period. In ed. Yasmin Haskell, xv–xxiii. Brepols Publishers.

Biegler, P. 2011. *The Ethical Treatment of Depression: Autonomy Through Psychotherapy.* Cambridge, MA: MIT Press.

Borsboom, D., and A.O. Cramer. 2013. Network Analysis: An Integrative Approach to the Structure of Psychopathology. *Annual Review of Clinical Psychology* 9: 91–121.

Brown, W. 1998. The Placebo Effect. *Scientific American* 278: 90–95.

Burton, R. 1989–2000. The Anatomy of Melancholy. 6 vols. In Vols I-III: Text; vols IV-VI: Commentary by Bamborough JB and Dodsworth M, ed. T. Faulkner, N. Kiessling, R. Blair. Oxford: Oxford University Press.

Cramer, A.O.J., D. Borsboom, S.H. Aggen, and K.S. Kendler. 2011. The Pathoplasticity of Dysphoric Episodes: Differential Impact of Stressful Life Events on the Pattern of Depressive Symptom Inter-correlations. *Psychological Medicine* 42: 957–965.

Cuijpers, P. 1997. Bibliotherapy in Unipolar Depression. *Journal of Behavior Therapy and Experimental Psychiatry* 28: 139–147.

Culpeper, N. 1985. *Culpeper's Compete Herbal*. Ware: Omega Books.

Dobos, G., and I. Tao. 2011. The Model of Western Integrative Medicine: The Role of Chinese Medicine. *Chinese Journal of Integrative Medicine* 17: 11–20.

Eardley, S., F.L. Bishop, P. Prescott, et al. 2012. A Systematic Literature Review of Complementary and Alternative Medicine Prevalence in EU. *Research in Complementary Medicine* 19(Suppl 2): 1661–4119.

Eherenberg, A. 2010. *The Weariness of the Self: Diagnosing the History of Depression in the Contemporary Age*. Montreal: McGill-Queens University Press.

Embry, D.D. 2011. Behavioral Vaccines and Evidence-Based Kernels: Nonpharmaceutical Approaches for the Prevention of Mental, Emotional, and Behavioural Disorders. *Psychiatric Clinics of North America* 34: 1–24.

Finnis, D., T. Kaptchuk, F. Miller, and F. Benedetti. 2010. Biological, Clinical and Ethical Advances of Placebo Effects. *Lancet* 375: 686–695.

Foucault, M. 2006. *History of Madness*. Trans. Jonathan Murphy and Jean Khalfa. London/New York: Routledge.

Frass, M., R.P. Strassl, H. Friehs, et al. 2012. Use and Acceptance of Complementary and Alternative Medicine Among the General Population and Medical Personnel: A Systematic Review. *The Ochsner Journal* 12: 45–56.

Gowland, A. 2006. *The Worlds of Renaissance Melancholy: Robert Burton in Context*. Cambridge: Cambridge University Press.

Grimaldi-Bensouda, L., P. Engel, J. Massol, et al. 2012. Who Seeks Primary Care for Sleep, Anxiety and Depressive Disorders from Physicians Prescribing Homeopathic and Other Complementary Medicine? Results from the EP13 population survey. *British Medical Journal Open* 2: 001498.

Hahn, R.A. 1999. The Nocebo Phenomenon: Scope and Foundations. In *The Placebo Effect: An Interdisciplinary Exploration*, ed. A. Harrington, 56–76. Cambridge: Harvard University Press.

Harrington, A., ed. 1999. *The Placebo Effect: An Interdisciplinary Exploration*. Cambridge: Harvard University Press.

Haskell, Y., ed. 2011. *Diseases of the Imagination and Imaginary Disease in the Early Modern Period*. Turnhout: Brepols Publishers.

Healy, D. 1997. *The Anti-Depressant Era*. Cambridge: Harvard University Press.

Hollandare, F., S. Johnsson, M. Randestad, M. Tilfors, et al. 2011. Randomized Trial of Internet-based Relapse Prevention for Partially Remitted Depression. *Acta Psychiatrica Scandinavica* 124: 285–294.

Jacob, K.S. 2012. Depression: A Major Public Health Problem in Need of a Multi-sectorial Response. *The Indian Journal of Medical Research* 136(4): 537–539.

Jacobson, N., and D. Greenley. 2001. What is Recovery? A Conceptual Model and Explication. *Psychiatric Services* 52: 482–485.

Jopling, D. 2008. *Placebos and Placebo Effects*. Oxford: Oxford University Press.

Kendler, K.S., P. Zachar, and C. Craver. 2011. What Kinds of Things are Psychiatric Disorders? *Psychological Medicine* 41: 1143–1150.

Kirsch, I., ed. 1999. *How Expectancies Shape Experience*. Washington, DC: American Psychological Press.

Kirsch, I., T.J. Moore, A. Scoboria, and S.S. Nicholls. 2002. The Emperor's New Drugs: An Analysis of Antidepressant Medication Data Submitted to the U.S. Food and Drug Administration. Prevention and Treatment 5, Article 23.

Klein, D.F. 1974. Endogenomorphic Depression: A Conceptual and Terminological Revision. *Archives of General Psychiatry* 31: 447–454.

Kramer, P. 2005. *Against Depression*. New York: Viking.

Kupfer, D.J., E. Frank, and J.M. Perel. 1989. The Advantage of Early Treatment Intervention in Recurrent Depression. *Archives of General Psychiatry* 46: 771–775.

Long, A.A. 1986. *Hellenistic Philosophy: Stoics, Epicureans, Sceptics*. 2nd ed. Berkeley/Los Angeles: University of California Press.

Lund, M. 2010. *Melancholy, Medicine and Religion in Early Modern England: Reading The Anatomy of Melancholy*. Cambridge: Cambridge University Press.

Mathers, C., D.M. Fat, and J.T. Boerma. 2008. *The Global Burden of Disease: 2004 Update*. Geneva: World Health Organization.

McLaughlin, K.A. 2011. The Public Health Impact of Major Depression: A Call for Interdisciplinary Prevention Efforts. *Prevention Science* 12: 361–271.

Merry, S., S.E. Hetrick, G.R. Cox, et al. 2012. Evidence-Based Child Health. *Cochrane Review Journal* 7: 1409–1685.

Moerman, D. 2002. *Meaning, Medicine, and the 'Placebo Effect'*. Cambridge: Cambridge University Press.

Moncrieff, J. 2009. *The Myth of the Chemical Cure: A Critique of Psychiatric Drug Treatment*. New York: Palgrave MacMillan.

Moncrieff, J., S. Wessely, and R. Hardy. 2004. Active Placebos Versus Antidepressants for Depression. *Cochrane Database of Systematic Reviews* 1: CD003012.

Mora, M.S., Y. Nestoriuc, and W. Rief. 2011. Lessons Learned from Placebo Groups and Antidepressant Trials. *Philosophy Transactions of the Royal Society of London: Series B* 366: 1879–1888.

Mrazek, P.J., and R.J. Haggerty, ed. 1994. *Reducing Risks for Mental Disorders: Frontiers for Preventive Intervention Research*. Washington, D.C.: National Academies Press.

Murray, C.J., and A.D. Lopez. 1997. Alternative Projections of Mortality and Disability and the Contribution of Risk Factors: Global Burden of Disease Study. *Lancet* 349(9063): 1436–1442.

Nayak, D., and P. Patel. 2014. Neural Mechanisms of Placebo Effects. *Psychiatric Annals* 44(2): 74–78.

O'Connell, M.E., T. Boat, and K.E. Warner (eds). 2009. *Preventing Mental, Emotional and Behavioral Disorders Among Young People: Progress and Possibilities*. National Research Council (US) and Institute of Medicine (US) Committee on the Prevention of Mental Disorders and Substance Abuse Among Children, Youth, and Young Adults: Research Advances and Promising Intervention. Washington, DC: National Academies Press.

Palazidou, E. 2012. The Neurobiology of Depression. *British Medical Bulletin* 101: 127–145.

Pampallona, S., S. Bollini, G. Tibaldi, B. Kupelnick, and C. Munizza. 2004. Combined Pharmacotherapy and Psychological Treatment for Depression: A Systematic Review. *Archives of General Psychiatry* 61: 714–719.

Parker, G., and D. Hadzi-Pavlovic, ed. 1996. *Melancholia: A disorder of Movement and Mood: A Phenomenological and Neurobiological Review*. Cambridge: Cambridge University Press.

Radden, J. 2003. Is This Dame Melancholy? Equating Depression and Melancholia. *Psychiatry, Philosophy and Psychology* 10(1): 37–52.

———. 2009. *Moody Minds Distempered: Essays on Melancholy and Depression*. New York: Oxford University Press.

———. 2013. Shared Descriptions: What Can be Concluded? Commentary. *Philosophy, Psychiatry and Psychology* 20: 157–159.

Rhead, J.D. 2014. The Deeper Significance of Integrative Medicine. *Journal of Alternative and Complementary Medicine* 20: 329–329.

Schmidt, J. 2007. *Melancholy and the Care of the Soul: Religious, Moral Philosophy and Madness in Early Modern England*. Aldershot: Ashgate Publishing Company.

Shorter, A. 2011. Still Tilting at Windmills Commentary on: The Myth of Mental Illness. *The Psychiatrist* 35(5): 183–184.

Shorter, A., and M. Fink. 2010. *Endocrine Psychiatry: Solving the Riddle of Melancholia*. New York: Oxford University Press.

Stahl, S.T., et al. 2014. Coaching in Healthy Dietary Practices in At-risk Older Adults: A Case of Indicated Depression Prevention. *The American Journal of Psychiatry* 171: 499–505.

Taylor, M.A., and M. Fink. 2006. *Melancholia: The Diagnosis, Pathophysiology, and Treatment of Depressive Illness*. Cambridge: Cambridge University Press.

Turner, E.H., et al. 2008. Selective Publication of Antidepressant Trials and Its Influence on Apparent Efficacy. *The New England Journal of Medicine* 358: 252–260.

Van Straten, A., P. Cuijpers, and N. Smits. 2008. Effectiveness of a Web-Based Self-help Intervention for the Symptoms of Depression, Anxiety, and Stress: Randomized Controlled Trial. *Journal of Medical Internet Research* 10(1): e7.

Van Voorhees, B.W., et al. 2011. Internet-Based Depression Prevention over the Life Course: A Call for Behavioral Vaccines. *Psychiatric Clinics of North America* 34(1): 167–183.

Varga, S. 2013. From Melancholia to Depression: Ideas on a Possible Continuity. *Philosophy, Psychology and Psychiatry* 20(2): 141–155.

Walsh, R. 2011. Lifestyle and Mental Health. *The American Psychologist* 66: 579–592.

Wichers, M. 2014. The Dynamic Nature of Depression: A New Micro-Level Perspective of Mental Disorder that Meets Current Challenges. *Psychological Medicine* 44: 1349–1360.

World Medical Association. 2001. World Medical Association Declaration of Helsinki. Ethical Principles for Medical Research Involving Human Subjects. *Bulletin of the World Health Organization* 79(4): 373.

Zachar, P. 2014. *A Metaphysics of Psychopathology*. Cambridge, MA: MIT Press.

Love in the Time of Antibiotic Resistance: How Altruism Might Be Our Best Hope

Dien Ho

The emergence of antibiotic-resistant bacteria is nothing new. In 1921, Alexander Fleming discovered the antibacterial property of lysozyme. Merely 2 weeks later, Fleming et al. identified strains of bacteria in their culture that had been exposed to lysozyme and survived the lytic action. The cultures grown using these strains retained their lysozyme resistance 9 months after their initial discovery (Fleming and Allison 1927). Fleming, of course, went on to discover penicillin in 1928. Penicillin transformed modern medicine: it not only cured previously fatal infectious diseases, but it also made possible invasive surgical procedures that would have otherwise exposed patients to dangerous risks of infections. Clive Butler recounts his experience on the staff at London Hospital before and after the introduction of penicillin in his treatment of more than 500 cases of acute osteomyelitis between 1919 and 1937. He writes,

> The overall mortality rate [before penicillin] was 25%—due, in almost every case, to staphylococcal bacteraemia…[S]urgical treatment consisted in early relief of tension by drilling the affected bone and then prolonged immobilisation…I particularly recall one youth who used to pass the time by killing the maggots when they emerged from his plaster by squirting them from an ethyl chloride spray…Towards the end of the war penicillin became available in Britain and I had the opportunity of treating 21 patients—none died…What a transformation from killing maggots, and all this achieved in under 10 years: it seemed miraculous to me and others of my generation. (Butler 1979: 482)

In his Nobel Lecture of 1945, Fleming highlighted the danger of antibiotic-resistant microbes. "The time may come," he warned, "when penicillin can be bought by anyone in the shops. Then there is the danger that the ignorant man may easily underdose himself and by exposing his microbes to non-lethal quantities of the drug make them resistant" (Fleming 1945: 93). Yet, 70 years after his warning, we find

D. Ho (✉)
School of Arts and Sciences, MCPHS University, Boston, MA, USA
e-mail: dien.ho@mcphs.edu

© Springer Science+Business Media B.V. 2017

D. Ho (ed.), *Philosophical Issues in Pharmaceutics*, Philosophy and Medicine 122,
DOI 10.1007/978-94-024-0979-6_15

our arsenal of antimicrobials becoming less effective in the face of growing resistance. In this essay, I wish to explore the causes of antibiotic resistance. The solution will require not merely a change in the development, distribution, and usage of antibiotics but a willingness to put aside our rational self-interest. If one is pessimistic of our chances to place other's interests before our own, then one should be pessimistic of our chances of winning the war against microbes.

The Dreaded Reality

Antibiotic-resistant pathogens have been identified in virtually every corner of the world. Nevertheless, the exact scope of the problem remains unclear. The World Health Organization (WHO) estimates that of the 6.1 million cases of tuberculosis diagnosed worldwide in 2013, approximately 300,000 of them are multidrug-resistant tuberculosis (MDR-TB) defined as being not responsive to two of the standard treatments (isoniazid and rifampicin) (WHO 2014a). MDR-TB has been identified in all six WHO global regions: Africa, the Americas, Eastern Mediterranean, Europe, Southeast Asia, and the Western Pacific. Although treatment success rate for TB is about 86%, it drops to 48% with MDR-TB. More worrisome is the emergence of extensively drug-resistant TB (XDR-TB): TB that is resistant to standard treatments including the most effective second-line treatments. Nine percent of all patients with MDR-TB were also diagnosed with XDR-TB, and more than 100 countries reported cases of XDR-TB in 2013 (WHO 2014b). The data collected by WHO, however, do not paint a complete picture. Although the number of reported cases of MDR-TB in Southeast Asia, for instance, has grown from 68 in 2005 to 28,618 in 2013, it is not entirely clear if the dramatic change was a result of significant increase of antibiotic resistance or better diagnostic and reporting efforts. Indeed, data collection has remained a major hurdle in our attempt to gain a full understanding of the scope of the problem. Zumla et al. report that among 27 countries with a high burden of MDR-TB, "only 1% of new tuberculosis cases and 3% of previously treated cases are screened for [drug-resistant] tuberculosis by a laboratory" (Zumla et al. 2012: S229). The actual number of individuals who suffer from MDR-TB is likely significantly higher.

In addition to a lower treatment success rate, patients who suffer from antibiotic-resistant infections are also more costly to treat. On average, treatment for MDR-TB cost 3–100 times more (Koenig 2008: 894), and they last three to four times longer than drug-susceptible TB (leading to a greater chance of the TB spreading). A broad study commissioned by the Prime Minister of the United Kingdom and conducted by Jim O'Neill and the Wellcome Trust warns that unless drastic measures are taken, by 2050, antimicrobial-resistant diseases will claim ten million lives per year (O'Neill and Wellcome Trust 2014).

What Stokes the Fire?

One commonly cited cause of the growth of antibiotic resistance is the unnecessary use of antibiotics at the clinical level. In a 2014 study, Schultz et al. identify redundant antimicrobial usage in 394 of the 505 hospitals examined in the United States, amounting to 32,507 cases (Schultz et al. 2014: 1231). Physicians, even those who are familiar with the danger of overprescribing antibiotics, often do not consider the broader public health implication in their prescription habits. Of the 400 generalists and 429 infectious disease specialists they surveyed, Metlay et al. note that "risk of the drug contributing to the problem of antibiotic resistance" ranked last among a list of seven considerations in determining what antibiotic to prescribe, well behind ease of use and cost to patients (Metlay et al. 2002).[1]

Similar failures to prescribe properly have been noted in other developed nations and often at community healthcare access points such as pharmacies. Marković-Peković et al. report that of the 131 pharmacies their "secret shoppers" visited in the Republic of Srpska in Bosnia and Herzegovina, 58% sold antibiotics to clients without prescriptions. Among these pharmacies, 25% did not offer any symptomatic therapies and only sold the antibiotics to the clients (Marković-Peković and Grubiša 2012). Without proper diagnostic indications for antibiotics and emphasis on the importance of correct usage (e.g., the completion of a treatment), pharmacists risk under- and overuse of antibiotics that contribute to the proliferation of antibiotic-resistant pathogens.

The dispensing and usage of antibiotics in developing nations often involve even less rigor and monitoring. Indira et al. examined the prescribing behaviors of clinicians in health centers in four Indian cities, and they report that 70% of antibiotics were given for viral conditions (e.g., diarrhea and upper respiratory infections) that are nonresponsive to antibiotics (Indira et al. 2004). In a 2010 study in Abu Dhabi, 68.4% of clients were able to obtain antibiotics in community pharmacies without prescriptions (Dameh et al. 2010). Not only is dispensing antibiotics over-the-counter illegal in UAE, 63% of them obtained antibiotics after complaining of respiratory problems, which are often viral in origin. Similar inappropriate dispensing of antibiotics has been identified in China as well. One study reports up to 79% of patients suffering from the common cold were given antibiotics in hospitals (Yip and Hsiao 2008: 462). A study by Means et al. shows that 40% of the 45,591 patients

[1] It is important to note that a number of factors contribute to well-informed physicians' failure to prescribe antibiotics properly. Often, physicians succumb to pressure from patients who explicitly seek antibiotics. These requests can be a result of misunderstanding of the therapeutic use of antibiotics (e.g., they do not work on common colds). They can also be a product of intense marketing campaigns by pharmaceutical manufacturers. Pfizer's marketing campaign for Zithromax (Z-Pak) geared heavily toward parents who wished to seek short-term antibiotics for their children's ear infections (Petersen 2000). The campaign included a donation of a zebra to the San Francisco Zoo, sponsoring of Sesame Street episodes, and large-scale distribution of Max the Zebra to various clinics. Max proved so popular that patients reported that they thought it was the mascot for Boston Children's Hospital.

in Uganda with malaria received antibiotic treatments, even though they lacked documented clinical indications (Means et al. 2014).

The cause of inappropriate antibiotic prescription is a complex phenomenon. In the case of China and India, often pharmacies and clinics profit from selling antibiotics, thus creating a perverse incentive to overprescribe. There is also the lack of oversight of healthcare providers' prescribing habits. Of the 47 member states constituting WHO's African Region, only eight countries had the means to report antimicrobial resistance data, and only one has a funded national action plan to combat antimicrobial resistance. In China, the rampant counterfeit drug market also increases the difficulties in controlling antibiotic distribution and usage. *Wired* magazine reported that in 2001, 192,000 Chinese patients died from fake drugs, and the Chinese authorities closed 1,300 factories and pursued 480,000 cases of counterfeit drugs (McKenna 2011).

Although questionable motives certainly led to the failure to prescribe antibiotics appropriately, often the reasons are far less nefarious. Consider the decision as to whether one should prescribe a broad-spectrum or narrow-spectrum antibiotic when a patient has a bacterial infection. Given that standard diagnostic cultures can take days to grow (and weeks for some bacteria), the use of a broad-spectrum antibiotic prior to securing an accurate diagnosis allows care providers to begin treatment for the patient.[2] Of course, broad-spectrum antibiotics also render more likely the development of antibiotic-resistant pathogens since they kill more than they need to. Ignoring for the moment the health risks a patient undertakes by using a broad-spectrum antibiotic (e.g., the destruction of possibly helpful bacteria), we can see a potential conflict of interest. To wit, it is in the best health interest of a patient to receive speedy and effective treatment including broad-spectrum antibiotics. At the same time, the treatment involves a cost (an externality) that is shouldered by everyone else in the form of an increased risk of an antibiotic-resistant pathogen emerging. By killing a wide gamut of bacteria (many of which pose no health risks to the patient), a broad-spectrum antibiotic leaves a bacterial vacuum in which the remaining resistant bacteria have ample opportunities to repopulate. A widespread use of broad-spectrum antibiotics thus exerts selection pressure against antibiotic-susceptible bacteria tilting the reproductive landscape in favor of antibiotic-resistant ones. In addition to vertical transmission of antibiotic resistance (via reproduction), antibiotic resistance can also transmit horizontally. Segments of a bacterium DNA that confer antibiotic resistance can break off and attach themselves to other bacteria including those not in the same species. Thus, a broad-spectrum antibiotic may select antibiotic resistance in a banal bacterium, but the resistance can be passed on to pathogens creating a new lethal antibiotic-resistant strain.[3]

Nevertheless, there are many immediate benefits to using broad-spectrum antibiotics. In addition to being able to start treatment before diagnostics are completed,

---

[2] To be sure, there are newer and faster diagnostic methods. But these methods often require resources including sophisticated laboratories that many health centers simply do not possess.

[3] See Morley et al. (2005) for an explication of the mechanisms that give rise to bacterial resistance.

the use of broad-spectrum antibiotics can also lower the cost of care for the patients and the clinics. Even if we assume that a liberal use of broad-spectrum antibiotics will shorten the effective lifespan of these drugs, it is not an obvious conclusion that we ought not to be aggressive with our use of broad-spectrum antibiotics. Diagnostics cost money and time. Given how cheap broad-spectrum antibiotics are, hospitals and clinics might be able to save money by prescribing them and using the savings to meet other healthcare needs. Such a practice would likely impact future patients by lowering the effectiveness of antibiotics, but it may provide better care for current patients.

In order to conclude that limiting broad-spectrum antibiotic usage allows us to provide optimal care for the long run, we need to answer two questions: (1) how long is the long run and (2) what is the cost to the patients who suffer, either because of the denial of broad-spectrum antibiotics or because of the decline in available medical resources due to the burden of more expensive diagnostic techniques? The first question is obviously a philosophical one that requires difficult trade-offs between the well-being of people in the future and those in the present. Although a number of scholars have attempted to balance the two, justifying a particular tipping point where the interests of nonexistent future people outweigh those of real sick people is hardly an easy task (Leibovici et al. 2012: 4).

Indeed, the very idea that a physician might provide less than optimal care (from a patient's point of view) for the interests of future individuals might erode the trust between patients and physicians.[4] Physicians are advocates for patients' healthcare interests. If patients believe that their physicians are actively limiting care in order to benefit others, patients would rightly question whether their healthcare needs are in fact the determinants of the treatments recommended.

One way we can avoid this problem is to enact regulations that restrict and guide physicians' clinical practices when patients' interests need to be compromised. Physicians sometimes make clinical decisions that place their patients' interests below the welfare of someone else. Quarantine orders, reporting of gunshot wounds, therapists' breaking confidentiality to report the abusing of minors, and emergency triage all can place the interests of a third party over that of the immediate patient. Clear and enforceable regulatory restrictions—be they from professional organizations or governmental agencies—allow physicians to remain full advocates of patients' healthcare interests while taking into consideration public health concerns. By "tying the hands" of clinicians, we preserve the trust essential for the proper working of a physician-patient relationship.

Strong regulatory guidelines for the prescribing of antibiotics can also limit unnecessary prescriptions of antibiotics that stem from liability worries on behalf of healthcare providers. In a 2009 study, Sakoulas et al. note that among 17 European countries and 41 states in the United States, there is a statistical correlation between the density of attorneys and cases of methicillin-resistant *Staphylococcus aureus* (MRSA). Sakoulas et al. hypothesize that the fear of malpractice lawsuits (e.g.,

[4] For a discussion of why trust is necessary for the proper functioning of medicine, see Ho (2008: 81).

undertreatment) causes physicians to overprescribe antibiotics, thus leading to a higher rate of MRSA (Sakoulas et al. 2009). Strong regulatory guidelines can alleviate some of the fear of litigation by providing institutional support for the proper usage of antibiotics.

National regulations that limit access to antibiotics can help eliminate a number of factors that contribute to the emergence of antibiotic resistance; for example:

1. Limit antibiotic usage for clinically indicated conditions (e.g., prohibiting prescription of antibiotics for viral infections).
2. Decrease antibiotic prescriptions for marginally indicated conditions such as ear infections.
3. Rigorous monitoring of antibiotic prescriptions by clinicians.
4. Reserve broad-spectrum antibiotics as second-line treatments.

Clear national policies restrict clinical decisions while preserving the trust between physicians and patients. In addition, they also lessen that pressure physicians feel when prescribing antibiotics out of fear of litigation.[5]

When Local Meets Global

Professional and governmental guidelines, however, run up against the fact that the emergence of antibiotic-resistant pathogens is a global problem that requires transnational-coordinated solutions. However, effective intranational agencies are at monitoring and regulating antibiotic resistance; the porous nature of international borders renders these efforts futile unless other nations also undertake similar efforts. Pathogens can spread easily and become endemic to their new environment. A recent genetic study of samples of typhoid in two dozen countries has revealed that in the past 30 years, a drug-resistant strain—H58—has spread from India to African and Pacific island nations. H58 is becoming the dominant strain of typhoid displacing those that are susceptible to traditional drug treatments in their new homes (Wong et al. 2015). Here the lack of any strong regulatory bodies renders controlling antibiotic resistance a far more daunting task. The United States, for

[5] It is important that we should retain a healthy dose of skepticism with regard to the political wisdom and will necessary to draft and implement rational antibiotic policies. When it comes to public health concerns, the United States has a troubling history. In 2014, for instance, Congress appropriated $5.2 billion to an emergency bill for combating the Ebola virus. The White House requested in the 2016 budget $1.2 billion for all antibiotic research. Although Ebola is a formidable infectious foe, its threat pales in comparison to antibiotic-resistant bacteria. Ebola killed approximately 11,000 individuals by 2015 (The Data Team 2015). Antibiotic-resistant bacteria kill about 700,000 people/year (O'Neill and Wellcome Trust 2014). Media attention on Ebola probably helped shift awareness and the subsequent funding to Ebola research. The creation of effective regulatory guidelines concerning antibiotic usage would depend heavily on an objective evaluation of public health independent of political biases. The fact that the United States' political system is designed to be influenced by political action committees and lobbying groups suggests that competing private interests might not make for ideal public policies.

instance, constitute only 5% of the world's population. Even with the most effective domestic policies guiding the use of antibiotics, its contribution to the curbing of antibiotic-resistant pathogens globally is minimal, especially if the consumption of antibiotics in the rest of the world increases.[6]

Good antibiotic stewardship requires supportive services that permit rational infectious control. Antibiotic usage is but one arm of a holistic approach to limiting infections. Indeed, if there were no infections, there would be no need for antibiotics. Access to clean water and nutrition, adequate sanitation, properly trained healthcare providers, patients who are educated about their treatment, affordable methods of traveling to and from clinics, concerted efforts to control comorbidities that exacerbate infections (e.g., HIV), and other public health efforts that limit the spread of infections (e.g., cheap or free condoms to combat STIs) are but parts of a coherent effort to combat infection and to lower the use of antibiotics. To institute a global policy on antibiotic usage without addressing these related issues could end up depriving those who need them most of antibiotics.

Prisoner's Dilemma

Ignoring for a moment obviously inappropriate antibiotic usage (e.g., using them to treat viral conditions) and putting aside logistical challenges intra- and international organizations confront in monitoring antibiotics usage in clinics and pharmacies, a global effort to slow the growth of antibiotic-resistant pathogens faces the classic problem of multi-person prisoner's dilemma or n-person prisoner's dilemma (NPD). An NPD arises when more than two parties have the choice of either cooperating or not with other parties. Moreover, a dominant strategy has to be present for each party; that is, no matter what the other party does, a given party is always better off not cooperating. Finally, if all parties fail to cooperate, the outcome is worse than if they cooperate. Voluntary recycling is a good example of NPD. Suppose I value a clean environment and I believe that a community with a high percentage of

[6] One area of antibiotic usage that we will not discuss is the sub-therapeutic use of antibiotic in raising animals for meat. In the Unites States, 80% of all antibiotic used for the raising of farm animals for meat consumption is given at a sub-therapeutic level in order to promote animal growth (Levy 2002: 152). In a 2007 survey, "the estimated annual antibiotics production in China was 210 million kg, and 46.1% were used in livestock industries"—four times higher than in the United States (Zhu et al. 2013: 3435). The sub-therapeutic use of antibiotics for growth promotion has been banned in European countries. Although the volume of antibiotics used for growth promotion is alarming, it is not entirely clear what the health consequences are to humans and animals in banning their use. A 2003 study shows that the banning of antibiotics for growth promotion in Europe has led to an increased use of antibiotics in animals for therapeutic purposes because of greater incidents of infections (Casewell et al. 2003). However, what is clearly worrisome is the lack of close monitoring of sub-therapeutic use of antibiotics for livestock farming in both United States and China. Given the potential enormous impact their usage has on animals' and humans' welfare, a careful collection of data on antibiotic uses in farming is a minimal first step we need to take to ensure good antibiotic stewardship.

recycling provides the best chance of a clean environment. Suppose further that I believe my contribution to the improvement of the environment via recycling makes a negligible difference. If everyone else recycles, I am better off not recycling since I can reap the environmental benefit without undertaking the hassle of recycling. If no one else recycles, I am better off not recycling since my effort would not make any difference. The dominant strategy is not to recycle because I would be better off no matter what other people do. Thus, the rational self-interested strategy is not to recycle.[7] Of course, if everyone else does the same thing, we would lose out on the environmental benefits of recycling (something most of us value).

There are various ways to encourage cooperation in a game of NPD. The most obvious way is to identify those who are not cooperating and punish them. This strategy essentially changes the cost-benefits of the player by moving the cost of noncooperation high enough such that it is no longer in the player's self-interest not to cooperate. A fine levied against those who do not recycle is a good way of realigning the self-interest of the non-recycler with the collective interest of a clean environment. Likewise, mandatory contribution in the form of taxation also provides us with the funds necessary for public goods like paved roads, sewer system, and clean water. From a rational self-interest point of view, although each citizen might desire to have these goods, their individual contributions make negligible differences. A tax to fund public goods essentially removes the choice a citizen has in terms of whether or not to cooperate; thus, it eliminates one of the prerequisites of an NPD.

Antibiotics as Global Climate: Kyoto vs. Montreal

Suppose we think of antibiotics like clean air. We all want to have access to them and we think future generations ought to have access as well. However, suppose a sustainable usage of antibiotics requires all nations to make minor sacrifices (from more expensive meat to not having quick access to broad-spectrum antibiotics). How do we motivate nations to cooperate (for instance, signing a treaty and being monitored for compliance)? One natural place to look for a possible solution to NPD at a global level is to adopt lessons learned from international treaties on pollution control. Looking to the Montreal and Kyoto Protocols as models of successful and unsuccessful international efforts, respectively, Jonny Anomaly outlines a number of features that make a treaty more likely to succeed (Anomaly 2010). They are:

1. Flexibility: goals must be adjustable on the fly and nations must be able to meet goals in a variety of ways (e.g., cap and trade, taxes, etc.).

[7] The most self-serving action would be to appear as if one recycles and convinces other people to recycle. One enjoys the benefit of a clean environment without putting any work into it while getting other people to do the hard work.

2. Signatories must perceive the burdens as being distributed in a fair way and beneficial to all participants by effective use of carrots and sticks.

The Kyoto Protocol places a significant burden on industrialized nations to cut emissions of greenhouse gases while allowing nations with smaller per capita GDP (including China) exemptions. The result was the creation of free-rider states and the withdrawal of large industrialized nations who felt they were shouldering most of the burden of cutting emissions. The Kyoto Protocol also lacked the necessary flexibility. As Asian countries began to absorb migration of heavy industries from developed nations, the Kyoto Protocol could not adjust and restrict new emitters (e.g., China and India) of greenhouse gases who had been exempted from the treaty. The cuts by industrialized nations that remained parties to the protocols have been dwarfed by emissions from China and other parts of Asia, South America, and Africa. Worldwide emissions of greenhouse gases have gone up by 50% since 1990 even though US contribution has decreased from 66% to 50% (Schiermeier 2012).

Contrast the failure of the Kyoto Protocol with the success of the Montreal Protocol—a treaty strongly advocated by conservative political leaders like Ronald Reagan and Margaret Thatcher. The Montreal Protocol aimed to limit production of ozone-depleting gases and does so by first creating incentives for small nations to sign on (Gillis 2013). When a non-developed country becomes a signatory, it immediately receives subsidies to help it create non-ozone-depleting alternatives. The treaty also imposes trade restrictions between those countries that have signed on and those that have not. However, the restriction does not kick in until a critical mass of nations has signed on to the treaty. The result was that after the initial surge of small nations signing on (motivated by subsidies), larger nations had a disincentive to remain on the sidelines. The Montreal Protocol has been held up as a model of an international treaty that provides flexible carrots and sticks to all nations to sign on while minimizing free riders in a game of NPD.

As Anomaly rightly points out, however, controlling ozone-depleting gases presents a set of challenges different from instituting a sustainable antibiotic global policy. For starters, there are currently no effective alternatives to antibiotics that are of similar costs. In addition, unlike the elimination of ozone-depleting gases, restricting access to, say, broad-spectrum antibiotics can have immediate and serious implications to individuals' welfare. Internal political pressure would be of a different magnitude: trade restrictions might not be sufficient to motivate a citizen to forgo beneficial antibiotics. Monitoring antibiotic usage (i.e., compliance with a treaty) might also present far more logistical problems. Unlike a blanket prohibition of ozone-depleting gases like chlorofluorocarbon (CFC), for example, detecting inappropriate use of antibiotics would require close monitoring at the clinical level to determine if a particular prescription of antibiotic is appropriate. Given the fact that even nations that have robust healthcare monitoring systems like the United Kingdom and the United States have had an exceedingly difficult time gathering data on antibiotic usage domestically, the mechanism necessary for a global monitoring of antibiotic usage would be tremendously complex and resource intensive.

Free Market Fails to Rescue

One might be tempted to appeal to the invisible hand of the free market to align national and individual's interests with the public health interest of having a global policy of sustainable antibiotics. The problem, as Michael Selgelid has argued, is that we have no reason to believe that a free market distribution of antibiotics would actually deliver the desired outcomes (Selgelid 2007). A major obstacle to the proper distribution of antibiotics is the lack of access to affordable antibiotics in developing nations. The slim profit margins undercut the financial incentive for pharmaceutical companies to develop antibiotics for poor nations. What we need is a coordinated effort to fund research that looks to create affordable and effective antibiotics and ensure that they are distributed and used properly. The solution "will require governmental intervention/funding. The fact that the problem of drug resistance is global in scope, while there is no global government, however, is troubling" (Selgelid 2007: 229).

Here lies, I believe, the most serious problem with a global solution to the antibiotic-resistant threat. In order for us to solve an NPD problem, we need to be able to identify, monitor, and punish those who do not cooperate (or provide incentive for those who do). The absence of any robust international organization that can undertake these arduous tasks means that a traditional solution to this NPD is unlikely to materialize. We cannot simply change the cost-benefit calculus of the parties involved hoping that we can steer them to do the right thing by appealing to their rational self-interests.[8]

Spreading Altruism as a Public Health Effort

The core conflict in a game of NPD is that if the participants decide to act on the basis of their rational self-interest, they would all be worse off than if they had not followed their rational self-interest. Real life examples of individuals not acting out of rational self-interest are plentiful. For instance, individuals often vote in general elections even if their self-interest dictates that they are better off not voting. For US presidential elections, an individual's vote would make minimal difference to the eventual outcome and casting votes often entail long waits. Yet, 59% of registered voters voted in 2012 presidential election (FairVote). Turnout in the 2005 Iraqi

[8] There are a lot of reasons why we would want to control the emergence of antibiotic-resistant pathogens. The most obvious is self-preservation. We care about ourselves, the people we love, and their offspring. But there are also moral reasons as well. The severe inequality of wealth in the world means that many individuals, in virtue of being born in the poorer parts of the world, will not have access to adequate healthcare. Healthcare, as Norman Daniels (1995) has argued, is special in that it is strategically important to one's ability to pursue the normal range of life's opportunities. Our failure to ensure that poor nations have adequate access to antibiotics not only runs contrary to our self-interests, it is also unjust.

parliamentary election—the first in the history of Iraq—was estimated at 70% even though insurgent groups threatened violence. From the point of view of rational self-interest, individuals should not be voting voluntarily. The same altruistic behavior occurs with regard to voluntary recycling. In 2012, 34.5% of all US households recycled. The majority of these households are in municipalities that do not mandate recycling (EPA). Rational self-interest does not always dictate our choices. We often pursue a course of action that is not in our best interests guided by considerations such as the well-being of others.

The global monitoring approach to solving the problem of antibiotic resistance depends on extensive surveillance of appropriate drug use at the clinical and pharmacy level, a combination of proper carrots and sticks to incentivize nations to participate, and a sufficiently objective and resourceful organization to implement the consequences of cooperation and defection by signatories. Given the slim chance that such a system can exist, it is wise for us to explore a possibility that has been largely ignored: that is, encouraging individuals to act not just in their rational self-interest but also the interests of other individuals including future and distant people. Dan Ariely has done extensive research into the psychological mechanisms at work when one decides whether or not to cheat (Ariely 2012). What he has uncovered is that most subjects in his studies do not base their actions on a simple cost-benefit analysis in the face of an opportunity to cheat. In one study, Ariely et al. provide their subjects with a matrix of simple mathematical questions. After a short period of time, the subjects report how many questions they have solved and they are paid accordingly. In one variation, the subjects' answer sheets are shredded by the subjects after they have had a chance to check their answers. There is no way the experimenters can tell if the subjects were lying about the number of correct answers they had on their sheets. In experiment after experiment, Ariely et al. observe that the majority of subjects lie about how many answers they get right by only a small margin when they could easily have lied to a greater degree and receive more money. (They determine the subjects regularly lie by comparing the subjects' reported number of right answers with their control group whose answers are checked by the experimenters.) One of the key conclusions that Ariely et al. draw is that the degree of cheating and propensity to cheat is determined by our perception of ourselves as morally decent people. Cheating a little allows us to benefit from our transgressions while letting us retain the sense that we are still good people. In other words, for most of us, the sense of moral self-worth trumps simple rational self-interest.

Indeed, moral decency plays a significant role in one's decision to cheat. Ariely et al. discovered that if subjects were exposed to a "moral reminder" prior to reporting their answers, their propensity to cheat decreases. Ariely et al., for instance, ask one group of subjects prior to an opportunity to cheat to try to list the Ten Commandments and another group to list ten books they read in high school. The Ten Commandments group cheated less while the other group did not behave differently.

Ariel's insights into the psychology of cheating can perhaps help us craft a global public health effort to encourage individuals to think about the well-being of future and distant people when it comes to the use of antibiotics. Such an effort might include an emphasis on the impact antibiotic usage has on other people. Perhaps we can use moral reminders in the form of signing an education form that highlights the externalities of antibiotics prior to receiving care (on a par with consent forms). Likewise, a public health campaign to educate the public on the virtues of good antibiotic stewardship can help create a culture of viewing the cavalier use of antibiotics as morally dubious. Such a campaign can model itself on other attempts to encourage altruistic behaviors. In the United States, for instance, we value and celebrate the virtues of democratic elections, and there are subtle and not so subtle efforts to instill in citizens a sense that voting is important and morally admirable. We teach elementary school children about the electoral process and the importance of voting. States and municipalities ensure that voting in an election is relatively easy, recent efforts by some states to limit participation in the form of voter ID laws notwithstanding. Voters get little "I Voted" stickers to show off their participation. The swearing in ceremonies for many immigrants often take place in historically significant places emphasizing the value of a liberal democracy. These subtle efforts all contribute to a climate in which individuals feel a sense of duty to participate in a democratic political system even when it is not in their rational self-interest to do so. Our value for voting is so ingrained that we hardly notice the mechanisms that went into its cultivation in our political and cultural psyche.

Antibiotic-resistant pathogens represent a serious threat to our well-being. The typical solutions take for granted that individuals will always act as rational self-interested agents. As such, like any NPD, cooperation can only come about by aligning individuals' interests with public health interests with the use of incentive and punishment. Nonetheless, given the scope of the problem and the level of logistical challenges involved, such a solution is unlikely to materialize. Perhaps a more promising approach is to encourage individuals to act not on the basis of rational self-interest via a broad educational campaign that stresses and celebrates the importance of altruism. Parents, for instance, might feel the moral nudge not to demand antibiotics for their children's ear infections (that would likely clear up on their own). This nudge might be on a par with our attitude toward littering: even though one can likely get away with doing so, there is a sense of internal shame that often leads one not to litter.[9] This internal policing is a product of successful public service campaigns that fosters the communal values. It is certainly a challenge to create a culture where thinking of our collective interests comes naturally but our success in urging one another to cooperate (from voting to recycling) gives us some hope that teaching altruism might not be entirely a pipe dream. Adding moral education as a component of public health effort has not been seriously explored.[10]

[9] One of the most startling experiences a traveler might encounter in a foreign country is the realization that our disdain toward littering is not universally shared. It is a stark reminder that our own attitude toward littering is the result of carefully crafted public campaigns.

[10] The evolutionary biologist Richard Dawkins first introduced the concept of a meme in his 1976 book *The Selfish Gene* (2016). Dawkins notes that natural selection can occur in an infosphere as well as a biosphere. The unit of transmission would be a meme (which is the information analogy to a gene). "Releasing" an altruism meme that combats antibiotic-resistant pathogens via the prop-

Given the gravity of antibiotic resistance and our limited options, perhaps it is time to explore altruism as a solution to our shared woes.

Conclusion

The emergence of antibiotic-resistant pathogens represents a steady and accelerating global existential threat. Unless we collectively develop a global strategy, piecemeal efforts will be of marginal benefits. To be sure, the development of new antibacterial might buy us more time. Nevertheless, the fact that vast populations of bacteria constantly mutate in ever-growing sites of severe selection pressure (e.g., dense factory farms that utilize sub-therapeutic antibiotics as growth promoters) means that we are confronting a relentless and strengthening foe. Past attempts to identify solutions to global crises have assumed that nations and individuals will always behave as rational self-interested agents. The lack of any viable means to shift the costs and benefits of using antibiotics to generate analogous global cooperation suggests that we ought to look for a different type of solution. Spreading altruism might not save humanity, but it might be the best hope we have to coexist with hostile pathogens.

References

Anomaly, J. 2010. Combating Resistance: The Case for a Global Antibiotics Treaty. *Public Health Ethics* 3(1): 13–22.

Ariely, D. 2012. *The (Honest) Truth about Dishonesty*. New York: HarperCollins.

Butler, C. 1979. Surgery – Before and After Penicillin. *British Medical Journal* 2(6188): 482–483.

Casewell, M., C. Friis, E. Marco, P. McMullin, and I. Phillips. 2003. The European Ban on Growth-promoting Antibiotics and Emerging Consequences for Human and Animal Health. *The Journal of Antimicrobial Chemotherapy* 52(2): 159–161.

Dameh, M., J. Green, and P. Norris. 2010. Over-the-counter Sales of Antibiotics from Community Pharmacies in Abu Dhabi. *Pharmacy World & Science* 32(5): 643–650.

Daniels, N. 1995. *Just Health Care*. Cambridge: Cambridge University Press.

Dawkins, R. 2016. *The Selfish Gene: 40th Anniversary Edition*. New York: Oxford University Press.

EPA. *Municipal Solid Waste*. Retrieved May 15, 2015, http://www.epa.gov/epawaste/nonhaz/municipal/.

Fair Vote. *Voter Turnout*. Retrieved May 15, 2015, http://www.fairvote.org/research-and-analysis/voter-turnout/.

Fleming, A. 1945. *Penicillin*. Nobel Lecture. http://www.nobelprize.org/nobel_prizes/medicine/laureates/1945/fleming-lecture.pdf.

agation of responsible antibiotic usage might present a promising line of research for public health advocates.

Fleming, A., and V. Allison. 1927. On the Development of Strains of Bacteria Resistant to Lysozyme Action and the Relation of Lysozyme Action to Intracellular Digestion. *British Journal of Experimental Pathology* 8(3): 214–218.

Gillis, J. 2013, December 9. The Montreal Protocol, a Little Treaty That Could. *New York Times*.

Ho, D. 2008. When Good Organs Go to Bad People. *Bioethics* 22(2): 77–83.

Indira, K., R. Devi, R. Pillai, L. Jeyaseelan, S. Chandy, Rashmi Kumar, S. Suresh, S. Kumar. 2004. Antibiotic Prescribing Pattern (APP) and Related Factors in Primary and Secondary Health Care Facilities of Government and Private Settings. http://www.inclentrust.org/inclen/uploadedbyfck/file/publication/theinclen/USAIDIIDI-finalversion.pdf. Accessed 11 May 2015.

Koenig, R. 2008. In South Africa, XDR TB and HIV Prove a Deadly Combination. *Science* 319(5865): 894–895, 897.

Leibovici, L., M. Paul, and O. Ezra. 2012. Ethical Dilemmas in Antibiotic Treatment. *The Journal of Antimicrobial Chemotherapy* 67(1): 12–16.

Levy, S. 2002. *The Antibiotic Paradox: How the Misuse of Antibiotics Destroys Their Curative Powers*. Cambridge, MA: Perseus Publishing.

Marković-Peković, V., and N. Grubiša. 2012. Self-Medication with Antibiotics in the Republic of Srpska Community Pharmacies: Pharmacy Staff Behavior. *Pharmacoepidemiology and Drug Safety* 21(10): 1130–1133.

McKenna, M. 2011, August 1. Are Counterfeit Drugs Driving Antibiotic Resistance? *Wired*. http://www.wired.com/2011/08/counterfeit-driving-resistance/. Accessed 11 May 2015.

Means, A.R., M.R. Weaver, S.M. Burnett, M.K. Mbonye, S. Naikoba, and R.S. McClelland. 2014. Correlates of Inappropriate Prescribing of Antibiotics to Patients with Malaria in Uganda. *PloS One* 9(2): e90179.

Metlay, J.P., J.A. Shea, L.B. Crossette, and D.A. Asch. 2002. Tensions in Antibiotic Prescribing: Pitting Social Concerns Against the Interests of Individual Patients. *Journal of General Internal Medicine* 17(2): 87–94.

Morley, P.S., M.D. Apley, T.E. Besser, D.P. Burney, P.J. Fedorka-Cray, M.G. Papich, J.L. Traub-Dargatz, and J.S. Weese. 2005. Antimicrobial Drug Use in Veterinary Medicine. *Journal of Veterinary Internal Medicine* 19(4): 617–629.

O'Neill, J., and Wellcome Trust. 2014. *Review on Antimicrobial Resistance. Antimicrobial Resistance: Tackling a Crisis for the Health and Wealth of Nations.*

Petersen, M. 2000, August 27. What's Black and White and Sells Medicine? *New York Times*. Retrieved from http://www.nytimes.com/2000/08/27/business/what-s-black-and-white-and-sells-medicine.html?pagewanted=1.

Sakoulas, G., G.P. Wormser, P. Visintainer, W.S. Aronow, and R.B. Nadelman. 2009. Relationship Between Population Density of Attorneys and Prevalence of Methicillin-Resistant *Staphylococcus aureus*: Is Medical-Legal Pressure on Physicians a Driving Force Behind the Development of Antibiotic Resistance? *American Journal of Therapeutics* 16(5): e1–e6.

Schiermeier, Q. 2012. The Kyoto Protocol: Hot Air. *Nature* 491(7426): 656–658.

Schultz, L., T.J. Lowe, A. Srinivasan, D. Neilson, and G. Pugliese. 2014. Economic Impact of Redundant Antimicrobial Therapy in US Hospitals. *Infection Control and Hospital Epidemiology* 35(10): 1229–1235.

Selgelid, M.J. 2007. Ethics and Drug Resistance. *Bioethics* 21(4): 218–229.

The Data Team. 2015, May 10. The Toll of a Tragedy. *The Economist*. Accessed 20 May 2015.

WHO. 2014a. *Global Tuberculosis Report 2014*. http://apps.who.int/iris/bitstream/10665/137094/1/9789241564809_eng.pdf?ua=1. Accessed 11 May 2015.

———. 2014b. *Multidrug-Resistant Tuberculosis (MDR-TB) 2014 Update*. http://www.who.int/tb/challenges/mdr/mdr_tb_factsheet.pdf?ua=1. Accessed 11 May 2015.

Wong, V.K., S. Baker, D.J. Pickard, J. Parkhill, A.J. Page, N.A. Feasey, R.A. Kingsley, N.R. Thomson, J.A. Keane, F.-X. Weill, D.J. Edwards, J. Hawkey, S.R. Harris, A.E. Mather, A.K. Cain, J. Hadfield, P.J. Hart, N.T.V. Thieu, E.J. Klemm, D.A. Glinos, R.F. Breiman, C.H. Watson, S. Kariuki, M.A. Gordon, R.S. Heyderman, C. Okoro, J. Jacobs, O. Lunguya, W.J. Edmunds, C. Msefula, J.A. Chabalgoity, M. Kama, K. Jenkins, S. Dutta, F. Marks,

J. Campos, C. Thompson, S. Obaro, C.A. MacLennan, C. Dolecek, K.H. Keddy, A.M. Smith, C.M. Parry, A. Karkey, E.K. Mulholland, J.I. Campbell, S. Dongol, B. Basnyat, M. Dufour, D. Bandaranayake, T.T. Naseri, S.P. Singh, M. Hatta, P. Newton, R.S. Onsare, L. Isaia, D. Dance, V. Davong, G. Thwaites, L. Wijedoru, J.A. Crump, E. De Pinna, S. Nair, E.J. Nilles, D.P. Thanh, P. Turner, S. Soeng, M. Valcanis, J. Powling, K. Dimovski, G. Hogg, J. Farrar, K.E. Holt, and G. Dougan. 2015. Phylogeographical Analysis of the Dominant Multidrug-Resistant H58 Clade of Salmonella Typhi Identifies Inter- and Intracontinental Transmission Events. *Nature Genetics* 47(6): 632–639.

Yip, W., and W.C. Hsiao. 2008. The Chinese Health System at a Crossroads. *Health Affairs (Millwood)* 27(2): 460–468.

Zhu, Y.-G., T.A. Johnson, J.-Q. Su, M. Qiao, G.-X. Guo, R.D. Stedtfeld, S.A. Hashsham, and J.M. Tiedje. 2013. Diverse and Abundant Antibiotic Resistance Genes in Chinese Swine Farms. *Proceedings of the National Academy of Sciences of the United States of America* 110(9): 3435–3440.

Zumla, A., I. Abubakar, M. Raviglione, M. Hoelscher, L. Ditiu, T.D. Mchugh, S.B. Squire, H. Cox, N. Ford, R. McNerney, B. Marais, M. Grobusch, S.D. Lawn, G.-B. Migliori, P. Mwaba, J. O'Grady, M. Pletschette, A. Ramsay, J. Chakaya, M. Schito, S. Swaminathan, Z. Memish, M. Maeurer, and R. Atun. 2012. Drug-Resistant Tuberculosis—Current Dilemmas, Unanswered Questions, Challenges, and Priority Needs. *Journal of Infectious Diseases* 205(Suppl 2): S228–S240.

